Annual Review of Wireless Communications

IEC
Chicago, Illinois

About the International Engineering Consortium

The International Engineering Consortium (IEC) is a non-profit organization dedicated to catalyzing technology and business progress worldwide in a range of high technology industries and their university communities. Since 1944, the IEC has provided high-quality educational opportunities for industry professionals, academics, and students. In conjunction with industry-leading companies, the IEC has developed an extensive, free on-line educational program. The IEC conducts industry-university programs that have substantial impact on curricula. It also conducts research and develops publications, conferences, and technological exhibits that address major opportunities and challenges of the information age. More than 70 leading high-technology universities are IEC affiliates, and the IEC handles the affairs of the Electrical and Computer Engineering Department Heads Association and Eta Kappa Nu, the honor society for electrical and computer engineers. The IEC also manages the activities of the Enterprise Communications Consortium.

Other Quality Publications from the International Engineering Consortium

- *Achieving the Triple Play: Technologies and Business Models for Success*
- *Business Models and Drivers for Next-Generation IMS Services*
- *Delivering the Promise of IPTV*
- *Evolving the Access Network*
- *The Basics of IPTV*
- *The Basics of Satellite Communications, Second Edition*
- *The Basics of Telecommunications, Fifth Edition*

For more information on any of these titles, please contact the IEC publications department at +1-312-559-3730 (phone), +1-312-559-4111 (fax), *publications@iec.org*, or via our Web site (http://www.iec.org).

ISBN: 978-1-931695-69-5

International Engineering Consortium
300 West Adams Street, Suite 1210
Chicago, Illinois 60606-5114 USA
+1-312-559-3730 phone
+1-312-559-4111 fax

Contents

Current Trends and Business Strategies

Technologies and Networks

Contents by Author

University Program Sponsors

The IEC's University Program, which provides grants for full-time faculty members and their students to attend IEC Forums, is made possible through the generous contributions of its Corporate Members. For more information on Corporate Membership or the University Program, please call +1-312-559-4625 or send an e-mail to *cmp@iec.org*.

Based on knowledge gained at IEC Forums, professors create and update university courses and improve laboratories. Students directly benefit from these advances in university curricula. Since its inception in 1984, the University Program has enhanced the education of more than 500,000 students worldwide.

IEC–Affiliated Universities

The University of Arizona
Arizona State University
Auburn University
University of California at Berkeley
University of California, Davis
University of California, Santa Barbara
Carnegie Mellon University
Case Western Reserve University
Clemson University
University of Colorado at Boulder
Columbia University
Cornell University
Drexel University
École Nationale Supérieure des Télécommunications de Bretagne
École Nationale Supérieure des Télécommunications de Paris
École Supérieure d'Électricité
University of Edinburgh
University of Florida
Georgia Institute of Technology
University of Glasgow
Howard University
Illinois Institute of Technology
University of Illinois at Chicago
University of Illinois at Urbana-Champaign
Imperial College of Science, Technology and Medicine
Institut National Polytechnique de Grenoble
Instituto Tecnológico y de Estudios Superiores de Monterrey
Iowa State University
KAIST
The University of Kansas
University of Kentucky
Lehigh University
University College London
Marquette University
University of Maryland at College Park
Massachusetts Institute of Technology
University of Massachusetts
McGill University
Michigan State University
The University of Michigan
University of Minnesota
Mississippi State University
The University of Mississippi
University of Missouri-Columbia
University of Missouri-Rolla
Technische Universität München
Universidad Nacional Autónoma de México
North Carolina State University at Raleigh
Northwestern University
University of Notre Dame
The Ohio State University
Oklahoma State University
The University of Oklahoma
Oregon State University
Université d'Ottawa
The Pennsylvania State University
University of Pennsylvania
University of Pittsburgh
Polytechnic University
Purdue University
The Queen's University of Belfast
Rensselaer Polytechnic Institute
University of Southampton
University of Southern California
Stanford University
Syracuse University
University of Tennessee, Knoxville
Texas A&M University
The University of Texas at Austin
University of Toronto
VA Polytechnic Institute and State University
University of Virginia
University of Washington
University of Wisconsin-Madison
Worcester Polytechnic Institute

Current Trends and Business Strategies

The Progression of WiMAX Toward a Peer-to-Peer Paradigm Shift

Bojan Angelov
Doctoral Student in Technology Management
Polytechnic University, New York

Bharat Rao
Associate Professor of Management
Polytechnic University, New York

Abstract

The standardization of worldwide interoperability for microwave access (WiMAX) represents an opportunity for a radical shift in viewing the Internet and its access systems. Considering the current constellation of the Internet service providers (ISPs), WiMAX can both improve its presence (by offering broadband connectivity where previously unavailable) and trim down its authority (if used as an instrument for ad hoc peer-to-peer networking). Therefore, the deployment of WiMAX is important for several stakeholders. It can be a source of substantial revenue for the equipment manufacturers and the ISPs, an alternative way of offering service for the content providers, and an opportunity for increased use of wireless commons in the broader society. In this paper, we look at a progressive model of WiMAX, starting from a wireless ISP (WISP) perspective, through social learning of the commons, and ending with a shift toward the peer-to-peer paradigm as applied to a network of interconnected wireless fidelity (Wi-Fi) users.

Introduction

Since the commercial deployment of the World Wide Web, the ways in which information is viewed, gathered, shared, and analyzed has changed drastically. As a result, many business models (if not most of them) rely on or incorporate the Internet strategic service component into their plans. The Internet has thus become a widely used common resource, just like water, oil, or public highways. In addition, the emergence of Wi-Fi has enabled wireless connectivity to the Internet, thus improving and increasing its availability and presence among customers [1]. With the emergence of WiMAX, there is an opportunity to establish wireless networking as a potentially new paradigm of Internet computing. WiMAX represents a technology that can enable customers to create their own manifestation of the Internet. By connecting individual (or corporate) Wi-Fi networks, it can go beyond the initial idea of extending the reach of wireless local-area networks (WLANs). It can create a notion of parallel Internet while still providing the conventional Internet connection as we now it.

WiMAX can alter current business models by enabling Wi-Fi owners to enter the circle of ISPs and mobile providers. The astounding success and acceptance rate of the peer-to-peer (PTP) Internet platforms (among Internet users) can be used as a case in point when providing an analogy to the potential use of WiMAX [2, 3]. Information is not only valued, but also shared among people like never before. With the emergence of WiMAX, the information shared in PTP networks can potentially evolve into a wireless network itself, providing access to various resources such as local databases, storage capacity, high bandwidth, etc. Learning from the lessons of the Wi-Fi deployment, it is necessary to explicitly evaluate the WiMAX environment (ISPs, mobile operators, content providers, current regulations in the telecommunications, social and economic trends, user needs and abilities, etc.) and the possible business models that can accompany it [26, 6]. The emergence of Wi-Fi hot spots clearly conveys the trend toward the idea of wireless commons as the center of the new style of collaboration. We use WiMAX technology as a possible enabler of the PTP paradigm shift in the broader society interactions. In addition, to address the current Internet and social developments, we describe an evolutionary model for WiMAX deployment with the premise of PTP networking as the foundation of its mature state.

Background and Current Trends

WiMAX represents a technology based on an evolving standard for point-to-multipoint wireless networking. It is being promoted by the WiMAX Forum, a wireless industry consortium with more than 100 members, including major vendors AT&T, British Telecommunications, France Telecom, Fujitsu, Intel, and Siemens Mobile [4]. Even though broadband wireless access has been used by enterprises and operators for several years, the standard developed by the Institute of Electrical and Electronics Engineers (IEEE), 802.16, is likely to accelerate the adoption of the technology and expand the scope of usage [5]. Basically, WiMAX is the common name associated to the IEEE 802.16 suite of standards. The first standard was officially published in late 2001 by the IEEE and was followed up by the 802.16a standard in early 2003. Both standards support peak data rates up to 75 Mbps and have a maximum range of 50 km. The frequency range is 10 to 66 GHz for line of sight and 2 to 11 GHz for non-line-of-sight standards. The WiMAX Forum expects to begin certifying equipment in early 2005 in the 3.4 to 3.6 GHz and 5.7 to 5.8 GHz ranges for both time division duplex (TDD) and frequency division duplex (FDD) systems [6]. For the time being, WiMAX has developed a single system profile with regards to the 5.8 GHz license exempt band and the 2.5 and 3.5 GHz licensed bands.

The Subtle Link between WiMAX and Wi-Fi

WiMAX has been referred to by many as an extension of Wi-Fi [7]. The main reason for such a relationship is the increase in range and bandwidth that WiMAX provides. Even though the technology behind WiMAX is not a result of incremental improvement of Wi-Fi [5], its emergence and market acceptance are based on the older standard, 802.11. Introduced in 1997 (when IEEE formally developed the 802.11 standard), Wi-Fi represents a technology that enables broadband Internet access via unlicensed spectrum in the 2.4 GHz and 5 GHz bands. In addition, the technology can also be used to wirelessly connect Internet protocol (IP)–based devices to each other or to a wireline network [8]. Wi-Fi can provide data rates up to 54 Mbps within a range of 20 to 100 meters. There are various applications of this technology, among which enterprise and campus networking, ad hoc networking, public access in "hot spots," home networking, and others. Wi-Fi has a number of 802.11 standards (also known as the "alphabet soup"), which also was one of the primary problems in its commercial acceptance. Among the major issues concerning Wi-Fi, the most important were (and, in many cases, still are) the security of the wireless networks, quality of service (QoS), total cost of ownership (TCO), compatibility of standards, and architecture of the networks [9]. Similar issues are at stake with WiMAX deployment, if we consider it to be the next stage of Wi-Fi evolution.

Market Trends

According to a report from In-Stat/MDR, the fixed wireless broadband (FWB) market will grow from $558.7 million in 2003 to more than $1.2 billion by the end of 2007 [10]. One of the primary reasons for such an increase is the introduction of standardized WiMAX technology. Similarly, ABI Research predicts that the combined equipment market for 802.16 and 802.20 (a standard that provides wireless broadband mobility) will reach $1.5 billion by 2008 [11]. The impact of WiMAX can be profound considering the broadband constellation in the United Sates. According to the research firm The Yankee Group, just 21 percent of homes and 51 percent of businesses in the United States had broadband access in 2003. Among the businesses, almost 90 percent of large enterprises and only 35 percent of small and medium-size businesses have broadband access [11]. This clearly identifies the small-to-medium enterprises (SMEs) as a near-term opportunity for WiMAX.

When considering these trends, it is inevitable to emphasize how truly ubiquitous the Internet has become as a means of global communication. Today, the total amount of packet-based network traffic surpasses traditional voice network traffic, and it is growing 125 percent per year, while voice is increasing at less than 10 percent. According to forecasts by market researcher Technology Futures Inc., at the given rate, voice traffic will be less than 1 percent of total traffic by 2007 [12]. Also, PTP file-sharing traffic volumes are at least double that of hypertext transfer protocol (HTTP) during the peak evening periods and as much as tenfold at other times [13]. More important, in the last mile, PTP makes up 80 percent or more of the traffic on the network. In October 2004, content provision became the leading on-line activity in the United States. More than 40 percent of the time spent on-line in the U.S. can be attributed to content distribution and acquisition activities [29]. As far as the nature of the content is considered, WinterGreen Research expects a $359.1 million worldwide on-line music market in 2003 to be $14.7 billion by 2009 [29]. In addition, emphasizing the social inclination toward the PTP collaboration platform, at any given instant, more than 7 million people are using PTP networks in the United States [29]. We use these developments as a case in point when trying to argue that PTP can be viewed as a dominating configuration in the future evolution of the communication and business models. If we add the recent trends and success of the IP telephony, particularly the case of Skype [14], it is clear that the WiMAX has a broad range of deployment opportunities leading toward possible success.

WISP as a Starting Point for WMAX

WISP can be considered as the starting point in the WiMAX evolution. Fixed wireless broadband is most likely to be the initial WiMAX target application. It appears that WiMAX is probably the most viable broadband alternative in markets where wired infrastructure is lacking [6]. The cost and complexity associated with traditional wired broadband infrastructure makes WiMAX the optimal solution for the significant broadband coverage gaps in the United States and the rest of the world (underserved areas). Even though market opportunities for fixed access in developed markets seem limited, newcomers in the ISP world can use the low deployment cost of WiMAX (as anticipated) to take market share of the incumbent broadband service providers such as digital subscriber line (DSL) and cable. Some estimates claim that WiMAX is likely to cost about one-fifth of a third-generation (3G) mobile network, but it is still unclear what volume levels are required to achieve this low pricing [6, 28]. According to *Wireless Watch,* it would cost $3 billion in equipment, towers, sites, labor, and setup costs to build a national U.S. WiMAX network reaching more than 90 percent of the mainland population [27]. Since WiMAX can operate both in unlicensed and licensed bands, it opens the possibility of independent companies deploying regional or nationwide wireless networks that will compete with the established cable and wireline service providers. However, to provide and guarantee QoS, it is necessary to operate in the licensed spectrum. Therefore, wireless service providers with available spectrum but without broadband service offerings, such as Sprint PCS and Nextel in the United States, can be among the first to successfully deploy and commercialize the technology [6]. In addition, WiMAX has enormous potential to lead the technological leapfrog in areas of the world with a less developed telecommunications infrastructure.

To support a profitable business model, ISPs and operators have to sustain a mix of business customers (for high-revenue services) and residential subscribers (high-volume users with lower QoS) [15, 16]. The advantages of the broadband wireless access when considering WiMAX can be summarized as follows [17]:

- Lower TCO than traditional leased lines and faster time to market
- Higher flexibility in service delivery using shared or dedicated bandwidth
- Scalability in handling rural and urban subscriber densities on the same infrastructure
- Standardization that creates a volume opportunity for chipset vendors/silicon suppliers
- Standards-based, common platform that fosters rapid innovation and the addition of new components and services
- Security, performance, and reliability at or above the standards for leased line networks

WiMAX Integration Phases

It has been generally accepted that WiMAX applications will have to transition through the following three major phases [8, 18]:

- Fixed location private line service or hot spot backhaul (providing services to fixed private locations and/or backhaul between the mesh/Wi-Fi networks and the points of presence)
- Broadband wireless access, including WiMAX as intercell transport (combination of Wi-Fi and WiMAX for optimal end-to-end performance)
- Mobile/nomadic client connection (with the emergence of the 802.16e standard)

In addition to these phases, we propose a PTP complementary integration process (where the peers are Wi-Fi networks) throughout each of the phases. At the beginning of this process, WiMAX will serve as a backbone for 802.11 WLAN hot spots, where roaming users can access carriers' Wi-Fi services [4]. According to the International Data Corporation (IDC), the number of Wi-Fi hot spots worldwide is set to grow from about 50,000 to about 190,000 by 2008 [6]. Through price adjustment and social integration of Wi-Fi commons [19], WiMAX can interconnect Wi-Fi owners' private networks (corporate or individual) and create a PTP-based network architecture. In this case, we will refer to the Wi-Fi network itself as a peer.

Since PTP has become popular through file-sharing platforms such as Kazaa and BitTorrent, WiMAX is supposed to provide similar capabilities, but in a more general sense. This means that users can chose their own path to a particular location (Wi-Fi network) through optimizing their peer relations. In addition, content providers can introduce their services in a more localized (and more secure) manner by personally selecting the networks of presence. The shift toward a PTP–driven paradigm lies in the power of the commons and its willingness to share resources.

An Untethered View of the WiMAX Integration

The previously mentioned integration phases of WiMAX are developed according to the "tethered" operator-centric–driven model of the telecommunications process. In addition, they do not explicitly account for the emerging PTP trends in the technological and social (business) constellation of interactions. As depicted in *Figure 1*, we consider a shift toward an untethered view of the wireless commons based on the promising PTP paradigm.

Basically, we see a transition from the current peer-to-peer client platforms toward an untethered collaborative effort of the wireless commons. Even though the first WISPs are supposed to provide services according to the conventional operators' view of telecommunications, the increased use of WiMAX as a Wi-Fi backhaul, along with the continuous proliferation of the community-based hot spots, is likely to have a profound influence on moving WiMAX toward enabling self-contained and self-initializing WiMAX commons.

FIGURE 1

The Untethered Positioning of WiMAX

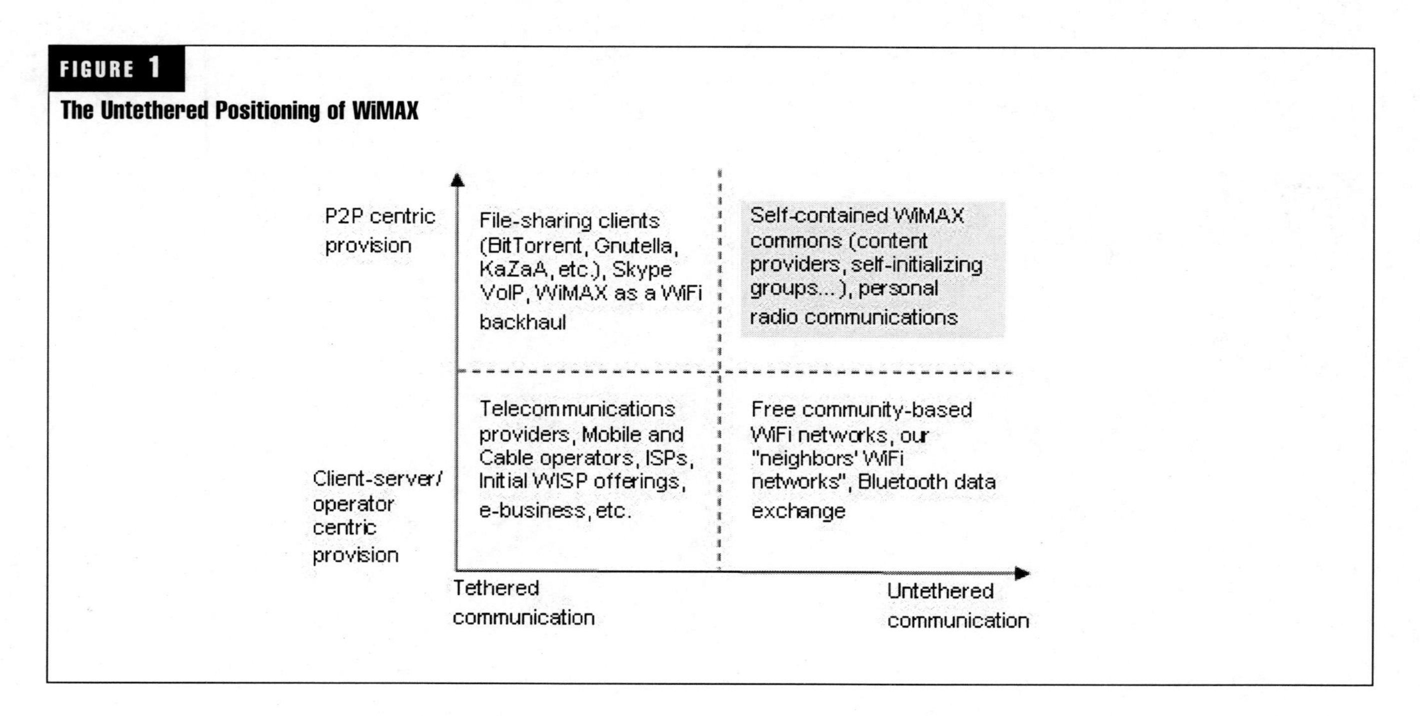

Wireless Commons and the Peer-to-Peer Paradigm

The emergence of WLAN infrastructure has allowed entrepreneurial organizations to engage in a dynamic process of clustering private and public WLANs to create pervasive wireless broadband networks [20]. Some of these initiatives have evolved into a community-oriented wireless commons, in which participants share their Wi-Fi networks with others. At the same time, through peer-to-peer networking, Internet users exhibited unprecedented willingness for information sharing (collaboration). They have allowed people from around the world access to their own files, and by borrowing their resources (bandwidth, processing power, storage), they facilitated the global effort toward the PTP paradigm shift. On the other hand, this initiative for information sharing comes in times when "new" knowledge is the essential builder of competitive advantage and the major driver of innovation [21]. Both individuals and corporations have hoisted the value of readily accessible information and PTP may seem to be not only the new technology paradigm, but the social and business one.

PTP represents a paradigmatic shift from the client-server model (both in the computing and business worlds) and assumes a dynamic interchange of goods and services, actively involving all participants. This PTP mechanism was formally defined as the action of mutually exchanging information and services directly between the producer and the consumer to achieve purposeful results [22]. A more precise definition was provided by the PTP Research Group, Internet Engineering Task Force (IETF)/ Internet Research Task Force (IRTF) [23]: "(PTP) is a way of structuring distributed applications such that the individual nodes have symmetric roles. Rather than being divided into clients and servers each with distinct roles, in PTP applications a node may act as both a client and a server."

This definition underlies our evolutionary model of WiMAX. Once the technology is accepted and commercially deployed, the Internet will no longer be the only access system for information servers. Instead, the Wi-Fi owners, both individual and corporate, will engage in a dynamic PTP participation, creating, offering, and sharing their own resources and services. This means that the need for a third-party enabler of the information (or resource) transfer (such as operators or ISPs) will be reduced to a minimum. In the new architectural paradigm, the traditional operators and telecommunications providers will have to adapt to the user-shaped and user-driven world of communications.

Peer-to-Peer WiMAX Commons

The evolutionary process of WiMAX is not going to be swift. It depends on many factors, both technological and social. The first step is acceptance from the current incumbents in the broadband arena and the new service operators. In addition, by resolving the mobility issues (advancement of 802.16e), WiMAX will create the opportunity to successfully penetrate the commercial barrier. Another important factor for our premise is the establishment of the wireless commons. Rao and Parikh [19] consider the bandwidth crunch, network integration, and group dynamics to be the major technological and social challenges regarding the future of the community-based Wi-Fi networks. Since WiMAX may upgrade the position of the users to the current level of ISPs and content providers, individual actions and regulations play a focal point in the establishment of the Wi-Fi PTP paradigm.

Since the foundation for our view of PTP WiMAX commons is the process of hot-spot interconnection and integration, we argue that instead of global Internet connectivity, many current applications and businesses can be better utilized if using the localized Wi-Fi constellation. *Figure 2* illustrates the possible integration of Wi-Fi commons using WiMAX. Each Wi-Fi network (hot spot) can be owned by individuals, corporations, or community institutions. In addition, each hot spot represents a peer entity that aggregates the genuine attributes of the local environment. Using WiMAX, these peers can be interconnected in such manner that their preferences are met most optimally. Generally, peers can be both individual users and groups of users or corporations.

It is important to notice that even though WiMAX networks are supposed to provide Wi-Fi interconnection, their initial deployment will be as an Internet backhaul. We argue that the WiMAX progression will lead toward increased use of inter–hot spot communication, thus virtually creating a par-

FIGURE 2
Wi-Fi Integration Using WiMAX

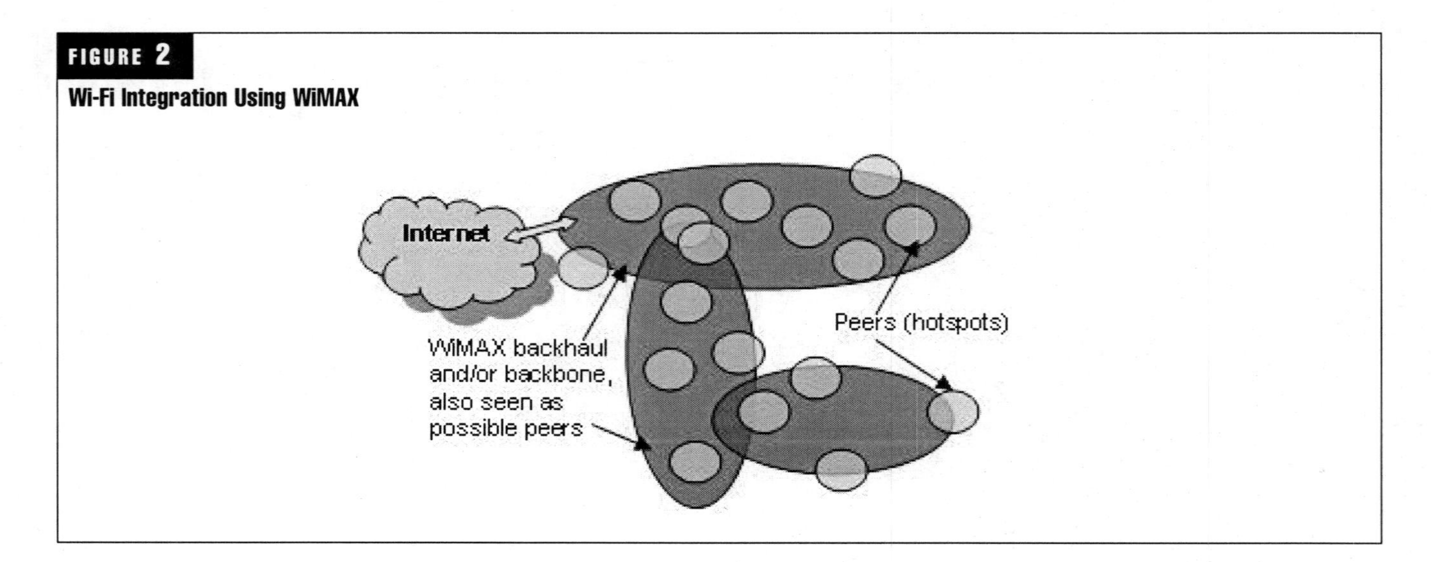

allel Internet. In addition, the WiMAX networks themselves can be considered as peers in the new architectural setup.

The peer networks enabled by WiMAX will differ in their needs and organizational structure. Since they will be characterized by high peer-selectiveness (meaning they/peers will participate based on their genuine properties and localized attributes), the WiMAX commons will range from family/friends type of peer relationships to business-to-business content distribution networks. In addition, it is our belief that separate WiMAX PTP networks will appear as new peers (integrated peers resulting into one aggregative peer) when dealing with larger, content-oriented business peers. For example, if all universities in New York are considered to be peers, and their local Wi-Fi commons are integrated via WiMAX, their aggregation can be considered as one peer. Furthermore, universities in New York and Pennsylvania universities can be seen as one peer when dealing with institutions such as NSF, ETC, and other similar to them.

Considering the wide range of possible WiMAX communities, we have selected the following structures as the most evident representatives of the previous:

- *Informal, local groups*: These groups consist of family members, friends, neighborhood members, and smaller, local service providers. The main purpose of these wireless commons is to share personal information and computer resources, such as bandwidth and storage. In addition, they may consolidate with each other and represent a single peer when dealing with other content-providing peers.

- *Scientific research communities*: These groups can use WiMAX technology to optimize their resources and level of collaboration. They can also integrate their research effort and allow secure content distribution.

- *Education and local government communities*: As in the previous example, one possible WiMAX community can be consisted of various educational institutions. In addition, local authorities can deploy their own WiMAX network and interconnect to other communities in the same environment.

- *Professional and business communities*: These communities refer to a specific group of users that share similar characteristics and needs (such as finance, stock news, entertainment, politics, sports, etc.).

- *Local ISPs and mobile operators*: Along with the previous communities, telecommunication providers can also use WiMAX to better utilize their resources (increase the mutual collaboration), and acquire and distribute user-specific content. They are supposed to be the first group to introduce WiMAX, mainly as a last-mile broadband access to underserved areas.

- *Corporate communities*: The enhanced QoS capabilities of WiMAX, along with the high security and bandwidth, can be used by corporations to build a very effective means of inter- and/or intra-collaboration with other corporations and stakeholders. They can further utilize their peer networks for high-quality voice applications and connection to the content providers.

- *Professional content distributors*: Content distributors can use the PTP design of the WiMAX communities to provide secure, timely, and localized access to various types of content (video on demand, financial information, etc.)

Conclusion

The latest developments in WiMAX technology may prove to be an important driver of overall wireless broadband evolution. After the introduction of Wi-Fi and its successful commercial adoption, WiMAX can also be viewed as its integrator and aggregator, and a potential bridge across the digital divide [5]. Because of competitive equipment prices and low operation costs, the scope of WiMAX deployment will broaden to cover markets where the low plain old telephone service (POTS) penetration, high DSL unbundling costs, or poor quality of conventional telephony lines have acted as a brake on extensive high-speed Internet and voice over broadband (VoB). In addition, the forecasts predict that WiMAX will reach its peak by introducing the portable Internet (WiMAX with embedded mobility). Such an outcome is possible when low-cost WiMAX chipsets are going to be integrated into laptops and other portable devices. The technology is supposed to provide high-speed data services on the move, extending the current limited coverage of public WLANs to metropolitan areas. The combination of these capabilities makes WiMAX an interesting technology for a wide diversity of stakeholders, such as fixed operators, mobile operators (2.5G, 3G, and 4G), WISPs, various vertical markets, and local authorities.

In this paper, we have tried to look beyond the technical capabilities of WiMAX and integrate the emerging technology with other existing business, technology and social trends. As PTP seems to be an inevitable paradigm of our current networking constellation—both from business and social perspectives—we argue that the same PTP architecture can be overlaid on Wi-Fi networks using WiMAX technology. In this case, peers are not just the individual users, but also the Wi-Fi networks themselves. This architecture is basically an evolution of the emerging wireless commons [19]. For wireless commons to survive and flourish, the society (businesses and individuals) has to accept and capitalize on the premise of communities of practice [24, 25]. In addition, there has to be a shift in the business practice allowing users (literally the Wi-Fi owners) to dynamically participate in the value creation. Such a shift assumes that instead of client-server–based business relations, the current Internet users need to evolve into entities that will actively participate in the configuration of the wireless networks (powered by WiMAX) and the trade of their resources and content. In a way, the users will have the authority (and responsibility) that now lies in the hands of the operators and ISPs. If the evolution is successfully realized, it should provide higher connection speed, increased reliability, lower operation costs, optimized interconnectivity, higher content control, and enhanced adaptability to the local environment. Basically, the PTP WiMAX architecture may prove to be the "natural" cure for the immense problem of regulating the digital content distribution.

We consider WiMAX to possess the capability to change the current communications paradigm and give users the power to shape and control the processes of accessing and utilizing the communication networks. The time has come for an integrating technology that simplifies wireless broadband.

References

[1] Werbach, Kevin, "Radio Revolution: The Coming Age of Unlicensed Wireless," New America Foundation, Public Knowledge, December 2003.

[2] K. P. Gummadi, R. J. Dunn, S. Saroiu, S. D. Gribble, H. M. Levy, and J. Zahorjan, "Measurement, Modeling, and Analysis of a Peer-to-Peer File-Sharing Workload," SOSP, Bolton Landing, New York, USA, 2003.

[3] E. C. Efstathiou and G. C. Polyzos, "A Peer-to-Peer Approach to Wireless LAN Roaming," WMASH, San Diego, USA, September 19, 2003.

[4] Steven J. Vaughan-Nichols, "Achieving Wireless Broadband with WiMax," Industry Trends, IEEE Computer, June 2004.

[5] Alcatel Strategy White Paper, "WiMAX, making ubiquitous high-speed data services a reality," Alcatel, June 2004.

[6] J. Osha, S. Pajjuri, S. D. Woo, M. Heller, "Semiconductors: WiMAX – Not your father's Wi-Fi," Merrill Lynch, Global Securities Research & Economics Group, June 10, 2004.

[7] WiMAX White Paper, "IEEE 802.16a Standard and WiMAX Igniting Broadband Wireless Access," WiMAX Forum, 2004.

[8] S. Panwar, J. Z. Tao, "More Than Just Wi-Fi: IEEE 802 Wireless Network Technologies," One-day short course, Polytechnic University, New York, February 18, 2005.

[9] C. J. Mathias, "Wireless LANs: The Big Issues," Farpoint Group, Network World Technology Tour, New York, 2004.

[10] D. Schoolar, K. Fischer, "WiMAX: 802.16 Brings Standards to Fixed Wireless Broadband," Reed business report, No. IN030845WN, January 2004.

[11] Internet article, "Wi-max: the (next) great wireless hope," Red Herring, The Business of Technology, www.redherring.com/PrintArticle.aspx?a=950§or=Industries, Jan. 12, 2004.

[12] K. Richards, (2000), "Terabit-scale traffic cops: Is your network equipped to handle billions of packets per second?" Technology & Profits, www.fiber-exchange.com/archives/technology/tech_0900.html, accessed January 28, 2005.

[13] A. Parker, (2004), "The True Picture of Peer-to-Peer Filesharing," CacheLogic.

[14] Bharat Rao and Bojan Angelov, (2005), "Skype: Leading the VoIP Revolution," Working Case Study, Department of Management Polytechnic University, New York.

[15] Intel White Paper, "IEEE 802.16* and WiMAX: Broadband Wireless Access for Everyone," Intel, 2003.

[16] Alvarion White Paper, "Introducing WiMAX: The next broadband wireless revolution," Alvarion, 2004.

[17] Proxim White Paper, "The Broadband Wireless Access Market: Evolution of the WiMAX Standard," Proxim Corporation, 2004.

[18] Tropos Technology White Paper, "Open Standards for Broadband Wireless Networks: Wi-Fi to WiMAX," Tropos Networks, October 2004.

[19] B. Rao and M. A. Parikh, "Wireless broadband drivers and their social implications," Technology in Society, vol. 25, pp. 477-489, 2003.

[20] J. Damsgaard, M. A. Parikh, B. Rao, "Wireless Commons: Perils in the Common-Good," Communications of the ACM, February 2006, Vol. 49, No. 2.

[21] M. Alavi and D. Leidner, "Review: knowledge management and knowledge management systems: conceptual foundations and research issues," MIS Quarterly 25 (1) 107-132, 2001.

[22] Moore, D. and Hebeler, J. (2002), Peer-to-Peer: Building Secure, Scalable and Manageable Networks (Osborne: McGraw-Hill), p. 4.

[23] IETF/IRTF (2004), Excerpt from the Chapter of PTP Research Group, www.irtf.org/charters/PTPrg.html, accessed September 26, 2004.

[24] E. Wenger and W. Snyder (2000). "Communities of Practice: The Organizational Frontier." Harvard Business Review 78 (1) 139-135.

[25] E. Wenger, (2000). "Communities of Practice and Social Learning Systems." Organization 7 (2) 225-246.

[26] Intel White Paper, "Understanding Wi-Fi and WiMAX as Metro-Access Solutions," Intel, 2004.

[27] Wireless Watch, "WiMAX turns the screw on 3G," The Register Internet article, www.theregister.co.uk/2005/02/14/wimax_versus_3g, February 14, 2005.

[28] WiMAX Forum White Paper, "Business Case Models for Fixed Broadband Wireless Access based on WiMAX Technology and the 802.16 Standard," WiMAX Forum, October 10, 2004.

[29] Telecordia Technologies, "The future of digital content distribution...enabling the controlled distribution of digital content," Conference Presentation, CCNC, 2005.

Affirming the Mobility Quotient: How Wireless Connectivity Is Shaping Worker Productivity

Frank J. Bernhard
Managing Principal, Supply Chain and Telecommunications
OMNI Consulting Group LLP

Abstract

Mobility in the past decade has brought about sweeping changes to the productivity of the global workforce as information becomes an increasingly vital link to job performance. We will explore the fundamental shifts in macroeconomic and microeconomic forces that face enterprise decisions regarding the adoption of wireless connectivity. And more relevant than ever before, the focus is on improving productivity for workers who capitalize on wireless advantage to actually do more in less time. This translates into not only a boost of efficiency, but also a source of competitive wealth for companies that place strategic value on information to succeed.

Of the roughly 180,000 jobs being created monthly in the United States, approximately 8.3 percent of these are a direct result of mobile workforce transformation. At the macroeconomic view of labor, this means that positive job growth is inherently linked to workers capable of performing their role in distinctly different way. From the insurance claims representative to law enforcement personnel, mobility appears to be changing the equation of getting more done in less time and at higher volumes than ever before. This trend line of productivity demonstrates an overall 13.37 percent improvement in total factor productivity (TFP), supporting the notion that employers can actually gain five additional hours per week of output from their workers by simply adding mobility to the mix.

An Economic Agenda for Mobility

Neoclassical economics tells us that adding technology to our mix of capital, labor, and assets should produce a positive return on the investment, or at least provide a distinguishable improvement in our ability to produce a market basket of goods and services. Through the Industrial Era of the 1900s, we saw technology concentrated squarely on changing production efficiencies (i.e., eliminating labor through automation and increasing output levels). However, as the global economy begins to morph into a knowledge-driven business fabric, the context of manufacturing steps aside in favor of information sharing. Now, competitive advantage hinges on rapidly executing bits of data rather than mastering the capacity angle of widget production. Lest we forget that widgets really still matter, the proper notion of an information economy still entwines exchanging data that makes commerce connect.

And that is exactly where mobility has made a quantum difference in enabling workers across the globe to enrich their productivity by having information at their fingertips and the freedom to execute wherever and whenever. The core of our research in looking at mobility's role today is nested inside an understanding of the behavioral and operational economics that face corporations and individuals choosing mobile solutions. Armed with real-world data and the concert of domestic and international reporting sources, we set out to define some of the critical patterns that make mobility a surging portion of economic growth in the past five years. By virtue of economic modeling and statistical sampling, this research highlights the ways in which mobile services infrastructure is improving the gross domestic product (GDP) outlook for countries and changing the game plan for organizations worldwide. Our premise that mobility does, in fact, equal productivity presents a number of insights at the macro- and microeconomic levels.

Macroeconomic Lessons in a Global Economy

Take the top-down approach to thinking about mobility as a larger input of change. It is of little wonder that technology and the effects of globalization have changed our business output schedules in recent years. As we said, a key concept that most executives accept involves productivity being derived from implementing technology; that is to say, technology adds benefits to the balance sheet across many modes, from higher per-customer profit to lower per-unit

FIGURE 1

Mobility Equals Economic Productivity

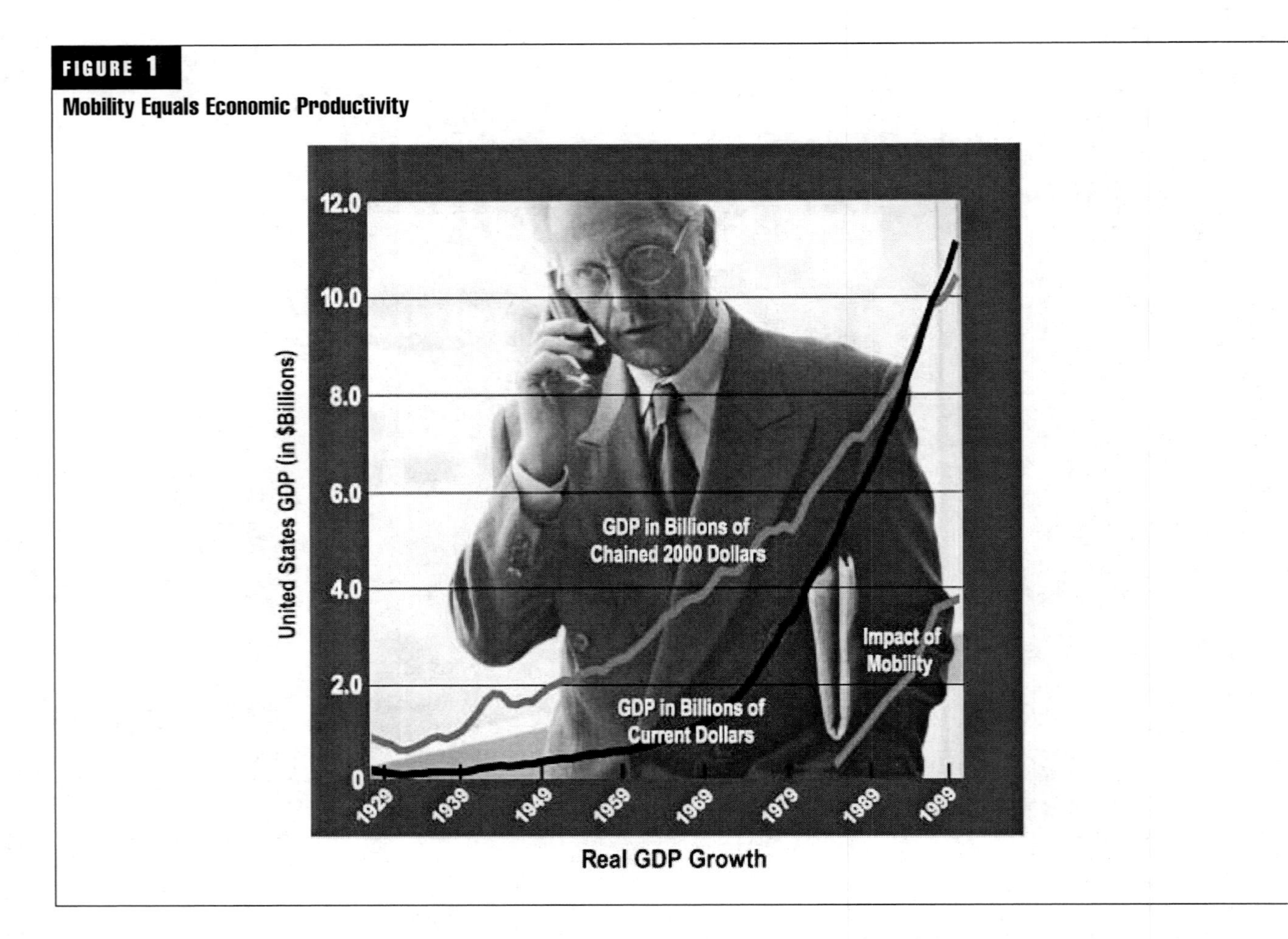

production costs. The point of contention, to some, is just how investments in mobile technology perform against other choices of spending. On average, we observe that the annualized productivity growth rates in the United States alone have steadied at a 4.5 percent year-over-year improvement. And of this climbing growth, data shows that mobile data services are beginning to account for a hefty 23.6 percent compound annual growth rate (CAGR) as the information services economy evolves. Whereas before the economy weighed heavily on durable goods manufacturing, the next leap ahead is clearly about service creation as a wealth-building cornerstone.

This indicates that the global economy is witnessing an era of measuring itself on knowledge resources and how information is playing a greater role in changing our productivity quotient. A number of observations in the data point toward why this phenomenon is happening. Among the most significant conclusions, we recognized the following:

- Network convergence is bringing users of information technology applications closer to enterprise data through wireless connectivity
- Declining costs for wireless access are driving up usage to data warehouses
- Reliability of mobile performance is improving to meet enterprise service expectations
- Decentralization of workforce operations is forcing the need for mobile communications infrastructure
- Adoption of mobility services are better accounted for and budgeted within corporate and individual spending
- Willingness to spend for mobility enhancement is sharply rising as priorities of discretionary income lean toward service access
- Regulatory barriers are softening to invite new service providers and thus drive competition among mobile data options

At the pinnacle of change, mobility is forging new ground because of the positive realization that companies can adjust their capability to produce goods and services without adding costlier inputs such as people, capital assets, and long-term liabilities. And as the early adopters of mobile services in the late 1990s became mainstream users in 2004, we noted a substantial increase in value recognition at both corporate and individual levels. Mobility is no longer a fashionable sidekick to the corporate world, but more so the propagation of technology is reaching down the employee chain to those on the front lines with an information need. To an obvious degree, this is being driven by corporate instincts to perform the following:

- Reorganize labor to create higher forms of efficiency
- Improve quality of service (QoS) to yield higher profits
- Lower the costs of business operations
- Increase marginal output of goods and services
- Maximize revenue opportunities

FIGURE 2

The Impact of Mobility on Economic Growth Rates

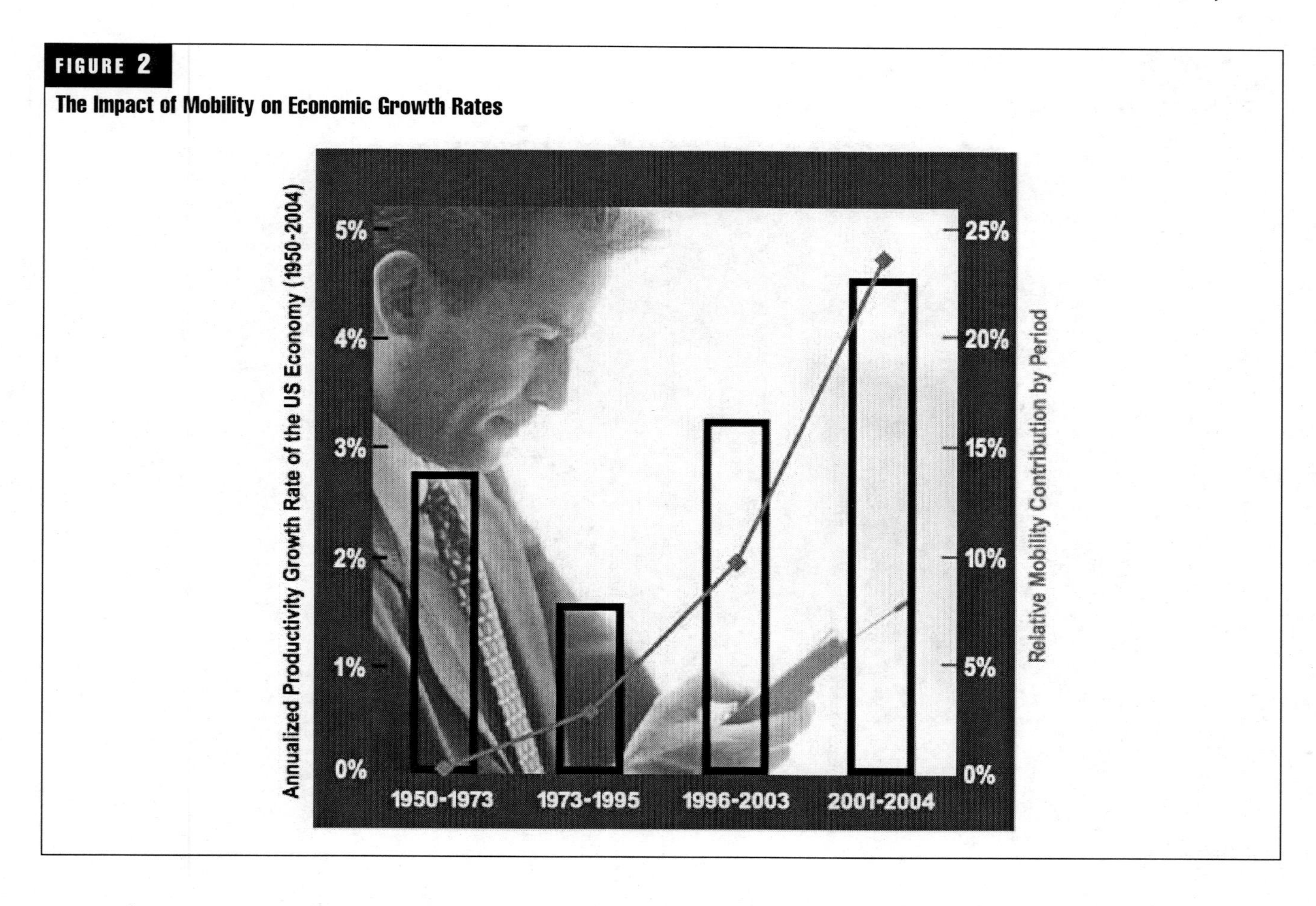

One of the conspicuous phases of mobile transformation has been the slant of technology influence on previously untouched technology domains. For instance, some of the power users of mobility are quickly appearing from the blue-collar sector, where simple, rote processes are being replaced with handheld automation. From the delivery driver to the parking garage attendant, wireless devices are realigning the use of data to better connect the enterprise, whether it be packages in transit or parking spaces for rent.

And as the costs of business operations dip slightly with better information about performance, organizations are left to choose how to allocate the savings offset and make plans. Over the next few years of seeing mobility change whole industries, our data reveals significant benefits for consumers as goods and services flow with greater velocity and competition rises to bear lower prices. A crucial part of the information era is the way in which pricing approaches have changed to reflect the knowledge of consumers in a given market. Wireless data networks are fundamentally accelerating the relationship of commerce and specific components of pricing goods and services.

In total, organizations and individuals are beginning to set themselves apart by using mobility as a strategic lever in adapting routine business processes to accomplish more through the execution of information resources. And mobility is finally delivering on its promise to enhance workers' quality of life—the subsequent productivity gains reflected in larger payrolls and spending speak volumes at the macroeconomic root of change.

A Microeconomic Perspective: Drilling Deeper into Vertical Industries

Not every industry views mobility in quite the same manner. Nor do those who use mobility gain the same benefits across a set of standardized measurements. In our research of several vertical industries that applied a cross section of mobile data devices, portable computing applications, and combination voice/data infrastructure investments, the net effect of mobility could be boiled down to the following primary factors:

- *Human capital improvement by the automation of constrained information*: Workers gain access to otherwise difficult information required to perform their job.

- *Point-of-sale transactions that occur beyond a traditional brick-and-mortar location*: Commerce happens in multiple locations apart from branch offices and physical locations.

- *Volume throughput as a function of job performance*: The measurable standard is how much mobility increases production goals and output of work products.

- *Self-reliance on information to accomplish job task objectives and increasing isolation from workgroups*: Workers

FIGURE 3

The Future: Mobility Contributes Heavily to New Job Creation

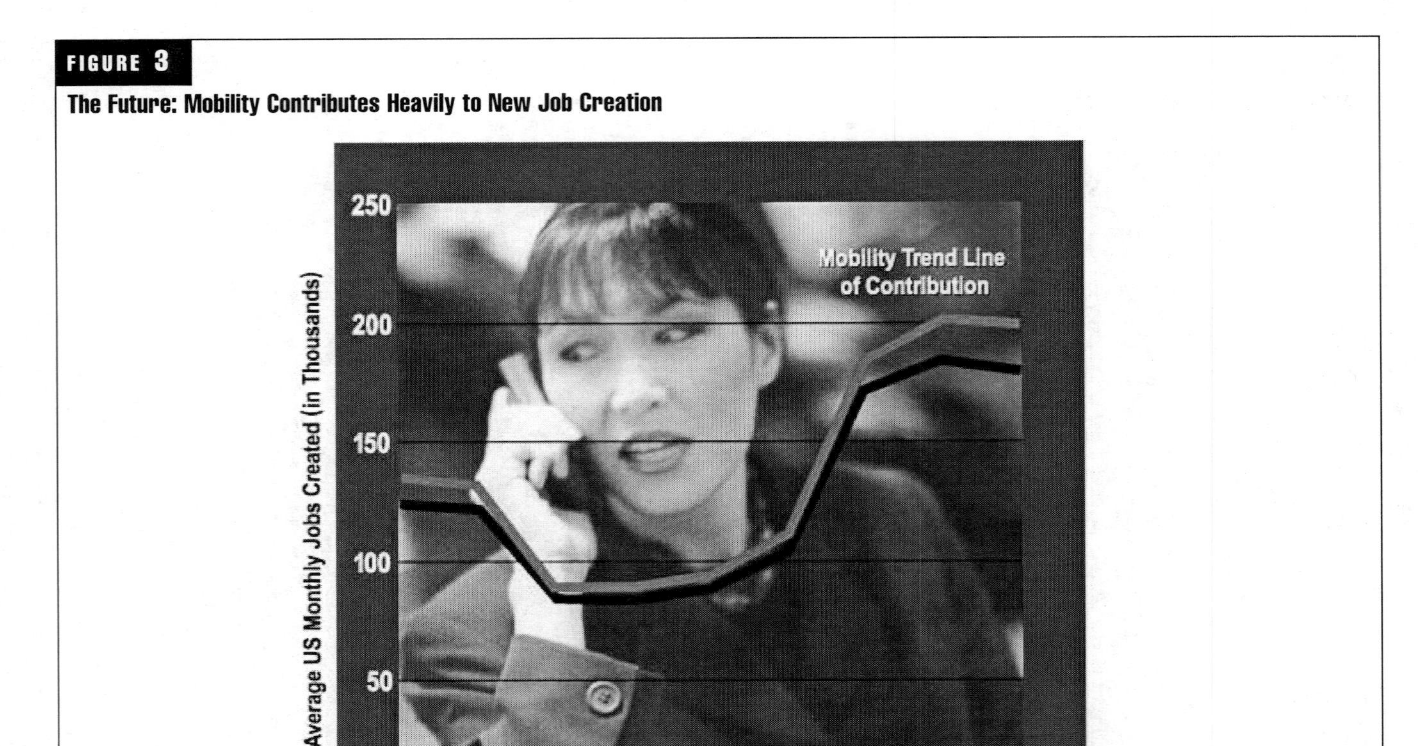

require a certain amount of information access to accomplish tasks remotely.

Specifically, we examined six industries—insurance services, hospitality, public safety, manufacturing and logistics, health care and pharmaceuticals, and financial services—to gain a perspective on why mobility changed the productivity quotient. The key statistics from each of these vertical segments relate to how companies can choose to improve their operations by the four primary factors in microeconomic circumstances.

Insurance Services

Insurance services at the consumer level delegate claims management to field representatives who must gather information and adjust loss payouts according to a policy basis. Key performance indicators include the following:

- Operational efficiency and field staffing of claims management
- Loss-to-policy limit ratios
- Resolution and expiration of claim cases
- Optimization of information resources and handling accuracy

Key Statistic: Field claims representatives can improve case management workloads by handling 7.4 additional claims per week and improve payout ratios by an annual savings of 6.35 percent per adjuster. This means that the workload of an adjuster can increase moderately by avoiding the manual input, search, and gathering of data processing as part of the claim. The accuracy of information regarding loss values coupled to real-time settlement has been shown to improve operational payouts by 6.35 percent of a given adjuster's payout performance.

Hospitality

Hospitality services run the full gamut of hotels, restaurants, and retail locations that provide managed access to patrons. More recently, travelers have begun to show a distinct bias toward selecting properties that offer broadband connectivity options—both wired and wireless. Key performance indicators include the following:

- Guest revenue yield per visit or stay
- Vacancy or loading factor per day
- Site improvement and introduction of service costs
- Elasticity of patron and guest spending

Key Statistic: Across the range of options for broadband access, guests or service patrons are more likely to select a location with Internet access—7.2 times out of 10—when given a comparable hospitality solution. Travelers to hotel properties will equivocally spend 6.85 percent more per room for suitably equipped broadband facilities. On an average $100 per night room, a guest will spend roughly $7 more on equipped property versus one without mobile high-speed connectivity.

Public Safety

Immediate access to information is critical for the performance of law enforcement duties in the public safety industry.

FIGURE 4

Industry Contrast: The Effect of Mobile Data within Financial Services

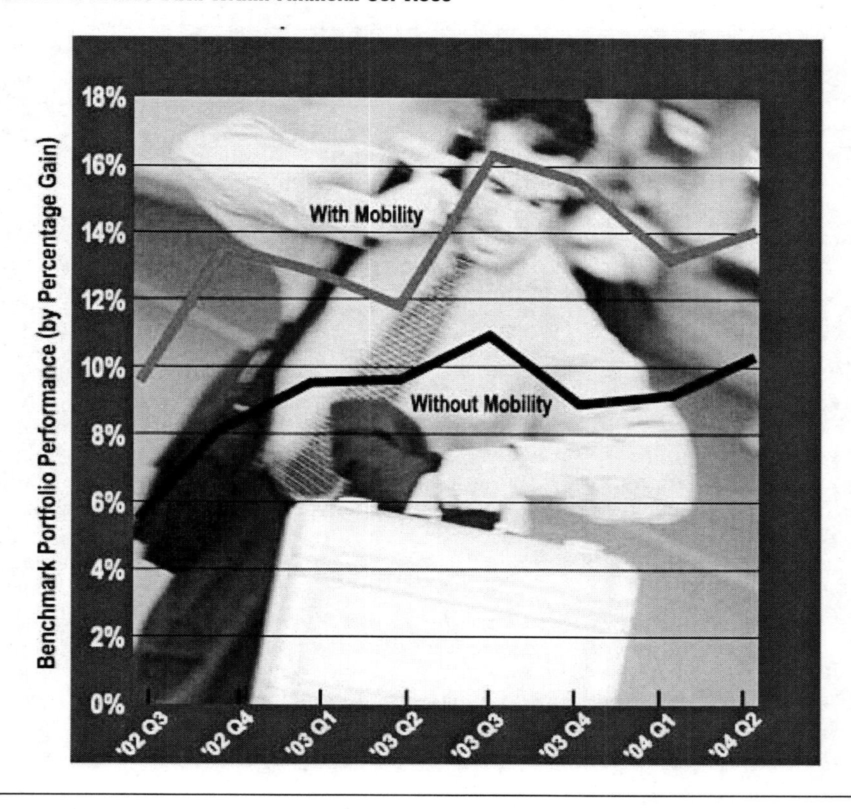

Peoples' lives depend on timely data to help accomplish the capture and prosecution of criminals at large, while vastly improving the chances of survival against attack. Key performance indicators include the following:

- Reduced loss of life and property
- Increased retention of criminal persons at large
- Reduced officer and public servant injury
- Accuracy and expediency of information from central database resources

Key Statistic: Mobile data services resulted in approximately 3.7 percent more arrests for wanted persons during the period of 2003-2004. The application of mobile data terminals inside police vehicles aided in the statistical measure of arrested persons because of information links.

Manufacturing and Logistics

The movement of goods and services within the economy requires a well-performing supply chain. Mobile data services in the warehouse environment help enhance the quality of logistics in getting products from one point to another. Key performance indicators include the following:

- Technical, operational, and economic efficiency improvements to move products at a lower cost
- Quality of shipments reaching end destinations in the least amount of time and on a least-cost basis
- Accuracy of data concerning the movement of high-value or volatile products
- Management of inventory thresholds to demand requirements

Key Statistic: Use of radio frequency information data (RFID) wireless technology eliminates 5.93 percent of logistics errors within any given warehouse operation. Based on the initial design of RFID networks, a warehouse shipper operation can effectively reduce the errors of miscalculated shipments, improperly picked loads, and the inventory handling process to achieve better spend performance on logistics services.

Health Care and Pharmaceutical

Most pharmaceutical sales organizations employ field sales representatives to directly affect the demand fulfillment and marketing of prescription drug products. The individual field representative relies on access to physician networks and information pertinent to recommending certain solutions. Key performance indicators include the following:

- Prescription fulfillment rates by physicians and medical practice groups
- Adoption and conversion of sales leads
- Order fulfillment and demand planning for manufacturer operations

Key Statistic: Field sales representatives may increase incremental physician briefings by an average of 8.3 visits per week because of mobile data and voice access. Depending on the size of a given sales territory, a single representative can maximize his or her time within a given market at greater efficiency because of sales relationship and supply chain management applications.

Financial Services
Markets fluctuate in real time as stocks, bonds, and other currencies move within trade circles. To place trade orders, financial services professionals have come to rely on data at their fingertips as they manage investment portfolios. Key performance indicators include:

- Execution of trade options to manage client wealth portfolios
- Profitability of fund and portfolio yields
- Data synchronization with field agents, brokers, and consultants

Key Statistic: Financial services agents can execute approximately 11.4 percent more trade options with wireless data services and achieve an average nominal improvement of 3.1 percent portfolio performance. Comparing two series of data on brokers with mobility against those without, the average volume of increased trade activity was marked, and the performance of the overall portfolio increased notably.

In Summary: Mobility Does Equal Productivity

The business logic of investing in mobility rests upon changing the productivity quotient. Workers that apply mobile data solutions to effectively increase their output do so because of scaled returns on labor utility. Time is essentially harnessed to produce greater business value with the result being more output over a shorter period. And the elimination of bottleneck constraints to information translates to improved execution on the part of employees. Empowered employees can then achieve higher value returns on the information technology investments already made by their organization.

Total factor productivity (TFP) is shaping up to be one of the landmark accomplishments of the economics discipline when it comes to measuring the real effects of technology. In the circle of mobility, TFP rates are steadily rising because of the positive correlation between mobile data and voice access becoming ingrained in the job requirements set forth by demands of the information economy. Looking ahead at labor economics of a future global society, we expect to see TFP continue to rise from 3.2 to 4.1 percent annually as information resource requirements transform the landscape. And as part of GDP estimates, this implies a widening in the respective size that electronic information and mobility shares in growing world economies (see *Figure 1*).

On the supply side of the equation, mobile service operators are poised to see the greatest boost in true average revenue per user (ARPU) as wireless data takes the stage in advance of pure voice networks. The sustained convergence of IP networks and mobile data will be the ultimate catalyst in bringing along the value of devices that can shed the four walls to roam freely with purpose. Inhibitors to spending

FIGURE 5

Total Factor Productivity Soars with Mobile Worker Technology

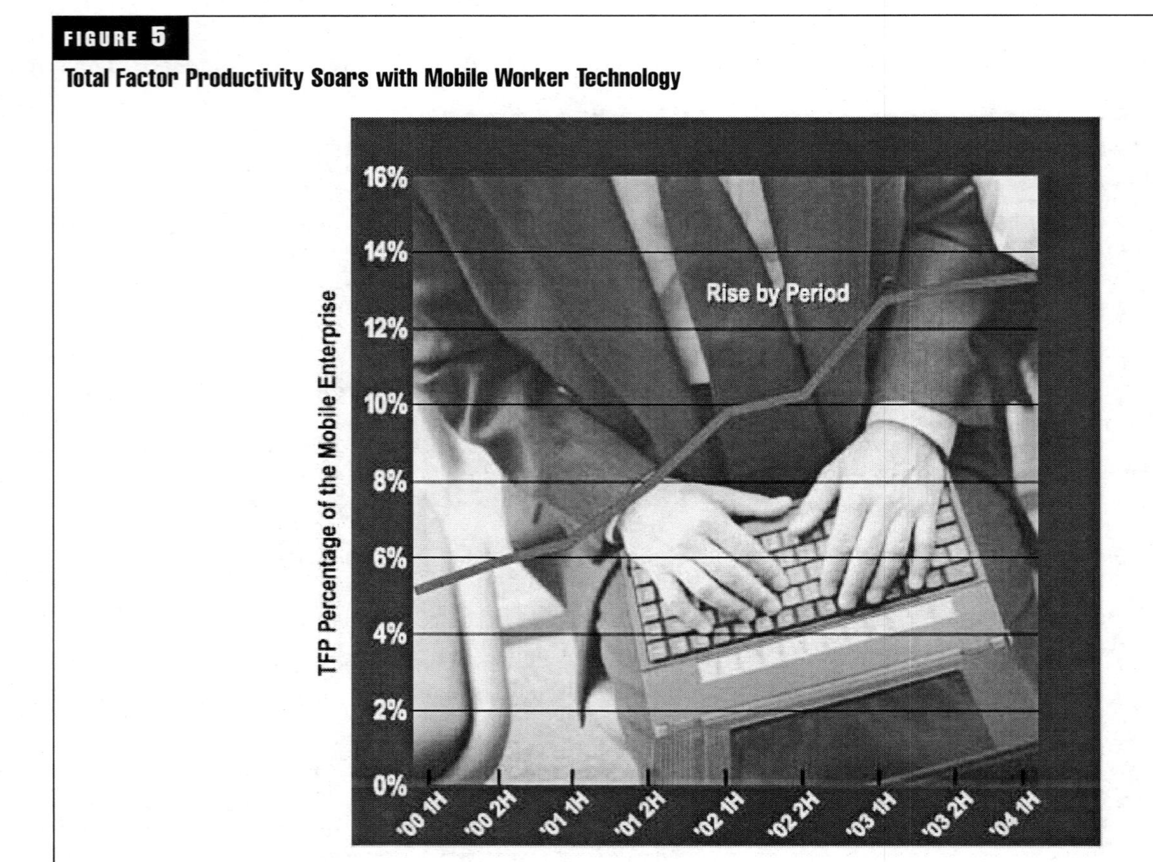

for premium service will eventually disappear, and operators may finally realize a market condition that supports inelastic pricing models. Higher degrees of inelasticity imply greater demand for the service and less focus on price determination at the consumer level.

As the data reflects today, we anticipate the persistence of a strong bias among consumers for mobile technologies and service offerings that complement not only their information-based lifestyles but also their conduit to business applications. A steady uptick in capital investment for wireless applications is proving that investor confidence remains high for where access demand is moving. And eventually this ties back to currency being well-spent on ensuring that mobile workers continue to reap the tangible benefits of unfettered productivity.

Methodology Endnote

Conclusions and various perspectives contained within this appraisal integrate the latest economic data related to users and organizations that apply mobility solutions around the globe—covering metropolitan statistical areas in the domestic United States, Europe, and Asia-Pacific. Our sample data populations were derived from sources such as the OMNI Network Economics Study, mobile operator networks, enterprise transactional databases, U.S. Bureau of Labor Statistics, U.S. Bureau of Economic Analysis, the U.S. Federal Reserve Bank, World Bank, and various university consortia studying human factor relationships with mobile technology.

We launched our modeling exercise with the compilation of this data concerning mobile data usage inside the framework of how organizations and individuals interface to specific economic patterns. The initial phase involved applying multivariate analyses for logistic regression and chi-square detection (CHAID) relationships. We expanded the scope of the modeling exercises to concentrate an understanding of the relationship between factors influencing the aspects of labor productivity and the economic output of firms within given metropolitan statistical area (MSA) populations. The resulting statistics and measurements were computed using the best available data and commonly applied stochastic methods.

How Small Can It Get?

Problems and Opportunities with Miniaturization and the Wireless Device

Michael S. Irizarry

Executive Vice President and Chief Technology Officer
U.S. Cellular

Abstract

Technology is advancing at a pace never before seen in the history of civilization. One of the main factors of this advancement is the continued miniaturization of integrated circuit (IC) technology. This is enabling the integration of different technologies at the chip level, which is, in turn, enabling the integration of different technologies at the board level and, more important, at the user level in terms of service and feature functionality. This research paper will explore the advancement of miniaturization and its impact on wireless device functionality. It will focus on the integration of camera phone technology with wireless device technology, specifically cellular phone technology. The impact of these technologies on individual privacy will also be discussed. Lastly, this research paper will explore how technology might solve some of the privacy problems it creates.

Introduction

Microelectronics is a trillion-dollar industry and the key enabler of service and feature integration in consumer electronics (Role, 2004). Many of today's electronic systems are discrete systems consisting of different functional modules on a common printed circuit board. This method of architecting an electronic system does not lend itself well to the bundling of different consumer functions or consumer applications. As an example, the discrete method is fine for the engineering of a dedicated camera or a cellular phone where each of these devices will only be used for the purpose of taking pictures or making phone calls. A common printed circuit board with integrated circuit technology mounted on the board does not pose a barrier to the development of a form factor that will appeal to the consumer. However, if the consumer would like to have a single device that can take pictures, place phone calls, play music, and synchronize e-mail, the discrete circuit technology will not enable the device manufacturer to develop a multifunction device with a form factor that will appeal to the consumer. The device will be big, bulky, and not fashionable.

Consumers are demanding smaller form factors in the various electronic devices they use (Grant, 2004). Whether cameras, cell phones, or MP3 players, consumers are expecting the devices to be small, packed with features, and have full portability and mobility (Grant). This demand is driving the development of advancements in the core technologies that are used to design and engineer integrated circuit technology. The result of these advancements over the past few years has been a phenomenal reduction in the size of integrated circuits and a phenomenal increase in the number of transistors that can be included on a single integrated circuit (Pinto, 2004). An increase in processing power and the capability to integrate more functionality onto a single integrated circuit wafer is a direct result of the increase in the number of transistors (Pinto). These means camera technology, MP3 capability, and e-mail synchronization technology can now be put into a single device and yet maintain a form factor and size that will appeal to the consumer fashion and usability demands. As an example of the progress in this area, today's cellular phones have storage capacity, memory capacity, and processing power that is greater than that of a PC from 1998 but in a form factor that is significantly smaller (Landon, 2004).

While the evolution of functional integration continues, the societal implications of this type of integration are only now being understood. This integration has major implications for the individual's notion of privacy, for example. The reason for these implications is because the multifunction devices can now be used in places and at times that were previously not practical because of the size of the devices. While integration of features and functionality such as cameras, MP3 audio, and e-mail synchronization are at the forefront of this integration trend, streaming video and personal location services are very close to full commercialization. This integration trend will continue, with the pace of integration accelerating as advances in integrated circuit miniaturization continues.

Wireless Service Integration and Microelectronics Integration

Cellular phones are now offering much more than just voice communication. Because of advances in technology, cellular phones can now offer access to a vast array of data and services over the Internet that was once only possible from the home computer (Landon, 2004). However, small screens, poor battery life, and inadequate keypads limit the practical benefits to users. Breakthroughs in silicon and process technologies provide the opportunity to minimize some of these constraints.

Handset manufacturers are combining technologies to offer features and functionalities to meet customer expectations. Today's handsets are built with the combination of separate wireless subcomponents on a common platform. A good example of this is Bluetooth functionality provided on one chip and GPS location technology on a different chip, but on the same platform within the cellular phone. While this system-level integration of different wireless capabilities makes sense for certain types of applications, it does not result in a solution that has the lowest cost or the smallest form-factor handsets. The ideal integration of multimode capabilities is at the component level for the radio frequency (RF) front end, the base band, the main processor, and the applications areas of the cellular phone platform (Fan, 2004). Innovation in wireless technology and electronics integration will continue to fuel further advances in these critical areas.

The complementary metallic oxide semiconductor (CMOS) process technology advances are showing no signs of slowing down, and it is one of the key technologies driving the integration of features and services in today's wireless devices (Marchini, 2004). CMOS is the technology of choice for digital electronics, mainly for its speed and low power consumption (Taur, 2002). Advances in semiconductor technologies are measured in the size of the channel width on the silicon substrate: the smaller the channel width, the more integration possible because the density with which transistors can be packed together is higher (Taur, 2002). In 1983, the channel width was measured at 2 microns, whereas today, the channel width is measured at 0.25 microns—truly a significant improvement (Taur, 2002). This improvement is, nevertheless, expected to continue. This is because the improvements are being made in the manufacturing process of CMOS wafers, the innovations in the configurations of the transistors are cut onto the silicon wafer, and, more important, the improvements are being made in the lithography process. Lithography is the process by which transistor patterns are delineated on the silicon wafer. Improvements in the size of the patterns that can be placed on the wafer in the lithography process come primarily through the light source used. Smaller wavelengths of light enable smaller patterns. Much research is under way to explore how light wavelength changes can improve the lithography process. However, there is concern that optical lithography will not be able to produce channel widths below 0.1 microns (Mazumdar, 2005).

While CMOS fabrication technology is a key driver of further miniaturization, and advances in this area are critical to further integration for wireless technologies, miniaturization and size improvements will need to occur in other areas too. For instance, battery technology will need to improve, not only to keep up with the increased power requirements of the highly complex integrated circuits being used in wireless devices, but also to help with the new, sleek form factors consumers are demanding from their wireless devices. Moreover, consumers want to have a full and rich user experience when using the wireless device to watch or listen to their multimedia content that is streaming to the wireless device. This will require major advances in device display technology and audio technology to make the most of the precious resources of the wireless device. Display requirements could grow faster than the advances in display technology, which could further burden form-factor real estate. With the increased integration of features and functionality, new methods of interfacing with the wireless device will need to be developed. The basic keyboard that has served as the primary interface for wireless devices will not be able to adequately address the type of navigation required of the user to get through the many and varied menus and feature selections on future wireless devices (Vesling, 2005). Today, current technologies enable the production of mobile wireless devices as small as 60 cubic centimeters (Martin, 2004). Breaking out some of the key components of the wireless device, it is easy to see that electronics, display, and battery consume a significant piece of form-factor real estate. For current wireless devices, electronics consumes 15 percent, battery consumes 25 percent, and display consumes 30 percent (Martin, 2004).

Types of Wireless Service Integration

There are different methods to achieve wireless service integration. Some of them are very simple, while others are very sophisticated. An example of the simple type is the bundling of services (wireline and wireless) on the subscriber's bill. Bundling is the most common form of integration (Duncan, 2001), in which a wireless carrier offers a suite of wireline and wireless services to a subscriber. This type of integration has been available for a number of years (Duncan, 2001), but does not have service integration on a common device or a common network. Transparent service integration is an elegant type of integration, which enables multiple types of services to be delivered to the end user across the same device and network. The main benefits to the customer are simplification through one-stop shopping and reduced cost, since the bundle is cheaper than ordering individual services. In addition, bundling of services provides the user with a single bill.

Fixed network mobility, where a landline network operator adds services that enable subscriber mobility using a fixed network, is another type of integration. A good example of this is the calling card and personal 800 services (Duncan, 2001). This is a solution for landline service providers that are restricted from or unable to offer mobile services in certain geographic areas. Network service providers can augment their offerings with features that are independent of access method, such as unified messaging (Duncan, 2001).

Network integration is most valued by customers who need to receive consistent services across different types of networks. This is not usually visible to the customer. It can be a source of cost reduction to the network provider, regardless of whether the customer receives any integrated services as a result. Network integration is typically based on using a common resource across many networks such as feature logic, which include intelligent network components, transport, and billing systems.

The final category of integration, called network integration, promises to simplify the user's experience by reducing the number of phone numbers, voice-mail boxes, and contact addresses. Current surveys indicate users have an average of five contact addresses, and early adopters have up to 10 contact addresses or phone numbers (Buckley, 2004). This makes it very difficult for the user to have a seamless service experience. The reason for this is that the underlying transport technologies are not integrated. When a user is in an area that has Wi-Fi or wireless local-area network (WLAN) service, the device and home network are not aware of this. Because of this lack of location information, the home network attempts to deliver services on a network that may or may not be the most efficient network to deliver the service relative to the Wi-Fi or WLAN.

Network integration can only be enabled by advances in the capabilities of both the underlying transport technologies and the end user device. The Internet protocol multimedia subsystem (IMS) is one such advancement of the network that will enable the delivery of integrated devices (Ellis, 2004). On the user device side, multimode integrated circuits capable of supporting different types of wireless and wireline access networks are critical enablers of the service integration discussed above. Both IMS and multimode integrated circuits are required for full service integration. IMS is in the very early stages of deployment (Ellis, 2004). Most network carriers will deploy this technology incrementally, with a key focus on the cost savings and revenue enhancements it may bring. Multimode chips are becoming more widespread, and this is expected to continue.

The Path to Miniaturization

In wireless devices such as cellular phones, there are several major subsystems required to have a fully functional phone (Keutzer, 2000). The first section is the radio frequency or analog section. This section of the device is basically the transceiver section, made up of a receiver and a transmitter. The analog section of the phone handles decoding or demodulating of the radio signal so the user can hear the information being sent. This also converts the subscriber's voice to digital so it can be impressed on a radio frequency signal for emission into space. This is later used for reception on the other end of the communication connection (Keutzer, 2000).

The second section of the phone is the digital section (Keutzer, 2000). This section of the phone provides the memory and storage, which is used to store subscriber information and application data, as well as for use by the device when processing the more complex telecommunication algorithms (Keutzer, 2000). The digital section is also responsible for processing keypad information and for sending and formatting information for the display of the mobile device. The digital section typically consists of several discrete integrated circuits. The third major section is the digital signal processor (DSP) and the main processor core of the wireless device. The processor core is responsible for coordinating and synchronizing the different sections of the wireless device (Keutzer, 2000). The processor core ensures all the subsystems are working in unison and not at cross-purposes. The DSP handles complex algorithms used to generate sounds, take pictures, render an image on the device display, and perform other important functions.

Each of these sections of the wireless device has evolved and will continue to do so. In the first implementations of cellular radio systems and wireless networks, the analog section used discrete devices for the low noise amplifiers, and the power amplifier was largely a hybrid module. The low-frequency functions were integrated as a single chip (Dulongpont, 1999). The above design reduced the number of devices drastically versus earlier radio implementations. Combining low-frequency functions into a single 5-V CMOS device was the next step in the progress toward more integration (Dulongpont). Integrating the phase-locked loop proved to be the most difficult function (Dulongpont), though integrating the filters, amplifiers, the analog-to-digital converter, and a few of the digital control functions was relatively easy. Nevertheless, manufacturers were eventually able to integrate the phase-locked loop. The phase-locked loop provides the basis of frequency generation for the entire wireless device.

The current focus of research and development activity is centered on a single-chip radio (Dulongpont, 1999). Merging the analog baseband chip and the RF circuit can take advantage of dedicated CMOS technologies. Analog function density or power consumption is not sacrificed in this solution. A second solution that uses a digital CMOS chip is more cost-effective but has performance issues. Another area of intense focus is consumption of power resources. To improve operational time and standby time with small batteries, IC power supplies are being reduced to 3V and lower for digital functions. This results in a much lower supply current with the same processing capacity. Despite all the progress being achieved in integrating functions, the power amplifier will be challenging. This is due to the high power requirements of the power amplifier.

The digital section—which performs many functions such as handling the display, keyboard, speech coding and decoding, channel coding and decoding, synchronization, demodulation, and ciphering—is a critical component of the wireless device. First-generation digital sections were comprised of several application-specific integrated circuits (ASICs) and several generic ICs. Current implementations of digital sections use a single CMOS ASIC and two standard ICs. This is primarily due to the improvements in silicon technology die sizes and improvements in power consumption. Static random-access memory and flash memories are used for the two ICs.

The brain of the wireless device, along with the section that pulls it all together, is the DSP and processor section. Up to this point, the DSPs used in wireless devices were based on the von Neumann architecture (Keutzer, 2000). Now, DSPs in wireless devices are based on the Harvard architecture. The difference between a Harvard machine and a von Neumann machine is how the processor accesses memory for data information and instructions. The Harvard processor has two separate data buses (Keutzer). One bus is used for accessing application data; the other bus is used to access instruction data. The von Neumann machine has only one data bus, which is used to access data information and instruction information (Keutzer). Although the Harvard architecture has more computational power than the von Neumann design, telecommunication systems require more computational horsepower than the basic Harvard machines can provide. Because of this, the new DSPs include multiple arithmetic logic units and dual instruction sets. This helps speed up the very complex telecommunication algorithms needed for today's wireless devices.

This is a brief walk through the evolutionary changes the subsystems that comprise a wireless device have undergone since the introduction of wireless services. There will no doubt be additional evolutionary changes as research continues in the critical areas of wireless device technology. It is important this evolutionary progress continues if wireless technology is to stay on the path to greater integration. One of the major drivers and enablers of continued evolution of microelectronics is Moore's law.

Moore's Law

ICs are made up of transistors packed onto an integrated circuit wafer. The first ICs, developed in the late 1960s and early 1970s, had about 2,200 transistors. Today's ICs, the most complex of which are processors, contain almost 1 billion transistors per chip area. This incredible increase in transistor density is the primary reason IC technology is getting smaller and smaller in terms of chip size to processing power. The fundamental reason for the increase in transistor density is described by what is called Moore's law (Chang, 2003).

Moore's law basically states that transistor density per integrated circuit area doubles every two years (Chang, 2003). The key driver behind Moore's law is the process used to manufacture the ICs. Lithography is the technique used to create the circuit mask on the circuit wafer. This core technology is improving in its ability to create transistor masks on the circuit wafer of incredibly small size. Smaller size masks mean more transistors can be packed onto the circuit wafer (Polyviou, 2003). This also drives down the cost to manufacture ICs because less wafer material is needed.

Will Moore's law continue? There are those who believe continued innovation in the manufacturing process will keep Moore's law alive and strong. There are others who believe the end is near and different means will need to be developed to keep the processing capabilities of ICs improving. The future of Moore's law rests with the improvements that can be made in the lithography process, the materials used to manufacture the chips, the way integrated circuits are powered, and the fundamental design of the ICs (Polyviou, 2003). To the extent improvements continue in the manufacturing process for ICs, and improvements are made in the other areas mentioned, Moore's law will continue to predict the transistor density improvements for IC technology.

Future Enablers of Integration and Miniaturization

Clearly, the primary driver of integration and miniaturization will be further advances in lithography technology. With current optical lithography technologies hitting the 0.1-micron barrier, new methods are expected to not only breach the 0.1-micron barrier, but also to take it to a new plane (Taur, 2002). Several of the new technologies being explored are x-ray, scattering with angular limitation in projection e-beam lithography (SCALPEL), ion beam projection, extreme ultraviolet, and electron beam direct write (Mazumdar, 2005). In addition to exploring new light sources, types of semiconductor materials are being explored that could yield further reductions in channel bandwidths (Daneshrad, 2004). Single-electron transistors are being discussed as a possible successor to metal-oxide semiconductor field-effect transistor (MOSFET) technology, which is the basic building block for CMOS technology (Pinto, 2004).

Improvements in other areas are necessary to drive further integration. The basic IC interconnect can potentially bottleneck high degrees of scaling, where resistance-capacitance delays begin to distort the waveforms moving in and out of the IC package (Pinto, 2004). This phenomenon can be so pronounced that its effects can exceed those caused by transistor gate transition delays (Pinto, 2004). Metals with lower permittivity can be used to mitigate this issue (Pinto, 2004). Today's circuit packages predominately use copper for the wire connections.

Some researchers are looking at other ways to get electrons to travel across materials. As mentioned earlier, today's IC technology uses tiny channels carved on the silicon substrate through which electrons pass, constituting an electric current. Nanotechnology may provide an alternative method for controlling electric currents by manipulating the atom, the most fundamental molecular building block. To date, researchers have been able to line up molecules of certain materials that have free electrons in the atomic bonds to create molecular structures called nanostructures, or nanowires, which conduct current when stimulated with an external source (Beebe, 2005). An electric field applied at the right point on the nanowire can control the flow of electrons in the same way the transistors in an IC can control electron flow across the silicon substrate. The difference is that the nanowires are molecular in size and could potentially be used to create extremely small IC devices.

Significant Integration Points in Wireless Devices

Within the wireless device, there are many areas that still have opportunity for size reductions. The display, battery, keyboard, and storage system are such areas. Each of these integration points will benefit from the improvements made in IC technologies and the advances discussed above. However, other areas of improvement will be required to keep the evolution curve going strong. As an example, miniaturization can help make mobile phone keyboards small, but are ultra small keyboards really what the end user requires to have the best experience while using the wireless device? The answer is no. The reason is that very small keyboards might be difficult to use for users with large fingers. So, additional techniques will be required to achieve both improved mobile device form factors and improved user experience.

User Interface: Input and Output

Mobile phones and wireless devices would not be of much use if the users could not input information and read information out of the device. Today's wireless devices include a keyboard with all the keys the standard home phone has and, in some cases, all the keys a home computer might have. This latter component is generally called a smart phone. Even when the wireless device has a keyboard with all the keys a home computer may have, text input is very tedious at best. User displays on the wireless device can be very difficult to read because of poor resolution and wash-out in bright light. Several technologies are being researched to address the current limitations of wireless device keyboard technology and display technology.

Foldable keyboards are available today as accessory attachments to certain mobile phones and mobile wireless devices (Shaw, 2004). Today's foldable keyboards are full-size QWERTY keyboards with a docking port for the mobile phone or wireless device. These foldable keyboards can be folded down to one-third of the full size, making them easily transportable (Shaw, 2004). Nevertheless, they are still fairly large when folded down relative to the wireless device they are designed to support and are an additional device for the user to carry around, perhaps offsetting the benefit the device offers the user.

Projection keyboards are a promising technology that basically uses infrared and laser light to project a full-size QWERTY keyboard onto any surface (Alpern, 2003). The user then simply types onto the projected keyboard like a normal keyboard. The user's fingers interfere with the projected light in accordance with the keys pressed. The projection system not only projects light onto the surface, but also has light sensors that constantly monitor the reflected light from the projection surface that has been interfered with by using complex algorithms to determine which keys the user has pressed (Alpern, 2003). The beauty of the system is that the keyboard is completely virtual, so the only thing the user needs to carry around is a small projection device. The projection device is much smaller than the folded keyboard systems (Alpern, 2003). The two issues with projection systems are wash-out in bright light, making it hard for the user to see the keys, and the accuracy of the algorithms used to read the reflected light. As with most technologies, these issues will be addressed, or at least mitigated, with technological advancements.

Voice recognition is a technology that can completely free the user from keyboard dependency for data entry. Voice recognition systems have been around for a decade (Kirriemuir, 2003), but it is only in the past few years that recognition accuracy has achieved such a level that systems using voice recognition for input have gain widespread acceptance. Earlier systems were fraught with recognition accuracy below 80 percent, thus making it very difficult and frustrating for users to input information into the system. Voice recognition systems are now used to book airline reservations, call into customer care centers to get automated billing information, and purchase movie tickets. Wireless carriers have begun to deploy mobile-based and network-based voice recognition systems (Meisel, 2005). Network-based systems require no technology changes to the wireless devices. All the recognition technology is hosted within the carrier's mobile switching centers. Network-based systems have the advantage of not requiring any software or hardware on the mobile device. The significance of this is that users of older and newer phones can use voice recognition technology. The drawback to network-based solutions is that they use more network resources than phone-based recognition systems.

Phone- or device-based voice recognition systems depend on the mobile device to host the hardware and software that provides the voice recognition capability. Many of the newer-model phones include voice recognition as a standard feature. The level of accuracy in phone-based and network-based voice recognition is about the same. The mobile-based technology has the advantage of not requiring any changes to the wireless network. The disadvantage is that the host phone requires more memory and processor horsepower to handle the voice recognition function. This adds to the cost of the mobile device.

Both the network-based and mobile phone–based voice recognition systems provide good recognition accuracy for dialing numbers and other phonetically simple commands. Only in the past few years have phone- and network-based systems been rolled out to convert full voice to text to completely remove dependency on the keyboard (Kirriemuir, 2003). The issue is still one of accuracy. Voice recognition accuracy is still below 100 percent. Moreover, accuracy depends on the culture and geographic region. Special tuning is required to optimize the accuracy of recognition (Kirriemuir, 2003). Voice recognition accuracy will no doubt improve as the underlying recognition software improves.

For simple text and dialing functionality, display requirements for wireless devices are trivial (Levola, 2002). With the advancement of the capabilities of wireless networks and their ability to transport multimedia content, the display requirements for wireless devices have increased in complexity (Levola, 2002). The displays on wireless devices need to handle simple text, pictures, streaming video, and bundled content such as video or pictures combined with floating text. The liquid crystal display (LCD) used in the older model phones do not provide the content resolution, display responsiveness, and color richness needed to handle today's rich media content. One of the biggest challenges is fitting the content on the display of the device without compromising the quality of the content or increasing the size of the wireless device in terms of form factor. Several technologies are being explored to address these conflicting demands.

To meet the often conflicting demands of display quality and small size, display engineers are looking to organic light-emitting diodes (OLED) and organic transistors on plastic substrates to address the conflicting demands of size and quality (Keenan, 2002). The LCD, on which current display technologies are based, is described as "two thin plates of glass with a special liquid crystal solution between them. This liquid crystal solution has an interesting property. If there is no current going through the liquid crystal solution, the solution is clear. But when there is an electrical current going through the solution, the solution crystallizes and goes from clear to dark. The solution acts like a shutter when current is passed through it, changing from clear to dark. So, for instance, in the display of a calculator, by changing certain sections from clear to dark, the calculator can form different numbers" (Prescott, 2003). LCD displays are not self-illuminating. This means LCD displays require an external source of illumination, such as a backlight. This requirement means the mobile device battery has an additional load on it, decreasing phone talk time. Moreover, the illuminating source adds to the size of the display module, making it more difficult to reduce the overall size of the mobile phone.

OLEDs are self-illuminating, requiring no external illuminating source (Keenan, 2002). This reduces power consumption and enables the display to have a smaller form factor. OLEDs emit their own color, typically red, blue, or green. OLEDs are mixed to synthesize a range of colors, with very good contrast ratios. OLEDs also offer better power efficiency, wider viewing angles, and faster response times than LCDs (Keenan, 2002). All of these are advantages OLEDs have over LCDs, making them the choice for use in wireless mobile devices. Although at the present time the cost of OLEDs versus LCDs is still higher, this expected to change as OLEDs move up the manufacturing learning curve.

Storage and Memory

As more and more applications are developed for wireless mobile devices, the storage and memory requirements for wireless mobiles devices increase. The storage and memory requirements are increasing for the application code, as well as the data the application code uses to deliver the service to the user of the wireless device. Each one has different requirements. The application code requires a fast response for easy access by the application processor. This is important to ensure the user does not experience inordinately long response time when using the application. For the application data, especially with the capability on mobile phones to view videos and listen to music, storage for application data needs to be large and inexpensive. With good quality, storage would require about three and five megabytes in compressed audio for a typical song and 500 and 800 megabytes in compressed video, and thus future mobile wireless devices will require a lot of storage (Yokotsuka, 2004).

The storage and memory technology will also need to have a small cell size to ensure a high storage density to fit into the shrinking form factors of new wireless devices. Typically, cell-phone manufacturers now use static random access memory (SRAM) for data backup and pseudo-static random access memory (PSRAM) for the working area of the system, and North Royalton (NOR) flash for the bootable code storage on basic phone programs (Yokotsuka, 2004). Not and (NAND) flash memory is typically used for application software and the storage of huge data such as pictures and music. This is changing with improvements in NAND flash technology, however.

NAND flash technology offers higher storage density and therefore a lower cost than NOR memory (Yokotsuka, 2004). As a result, it is now being used for code and data storage. Coupling NAND technology with synchronized dynamic random access memory (SDRAM) ensures fast access times, which has been the primary advantage of NOR technology. The combination of NAND and SDRAM technology not only offers fast access times, but also lowers the storage cost per bit. NAND technology is becoming a pivotal storage technology in the ongoing evolution of wireless mobile devices (Yokotsuka, 2004). Removable memory cards such as secure digital cards, compact flash cards, and memory sticks, which are all based on NAND flash technology, are providing large removable storage devices. In fact, 1-gigabyte removable cards are quite common. Many cellular phones now come with a slot for one of these types of cards. Continued improvements in storage density and cost are expected with improvements in the manufacturing process used in the creation of NAND flash technology. Even with the improvements in NAND flash technology, additional storage technologies will be required to keep pace with the storage requirements of wireless mobile phones. One such device is the micro-hard drive or high-density hard drive.

Micro-hard drives have been used for digital cameras and camcorders for several years. Because of price reductions, micro-hard drive technology is making its way to wireless mobile devices. Hard-drive technology offers significantly lower storage costs per bit than semiconductor storage mechanisms. The cost advantages can be as much as one-tenth that of semiconductor memory. Disadvantages with hard disk storage for mobile devices have been power consumption and size, as well as their sensitivity to shock, such as that experienced when the mobile device is dropped on the floor (Schendel, 2004). Power and size are premiums on the mobile device platform, especially with ever-shrinking form factors.

Advances in the core technologies used to develop micro-hard drives have brought the devices within reach of mobile technology, especially along the dimensions of size and power consumption. Using special materials and advanced material deposition techniques, more tracks and higher storage densities are being achieved with today's micro-hard drives. One leader in developing and evolving this technology is Hitachi, which has been able to create a three-atom-thick layer of special material and sandwich it between magnetic materials, creating a hard drive with 4 gigabytes of storage capacity (Schendel, 2004). To reduce power consumption, micro-drive manufacturers are combining micro-drive technology with NAND storage technology. Data is written to the NAND storage or buffer before spinning up the hard drive. When the NAND buffer is full, then the data is written to the hard drive. Since the hard drive is not spinning all the time, power requirements are less (Schendel, 2004). Nevertheless, improvements in the power efficiency of hard-disk-drive technology and other core wireless technologies will not be enough. Research is focused on improving the power density of battery technologies used in wireless devices.

Battery

Some argue that the greatest limitation or constraint to further integration of mobile phones is the battery (Ohr, 2004). The power consumption loads placed on the battery of mobile phones have increased dramatically as new services and features have been integrated onto the mobile wireless device (Paulson, 2003). While advances in the processing capacity and storage capacity of the devices that support wireless technologies has continued, the power-sourcing capacity of batteries has not advanced at the same rate (Paulson, 2003). In fact, battery technology has effectively doubled in the past eight years (Ohr, 2004). This does not even come close to the progress made in IC technology, which has effectively doubled every 18 months. Even though battery technology has slightly improved, the size reductions of mobile phones have forced the size of the batteries to be smaller. The slow rate of improvement could be due to the incremental approach taken to enhance battery energy density which is, however, changing. Research focus is now being put on radically new types of battery technologies.

Batteries have evolved from those based on lead acid technology, then to nickel cadmium technology, and finally to lithium ion technology. Improvements to existing battery technology such as lithium ion technology are focused on finding new materials to use in the battery to increase the energy density. One such new material is organic polymer. Lithium ion batteries based on organic polymer technology offers higher energy density, greater flexibility in form factor design, and lighter weight (Buchmann, 2005). Moreover, lithium batteries that use polymers are safer because the electrolyte is a gel and inherently more stable than the standard electrolyte (Buchmann, 2005). Nevertheless, these incremental improvements are only yielding a 10 percent per year improvement in energy density (Paulson, 2003). Many believe this is not enough to keep up with the power-consumption needs of mobile wireless devices. As a result, research is focused on batteries based on fuel-cell technology.

Fuel-cell technology, which can increase energy density by a factor of 10, falls into a category of power technologies known as air-breathing technologies (Biancomano, 2001). Fuel cells work by passing a gas, such as hydrogen or methane, into a channel, where the gas reacts with a catalyst material. The reaction between the catalyst and the gas causes the gas to give up electrons. The electrons flow through the electric circuit or load that was designed to work with the fuel cell. As the electrons follow the path to complete the circuit path, they react with oxygen that is being drawn into a channel on the other side of the fuel cell. In addition to reacting with the oxygen, the electrons react with the hydrogen protons—what is usually left from the hydrogen gas that has given up its electrons—forming simple water vapor and heat.

One of the advantages of fuel cells is a very high energy density per volume, making them very attractive for mobile wireless devices. For the most part, fuels generate safe byproducts through simple water vapor and heat and make them ecologically safe. Despite the advantages, there are a number of issues that need to be addressed before fuel cells gain wide spread acceptance for use in mobile wireless devices. One of the main issues is the need for a fuel source, such as hydrogen gas or methane gas. A distribution system needs to be put in place for consumers to purchase fuel cartridge replacements or refills. This is similar to butane refills used for lighters. Since the fuel is a gas, special handling precautions need to be taken. Currently, gases are not legally transportable via commercial airplanes. A number of manufacturers are working to address these issues via regulatory changes and technological changes.

The Impact of Wireless Technology on Privacy

Wireless devices, specifically mobile wireless devices, have capabilities that enable people to stay connected in real time. Whether the communication is simple voice communication, e-mail, text messaging, video messaging, picture messaging, or some combination, the capabilities of the wireless device that enable people to communicate while mobile are also the reasons security and privacy protections need to be put in place. These privacy and protection measures will be of an unprecedented scale. The reasons for this are simple. Mobile wireless devices are getting smaller, making them inconspicuous. Wireless communication that goes beyond simple voice communication is inherently more difficult to secure. Lastly, the penetration of wireless mobile devices in society is increasing at a phenomenal rate. The current penetration in the United States is 65 percent (Landon, 2004). This is expected to rise to 85 percent by the year 2010 (Landon). This rise in penetration increases the number of potential wireless connections and, thus, the number of security and privacy concerns. There is an obvious tension between the use of mobile wireless technology and the need for privacy. Mechanisms designed to address security and privacy concerns in the mobile wireless ecosystem must be able to effectively balance the two. No other technology brings this issue to the forefront more than camera phone technology.

Camera Phone Technology

Most people take pictures during special occasions, mostly due to the inconvenience of carrying a camera (Dunphy, 2003). Camera phones have radically altered this behavior because they facilitate ubiquitous picture taking, anywhere and anytime. Moreover, camera phones enable instantaneous sharing of pictures by using multimedia-messaging services offered by most wireless service providers. In some countries, 70 percent of wireless subscribers have phones with an integrated camera (Kindberg, 2004). With this level of camera phone penetration, and with the very small form factors in use by wireless devices today, there is a high probability of privacy breaches.

All individuals—whether in public or in private—have the right to privacy. However, the definition of privacy varies from state to state in the United States, which truly complicates matters. For example, in the state of Washington, it is the physical location of the individual rather than the part of the person that determines if privacy has been breached (ABC News, 2004). Further, privacy is based on values, interests, and power, all of which have always confounded society (Westin, 1967). This, coupled with the conceivability of fully integrated wireless devices, complicates not only determining privacy breaches, but also regulating the use of camera phones.

Camera phone technology is fairly new. It has only been commercially available for about four years, but the penetration of camera phones is increasing very quickly. In fact, according to one report, more than half of cellular phones will have integrated camera phone capability by 2007 (Edwards, 2003). The current focus of privacy research, as it relates to camera phones, centers on voyeurism and identity theft. Identity theft is the unauthorized appropriation of an individual's Social Security number, credit card number, or bank account information, all of which is personal and private. Voyeurism is the act of watching unsuspecting individuals who may be naked or in the process of disrobing. They are usually strangers. The mobile cellular phone with camera phone capability is a powerful tool for people trying to perpetrate the acts of identity theft and voyeurism, which has led to the intense interest by the various regulatory organizations to come up with effective yet measured ways of protecting the privacy rights of individuals.

To protect the privacy of individuals from abuse by other individuals through the use of camera phones, societies can use legal, regulatory, and technical methods, voluntary bans, or a combination of all three. In regards to camera phones, what is happening is the institution of voluntary bands, followed by adoption of legal and regulatory polices and laws. This is occurring while technical methods are developed to strategically protect the privacy of individuals in a more precise way than is possible with a generalized ban or law. The Sports Club/LA in Washington, as a specific example, has instituted a policy that forbids the use of cell phones of any type in the locker rooms (Searing, 2003). Many other gyms have instituted similar bans due to feedback from patrons and concerns of lawsuits. Gym owners fear that patrons, upon seeing that pictures of themselves changing in the locker room have been posted on the Internet, might sue because the owners failed to take appropriate steps to protect their privacy (Searing, 2003). Bans of this type are a good thing, as they exhibit and reinforce social norms of privacy.

The legal and regulatory methods for addressing individual privacy has its best example in the Video Voyeurism Prevention Act of 2003, also known as S. 1301, signed into law by the president of the United States in 2004 (Mark, 2004). While the law only applies to federal land, it is a federal crime to photograph any part of a person's body or undergarments without their consent; this will undoubtedly serve as a model that states can follow to develop their own privacy laws at both the state and local level (Mark, 2004). At the core of this law, and presumably other laws of its type, is the basic premise that there are places where a reasonable person would believe that he or she could disrobe in privacy (Mark, 2004). This is the same premise that has driven the private sector to apply bans, such as those being implemented at gyms and spas, which forbid the use of recording devices in the locker rooms, specifically where a reasonable person would assume some modicum of privacy. More of these laws, at the state and city level, are expected. In fact, more bans similar to those implemented at gyms have been already put in place, such as at select schools and universities, to protect individual privacy.

Technical solutions promise to enable the negotiation of specific terms of privacy between individuals in real time. This is a much more flexible and powerful means for handling privacy, especially since each individual has a unique and specific definition of privacy that can only be served by flexibility. Bans and laws tend to only address the lowest common denominator. Technical solutions can enable the individual to regulate their personal privacy through privacy profiles. Nevertheless, even technical solutions can be coarse or fine in their approach to protect the privacy of the individual. As an example of a coarse type, there is a system called Safe Haven, which broadcasts a signal within a defined area that would disable all the camera phones in the area (Kotadia, 2003). A small transmitter is placed in the center of an area that is defined as a privacy zone. Cell phones must be equipped with the Safe Haven receiving technology (Kotadia, 2003). The Safe Haven receiver in the cell phone hears the signal transmitted from the Safe Haven transmitter unit and disables the imaging/recording functions in the phone until the phone is out of range of the Safe Haven transmitter. Once out of range of the signal transmitted from the Safe Haven transmitter, the imaging and recording functions in the cellular phone are enabled (Kotadia, 2003). While the solution is very effective at disabling camera phone functionality, it does so for all phones in the privacy area, even though there could be many individuals whose definition of privacy is such that they have no problem with a picture being taken of them. A more refined technical approach would enable the owner or user of the camera phone or image recording device to arrange the terms of privacy, in real time, on an individual basis.

A technical system that enables a real-time and negotiated approach to privacy requires an agreed protocol for the negotiation process. For instance, a technical system could be based on two principles (Felton, 2004). The first might be unanimous consent, where no event is recorded without the consent of all persons present. The second could be confidentiality of policy. To the fullest extent possible, a person's decision to grant or withhold consent will not be revealed to anyone else. This type of system does require all participants to carry a device that runs the same privacy protocol.

To execute the protocol, the devices must be able to discover other compatible devices in the area and communicate via short messages. Bluetooth is an example of a wireless access system that can be used for the exchange of messages. A single device will start the process by executing a discovery process to find other compatible devices. It basically does this by broadcasting a presence message, letting other devices know it is available to exchange information. Once the other devices in the area respond to the initial broadcast message, they exchange additional information about themselves and become privacy stakeholders. Information about the discovered devices and the information to be exchanged are encrypted to protect from eavesdropping (Felton, 2004). Each privacy stakeholder creates a privacy profile that is used by the privacy protocol to allow or disallow the recording of certain information.

While the above system enables privacy negotiation for those devices that are part of the stakeholder group, it does nothing to address those devices that do not support the technology but have recording capability, including camera phones. To fully address this issue, regulations are needed to push device manufacturers to include the privacy technology in their devices. This should not be an issue, however, because there is growing public anxiety over privacy and wireless devices such as camera phones. As device penetration increases, this will become less and less of an issue. Moreover, typical users replace their phone every 18 months, which will ensure over time that all wireless users have a device that supports the privacy protocol.

Ethics and Wireless Technology

As mobile phone penetration increases, clashes between groups who have different views of privacy, ethics, and etiquette in a wireless context will become more frequent and more intense.

The mobile phone brings about these clashes since it is a communicative instrument that brings together the opposing concepts of intimacy and privacy in the individual sphere and the sharing of norms and procedures in the public sphere (Katz, 2002). Nevertheless, at one time, phones were considered private. Even the 1967 Katz decision that prevented the FBI from listening in on a bookie's conversation in a phone booth reaffirmed the private nature of a phone conversation (Katz, 2003). New technologies such as camera phones have made the old laws based on phones with wires obsolete, forcing a battle over the definition of privacy rights. Studies show that cordless and cellular phones do not have the traditional expectation of privacy because they operate like radio stations sending signals by waves, which can be picked up with scanners (Scott, 1995). As more people use wireless devices, however, they will want and expect the privacy they were accustomed to in a pre-mobile phone age. The failure of government and society to help facilitate an environment that addresses the individual's notion of privacy will leave some disenfranchised. In sum, technology, coupled with the proper regulatory framework, can help address some of the privacy problems it creates.

The proper use of camera phones is prompting some handset manufacturers to come out with a code of ethics, which will serve as guidelines for the proper use and etiquette of the devices. Ethical guidelines, while softer than the laws and technology that govern privacy, can be very powerful since it is concerned with how a person, persons, or a society should behave (Joseph, 2002). Ethics deduces from deep values and beliefs held by both the individual and society. These deep-rooted values, which serve as powerful drivers of action, can ultimately drive the adoption of laws and technology that ultimately enforces and ensures privacy for all individuals. Ethics confronts the individual with his or her responsibility toward others (Joseph, 2002). In the case of the camera phone, many of the values underpinning privacy in a non–camera phone context transfer seamlessly to the camera phone context.

Several industry associations have instituted code of conduct guidelines for the proper use of camera phones. As an example, the Consumer Electronics Association (CEA) recently posted its camera phone rules of conduct. There are seven guidelines for camera phone use that the CEA has published. Reiter (2004) defines them as follows:

> Camera phones should not be used where photographic equipment is typically banned, for example: museums, movie theaters, and live performances, users should look for signs in public places that indicate whether photographic equipment is banned; Camera phones should not be used in public areas considered *private* by those who use them, for example: bathrooms, changing rooms, and gym locker rooms; Camera phones should not be used without authorization to record and/or transfer confidential information, whether in the corporate, government, or educational environment; Camera phones should not be used to take photos of individuals without their knowledge and consent; Discretion is advised to take photos of individuals under the age of 18; Safety is paramount when operating a motor vehicle, users should refrain from using the camera phone or video function of a wireless phone when driving; Camera phone users should always respect the privacy of others, when and where they have a reasonable expectation of privacy.

The aforementioned guidelines are based on common sense and norms that are already prevalent in society. They also flow from a perspective based on simple respect for the rights of others. Moreover, these guidelines require the individual to exhibit a certain amount of self-restraint. This manifests itself in the form of not doing what one may have the power to do, not doing what one may have the right to do, and not doing what one may want to do. It can all be summed up in the golden rule of do unto others as you would have them do unto you. Nevertheless, these guidelines will be tested as more people use these small multifunction devices in ways never before thought possible.

Conclusion

This research paper has explored the key technological advancements that have enabled the massive integration of new services, features, and functionality into the once humble cellular phone. Moore's law is the critical driver for many of the integration points of the mobile wireless device. While there is ongoing debate as to whether this law will continue, there will no doubt be improvements in the manufacturing process of the IC, the most fundamental building block of the mobile wireless device. Most of the improvements will likely come from improvements in the lithographical processes used to etch the tiny transistors on the semiconductor substrate. It is these improvements that will continue to ensure further advancements in the integration of new services, features, and functionality in the mobile wireless device.

The user interface (keyboard and display), internal and external storage, and radio components are critical building blocks for the mobile wireless device. The lowly battery is now a key component of the mobile wireless device. With all the new features, services, and functionality being integrated into the mobile phone, attention is now shifting to ensuring that there is adequate power to feed it. This is very important because all the new services and features are pretty much useless if there is no energy to power the device. Greater usage by the mobile wireless subscriber, coupled with all the new features, is putting a strain on today's battery technology. Some say this is the real roadblock to further device integration. Current state-of-the-art battery technology is chemical-based, using lithium-ion as its engine. To keep the battery from being a choke point, research is focused in two areas: squeezing more out of Lithium-ion technology and the so-called air-breathing technologies, also called fuel cells. While fuel cells offer huge energy gains over today's battery technology, they are still several years away from commercialization. This is mainly due to safety concerns, which may be addressed through advances in technology and new regulations.

The mobile wireless device has brought significant utility to society. Communication that was once only thought possible in science-fiction novels and TV series such as *Star Trek* is now routinely conducted in everyday situations. While it is still not possible to teleport individuals from one location to the next, it is very much possible for an individual to place a call, anywhere and anytime, for a simple ride from one point to another point. This was previously not possible until the widespread adoption of the mobile phone. Moreover, the simple act of sharing one's environment in real time, wherever that might be, with someone across the world, was simply not possible. The mobile wireless device enables this kind of powerful yet intimate communication.

However, this power and intimacy does not come without significant risk.

The rapid integration of features and functionality and continued advances in IC miniaturization have made the cellular phone incredibly small. This makes for a device that is stealthy and highly concealable. While these can be very valuable attributes for such a device to have, they also make for a device that can be abused. The very small form factor of devices with recording capabilities, such as camera phones coupled with video recording and voice recording functionality, would result in a device that can easily encroach on the privacy rights of others.

This research paper has explored the various options that can be used to protect the privacy rights of others in a camera phone world and, at the same time, balance the tremendous utility brought upon society. As is common with the introduction of any new technology, opposing views are discussed through public dialogue and individuals take steps on their own to solve the problem of, in this case, privacy protection. For example, gyms have taken steps to protect the privacy of their patrons by banning the use of camera phones in locker rooms. The same is occurring at universities and federal and government buildings. The government, feeling pressure from its citizens to protect their rights, institutes laws that make it a crime to violate the privacy of others through the use of a camera phone. This has occurred with the introduction of the Video Voyeurism Prevention Act of 2003. This law was introduced to address abuses occurring with the use of camera phones. Finally, technology catches up and offers more precise ways of protecting the privacy rights of others, which may differ in definition from one individual to the next.

The above progression from voluntary bans, to legal regulation, to technology solution, inevitably flows from the deep values that fuel the ethics of the society in question. At the core of this is the simple, common-sense perspective that each individual in society should be treated with respect and dignity. In a nutshell, the practice of the golden rule is the most powerful drive in the continual journey to protect the right to individual privacy and propel further improvement in the methods used to ensure privacy in practice. Wireless phone technology, with its incredible technological advances, will no doubt push the frontier of what is right and ethical, challenging the ethics, laws, and technology of society.

Bibliography

ABC News. (2004, October). *Skirting the law? Court says it's legal to videotape up woman's dresses*. Retrieved August 15, 2005, from abcnews.go.com/sections/GMA.

Alpern, M. (2003, May). *Projection keyboards*. Retrieved July 31, 2005, from www.alpern.org/weblog/stories/2003/01/09/projectionKeyboards.html.

Beebe, T. (2005, March). *Researchers break down barriers to miniaturization and revolutionize modern electronics*. Retrieved July 23, 2005, from www.azonano.com/news.asp?newsID=567.

Biancomano, V. (2001, September). *Fuel cells breathe new life into wireless*. Retrieved August 13, 2005, from www.eetimes.com/op/showArticle.jhtml?articleID=18305806.

Buchmann, I. (2005, April). *Is lithium-ion the ideal battery?* Retrieved August 7, 2005, from www.batteryuniversity.com/partone-5.htm.

Buckley, S. (2004, March). *Wireless/wireline integration: breaking the mode; what to do with multiple networks and a briefcase-full of devices?* Retrieved July 21, 2005, from www.findarticles.com/p/articles/mi_m0NUH/is_3_38/ai_n6081638#continue.

Chang, L. (2003, July). *Moore's law lives on*. Retrieved July 28, 2005, from www.eecs.berkeley.edu/%7Ejbokor/Full_text_pubs/1-198.pdf.

Daneshrad, B. (2004, September). *Integrated circuit technologies for wireless communications*. Retrieved July 10, 2005, from newport.eecs.uci.edu/~aeltawil/downloads/book.pdf.

Dulongpont, J. (1999, May). *Microelectronics in mobile communications: a key enabler*. Retrieved July 23, 2005, from ieeexplore.ieee.org/iel5/40/17341/00798110.pdf.

Duncan, R. (2001, May). *Wireless/wireline integration: simplicity and value-industry trend or event*. Retrieved July 24, 2005, from www.findarticles.com/p/articles/mi_m0TLC/is_5_35/ai_74940626.

Dunphy, J. (2003). The emergence of camera phones-exploratory study on ethical and legal issues. *Communications of the International Information Management Association, 3*(2), 23–25.

Edwards, L. (2003, November). *Camera phones: The next big thing*. Retrieved August 15, 2005, from www.sandiego.com/sdbusiness.jsp?id=188&searchText=camera%20phones.

Ellis, L. (2005, May). *Meet the IP multimedia subsystem*. Retrieved September 5, 2005, from www.multichannel.com/article/CA607831.html?display=Broadband+Week.

Fan, R. (2004, June). *CMOS enables increased RF integration in multimode mobiles*. Retrieved June 18, 2005, from www.berkanawireless.com/news/commsdesign.pdf.

Felton, E. (2004, October 28). Privacy management for portable recording devices. *ACM, 12*(3), 10–12.

Grant, D. (2003, June). *Playback apps push cell phones to edge*. Retrieved September 5, 2005, from www.commsdesign.com/design_corner/showArticle.jhtml?articleID=12802673.

Joseph, A. (2002, August). *Making ethical decisions*. Retrieved September 5, 2005, from www.josephsoninstitute.org/MED/MED-intro+toc.htm.

Katz, J. (2002). *Perpetual contact: mobile communication, private talk, public performance*. New York, New York: Cambridge University Press.

Keenan, R. (2002, January). *Advanced displays bring mobiles life*. Retrieved August 1, 2005, from www.commsdesign.com.

Keutzer, K. (2000, August). *Digital signal processors: applications and architectures*. Retrieved July 25, 2005, from bwrc.eecs.berkeley.edu/Classes/CS252/Notes/Lec09-DSP.pdf.

Kindberg, T. (2004, November). How and why people use camera phones. *ACM, 4*(2), 23–25.

Kirriemuir, J. (2003, June). *Speech recognition technologies*. Retrieved July 31, 2005, from www.jisc.ac.uk/uploaded_documents/tsw_03-03.pdf.

Kotadia, M. (2003, September). *Jamming device aims at camera phones*. Retrieved August 29, 2005, from news.com.com/2100-1009-5074852.html.

Landon, B. (2004, July). *Mobile phones could make your PC obsolete*. Retrieved September 5, 2005, from www.pdatoday.com/pdaviews_more/1647_0_4_0_C.

Levola, T. (2002, January). *Display technologies for portable communication devices*. Retrieved July 31, 2005, from www.nokia.com/downloads/aboutnokia/research/library/user_interfaces/UI1.pdf.

Marchini, L. (2004, January). *CMOS process*. Retrieved September 5, 2005, from www.electronicstalk.com/guides/cmos-process.html.

Mark, R. (2004, September). *Lawmakers ok video voyeurism privacy bill*. Retrieved August 28, 2005, from www.internetnews.com/bus-news/article.php/3411121.

Martin, N. (2004). High-density packaging for mobile terminals. *IEEE transactions on advanced packaging, 27*(3), 3–4.

Mazumdar, P. (2005, April). *Semiconductor lithography*. Retrieved July 10, 2005, from www.atp.nist.gov/focus/98wp-sl.htm.

Meisel, B. (2005, May). *Speech recognition technology and its varied market segments*. Retrieved July 31, 2005, from www.tmaa.com/Meisel%20commentary%20on%20market%20segments.pdf.

Ohr, S. (2004, February). *Nokia exec: new cell phone features need better batteries.* Retrieved August 7, 2005, from www.eet.com/showArticle.jhtml?articleID=18310970.

Paulson, L. (2003, November). *Will fuel cells replace batteries in mobile devices.* Retrieved August 7, 2005, from www.neahpower.com/news/newsfiles/IEEEComputer.pdf.

Pinto, M. (2004, February). *The transistor's discovery and what's ahead.* Retrieved July 27, 2005, from www.imec.be/essderc/papers-97/322.pdf.

Polyviou, S. (2003, October). *The future of Moore's law.* Retrieved July 23, 2005, from www.iis.ee.ic.ac.uk/~frank/surp98/report/sp24/index.html#1.

Prescott, R. (2003, August). *How LCDs work.* Retrieved August 1, 2005, from archive.chipcenter.com/circuitcellar/june01/c0601rr1.htm.

Reiter, A. (2004, October). *CEA posts "Camera Phone Code of Conduct."* Retrieved September 5, 2005, from www.wirelessmoment.com/2004/11/ces_posts_camer.html.

Role, T. (2004, January). *The role of packaging and its market.* Retrieved June 18, 2005, from www.erc-assoc.org/vision_value_added/GaTech%20PRC%2004.pdf.

Schendel, J. (2004, April). *Hard disk drives to be used in mobile phones.* Retrieved August 6, 2005, from www.mobilemag.com/content/100.

Scott, G. (1995). *Mind your own business: the battle for personal privacy.* Boston, MA: Insight Books.

Searing, L. (2003, September). *Some gyms ban camera phones.* Retrieved August 28, 2005, from www.detnews.com/2003/technology/0309/24/technology-279651.htm.

Shaw, K. (2004, April). *Adesso launches foldable illuminated keyboards.* Retrieved July 31, 2005, from www.networkworld.com/weblogs/cool/2004/006813.html.

Taur, Y. (2002, September). *CMOS design near the limit of scaling.* Retrieved September 5, 2005, from researchweb.watson.ibm.com/journal/rd/462/taur.html.

Vesling, R. (2005, February). *New technology makes feature-rich cell phones and media players easy to use.* Retrieved July 10, 2005, from www.interlink-elec.com/media/trade/pdfs/micronavfamilypr.pdf.

Westin, A. (1967). *Privacy and freedom.* Atheneum, New York.

Yokotsuka, M. (2004, April). *Memory motivates cell-phone growth.* Retrieved August 5, 2005, from wsdmag.com/Articles/Index.cfm?ArticleID=7916&pg=2.

Improving the Performance of Mobile Ad Hoc Networks Using Directional Antennas

Hetal Jasani

Assistant Professor, School of Technology
Michigan Technological University

Kang Yen, Ph.D., P.E.

Professor and Chairperson, Department of Electrical and Computer Engineering,
Florida International University

Abstract

A novel preventive link maintenance scheme based on directional antennas has been proposed to extend the life of the link by using the ability to orientate radio signals into the desired directions for multihop infrastructureless networks such as mobile ad hoc networks (MANETs). We investigate this preventive link maintenance scheme with on-demand routing algorithms. The scheme of creating directional link is proposed to extend the life of a link that is about to break. To be more specific, preventive warning is generated to a previous node in the path to create a directional antenna pattern. The link is considered in danger when received packet power is close to minimum detectable power. We call the process of changing the pattern of antenna from omnidirectional to the directional "orientation handoff." We see the performance improvement at the network layer by using the proposed scheme. We do a comparative performance study between omnidirectional and directional antennas for dynamic source routing (DSR, an on-demand routing protocol) using simulations. By using directional antennas, substantial gain is achieved in terms of end-to-end delay, aggregate throughput, average data packets dropped, packet delivery ratio, and routing overhead. The proposed scheme is general and can be used with any other on-demand routing algorithms.

Introduction

Typically, in ad hoc networks, omnidirectional antennas have been used to communicate with other nodes for transmission as well as for reception. Omnidirectional antennas may not be efficient due to interference caused by the transmission of packets in all directions (other than target direction) and limited range of communications. Directional antennas may be useful to increase network efficiency by directing the transmitted power in the desired direction toward the target location. Due to the mobile nature of ad hoc network nodes in their applications, it is important to observe the effect of directional antennas on the network layer. We propose a novel scheme of link life extension by using a directional radiation pattern, which helps to avoid or delay route rediscovery operation. Route rediscovery operation is an expensive process in terms of network resources and control overhead.

Recently, there has been increasing interest in developing protocols at the link layer and network layer for ad hoc networks where nodes are equipped with directional antennas [4, 6, 8, 12, 14, 16]. Previous research has shown that directional antennas–based communications increase throughput because of better spatial reuse of the spectrum [5, 7, 8, 13, 15]. By exploiting capabilities of directional antennas, a novel link maintenance scheme can yield better throughput, lower end-to-end delay, a decrease in the number of data packets dropped, an increase in packet delivery ratio, and a decrease in routing overhead.

In traditional routing algorithms of wired, wireless, and mobile networks, a change of path (route) occurs in one of two cases: a link along the path fails, or a shorter path is found. A link failure is very expensive, since multiple retransmissions/time-outs are required to detect the link failure and a new path has to be found and used (in on-demand routing), since a spare path may not be readily available. In wired networks, the cost of route rediscovery is not very high, since paths do not fail very frequently. Routing protocols in mobile and wireless networks also follow the same model, although they have a significantly higher frequency of path disconnections. For each path break (in the Institute of Electrical and Electronics Engineers [IEEE] 802.11 standard), three medium access control (MAC) layer retransmissions (a total of four time-outs, including the original transmission) are required before a link is considered broken.

The preventive link maintenance scheme proposed here initiates local recovery action early by detecting that a link is likely to break soon and uses a directional antenna pattern to prevent link failure, thus extending the life of the link and reducing the cost of link failure. The scheme maintains connectivity by proactively establishing a "higher-quality" link when the quality of a link in use becomes suspect. Note the similarity to on-demand protocols: we replace link failure with the likelihood of failure as the trigger mechanism for directional antenna orientation instead of the sending of a route error (RERR) packet to the source node, which initiates the costly operation of route rediscovery for a new path from the source to the destination. We study the effectiveness of proposed preventive link maintenance scheme by simulating with OPNET simulation software for DSR routing algorithms. Our scheme can be used for any other kind of routing algorithm.

We have organized our paper as follows: The second section reviews ad hoc routing algorithms and DSR in more detail. The third section discusses the related work in this area. The fourth section discusses the antennas model used in the proposed preventive link maintenance scheme. Then we discuss simulation model and performance evaluations. At the end, we make some concluding remarks.

Ad Hoc Routing Algorithms

For multihop wireless networks, traditional Internet routing protocols are no longer efficient, since the network topology is being changed dramatically due to node movement. It presents a great challenge for a routing protocol to keep up with the frequent and unpredictable topology changes. Existing multihop wireless routing protocols can be generally categorized into two classes: table-driven (such as destination sequence distance vector [DSDV] [19] and wireless routing protocol [WRP] [23]) and demand-driven (or source-initiated such as DSR [18], ad hoc on demand distance vector [AODV] [20], zone routing protocol [ZRP] [24], and location-aided routing [LAR] [25]).

Table-driven routing protocols of wireless networks are similar to the routing protocols used in wired networks. They attempt to maintain consistent, up-to-date routing information from each node to every other node, regardless of the need for such routes. They respond to topology changes by propagating updates throughout the network. In contrast, on-demand (reactive) routing protocols attempt to discover a route to a destination only when it is needed or when it senses some activities to forward to the destination. These discovered routes are maintained by a route maintenance procedure until either the destination becomes inaccessible along every path from the source or until the route is no longer desired. Most of the performance studies indicate that on-demand routing protocols perform better than table-driven routing protocols [21, 22]. The major advantage of the on-demand routing comes from the reduction of the routing overhead, as high routing overhead usually has a significant performance impact in low-bandwidth wireless networks.

Overview of DSR protocol is presented here briefly, since DSR is used to integrate with 802.11 and directional antennas as representative of on-demand routing protocols. DSR is a reactive (on-demand) routing protocol that is based on the well-known concept of source routing [2]. The protocol includes two major operational components—route discovery and route maintenance—and three types of route control messages—route request (RREQ), route reply (RREP), and RERR. When a source node in the MANET attempts to send a packet to a destination but does not have a route to that destination in its route cache, it initiates a route discovery process by broadcasting a RREQ packet. This packet contains the source node address, the destination node address, a unique sequence number, and an empty route record. Each intermediate node, upon receiving a RREQ for the first time, will check its own route cache. If it has no route to the destination, the intermediate node will add its own address to the route record and rebroadcast the RREQ. If it has a route to the destination in its route cache, the intermediate node will append the cached route to the route record and initiates a RREP back to the source node. The RREP contains the complete route record from the source to the destination. The intermediate node ignores the late arrival of the same RREQ by examining the sequence number. If the node receiving the RREQ is the destination node, it will copy the route record contained in the RREQ and send a RREP back to the source. In most simulation implementations, the destination node will reply to all the RREQs received, as DSR is capable of caching multiple paths to a certain destination and the replies from the destination most accurately reflect the up-to-date communication topology.

Due to the node movement, the routes discovered may no longer be valid over time. The route maintenance mechanism is accomplished by sending RERR packets. When a link is found broken, a RERR packet is sent from the node that detects the link failure back to the source node. Each node, upon receiving the RERR message, removes all the routes that contain the broken link from its cache.

In DSR, each node transmitting the packet is responsible for confirming that the packet has been received by the next hop along the source route. This can be done by link layer acknowledgment (as in IEEE 802.11), "passive acknowledgment" (in which the first transmitting node confirms the receipt at the second node by overhearing the second node transmitting the packet to the third node), or a DSR–specific software acknowledgment returned by the next hop. Thus, once a route is entered into the cache, the failure of the route can only be detected when it is actually used to transmit a packet but fails to confirm the receipt by the next hop. It is very critical to find a mechanism that helps avoid such link failure. We have proposed such a mechanism in this paper.

Related Work

Recently there have been several papers that have looked at the problem of data link layer and routing layer design for MANETs where nodes are equipped with directional antennas [4, 6, 8]. Most of the work toward the use of directional antennas has concentrated on the MAC layer. The directional antenna models used in various papers include switched-beam antennas (where the antenna is sectored and one of these sectors is used depending on the direction of the communicating node), multibeam antennas (where more than one beam can be used simultaneously), and adaptive antenna arrays.

Bao et al. [6] developed slotted scheduling-based MAC protocols for networks in which nodes are equipped with directional antennas. The directional antenna considered is a multibeam adaptive array antenna, which is capable of forming multiple beams. Assumption for the protocols was that nodes could engage in several simultaneous transmissions. Authors developed the neighbor-tracking scheme that is then used to schedule transmissions by each node in a distributed way.

Nasipuri et al. [4] proposed a directional carrier sense multiple access with collision detection (CSMA/CD) mechanism that utilizes a switched-beam antenna array and assumes that the gain of the directional antenna is equal to the gain of an omnidirectional antenna. The transmitters use omnidirectional antennas to transmit request to send (RTS) frames and the receiver antennas remain in omnidirectional mode. Assuming the receiver is idle, it receives the RTS and transmits clear to send (CTS), again using an omnidirectional antenna. The transmitter estimates the angle of arrival (AoA) of the CTS being received and transmits data using the directed antenna beam. Since the transmissions and receptions involving omnidirectional antenna patterns are susceptible to collisions, this mechanism suffers from high probability of packet error. The authors used a switched-beam antenna array that could only switch among a limited number of antenna patterns. Moreover, the assumption of having the same gain for both types of antennas is not to exploit the full capability of direction antennas, which give higher coverage (gain) and longer communication range. We exploit these benefits of directional antennas to extend the life of link.

Takai et al. [7] extended Nasipuri's work [1] by proposing the use of a caching mechanism to store information about angular location of neighboring nodes. This information is obtained from AoA estimation for frames received by each node. Whenever the MAC layer receives a packet from an upper layer, it would look in the cache to determine whether it has the information about the angular position of the destination node. If the angular position of the destination node is known, the packet is transmitted using the directional antenna after a directional exchange of RTS/CTS control packets. This helps to decrease the dependency on omnidirectional antennas. The scheme still reverts to omnidirectional mode if angular information is not obtained from the cache. In our work, we use the location information to decide the direction of communication.

Choudhury et al. [8] and Takai et al. [7] have suggested the use of directional virtual carrier sensing (DVCS), in which a directional network allocation vector (DNAV) is constructed. The DNAV table stores the angle of arrival of RTS packets along with the duration of data transmission in any given direction. Thus, when the MAC layer receives a packet from an upper layer, along with the angular profile of the destination node with respect to the source node, the DNAV table is consulted to determine whether the angle overlaps with any of the ongoing transmissions. If there are no overlaps, the packet is transmitted; otherwise, packet transmission is deferred in accordance with a back-off mechanism. We have not considered all such MAC layer issues in detail, since we concentrate on network layer design in our research.

In other work, Choudhury and Vaidya [26] have done rigorous analysis of using directional antennas and done performance evaluation for on-demand routing protocols such as DSR. They used only single switched-beam antennas. They have considerable sweeping delay due to the sequential transmission of the same packet with different antennas to cover the entire 360 degrees. We use multiple directional antennas to solve the problem of sweeping and deafness.

Ramanathan [13] proposed the scheme for considering higher transmission range using beam-forming antennas. He discussed the issues more related to MAC layer and has not discussed the issues related to the performance of the network layer. In this paper, we focus on a preventive link maintenance scheme for improving the performance of reactive routing protocols such as DSR, while Ramanathan [13] discussed proactive routing protocol over electronically steerable passive array radiator (ESPAR) antennas.

Very little work [11, 17] related to the work presented in this paper has been published on preventive maintenance using directional antennas. The pre-emptive routing [17] proposed keeps track of signal strengths and resorts to route repair procedures before a link breaks. Router handoff [11] is the strategy of finding another path pre-emptively before the link actually breaks and passing routing information to another suitably situated node after local recovery. However, neither work uses the ability of directional antennas for the above-mentioned purpose.

Antenna Model

A directional antenna module is implemented in an OPNET simulator [1]. There are two modes of operation for this model: omnidirectional and directional [8]. In normal operation, the omnidirectional mode is used, while the directional mode is used for both transmission as well as reception after a preventive warning has been generated due to decreasing signal strength in a received packet. Nodes can interchange the modes with negligible latency. We use multiple directional antennas (N) to avoid the sweeping operation and sweeping delay [26] that are present in the case of single switched-beam antenna. Using multiple directional antennas would also solve the deafness problem [26], since other antenna elements would be available to listen to other communication in other directions.

In omnidirectional mode, a node is capable of receiving signals from all directions with a gain of G_O. In the directional mode, a node can point its beam toward a specified direction with gain G_d (with G_d typically greater than G_O). Moreover, the gain is proportional to number of antenna beams (i.e., inversely proportional to the beam width). Since more energy can be focused toward a particular direction, this results in an increased coverage range. Though it is not feasible to have a completely non-overlapping pattern, we assume the non-overlapping pattern for directional antennas. To model antenna side lobes, we assume that the energy contributed to the side lobes is uniformly distributed in a circular area. For simulation purpose, we also assume that the side lobe gain is fixed and is set to a very small value.

Preventive Link Maintenance Algorithm

In this section, we propose a novel directional link maintenance scheme based on signal strength. We name it preventive link maintenance scheme, since it takes preventive action before a link actually breaks. We replace weak omnidirectional links with high-quality directional links to extend the life. More specifically, the scheme consists of two components: detecting that a link is likely to be broken soon, and establishing a directional link to it. Determining when link quality is no longer acceptable (which generates a preventive link maintenance warning) is a crucial component of the proposed scheme. The link quality can incorporate several criteria such as signal strength, age of a link, and rate of collisions. In this paper, we assume the link quality to be a function of the signal strength of received packets. Since most link breaks can be attributed to link failures due to node motion in a typical ad hoc scenario, the signal strength offers the most direct estimate of the ability of the nodes to reach each other. We should keep in mind that signal power variations due to fading and similar temporary disturbances do not generate erroneous preventive link maintenance warnings and, hence, creation of a directional link.

If a directional link is established successfully before the omnidirectional link breaks, the cost of overhead for detecting a likely broken omnidirectional link (the retransmit/time-out time) is eliminated using preventive link maintenance. In other words, the cost for creating a directional link is justified, since the route recovery is initiated before the current link actually breaks. Eventually, it is expected to improve the performance in terms of reduced latency (end-to-end delay), higher throughput, higher packet delivery ratio, and reduced routing overhead.

When the signal power of a received packet drops below an orientation threshold, the preventive link maintenance warning is generated. The value of this threshold is significant to the effectiveness of the algorithm—if the value is too low, there will not be sufficient time to create a directional pattern before the link breaks. However, if the value is too high, the warning is generated too early, with negative side effects of unnecessary computing for creating a directional link; the full life of the omnidirectional link currently in use is not exploited. Likewise, the moving nodes may change direction and the current link may never break, rendering the preventive action an unnecessary overhead. Moreover, if the threshold is too high, false disconnections may occur. Generating the preventive link maintenance warning is complicated due to fading that can cause sudden variations in the received signal power.

The decision to create a directional link is made by a node based on measured signal strengths of its neighbors with which it forms part of an active route. We maintain power information at nodes in terms of received power. The decision to create a high-quality directional link because of weak signal strength is made when one end of the link senses that the received power has dropped below an orientation threshold and a preventive link maintenance warning is generated. The operation of switching to the directional link from an omnidirectional link is called an orientation handoff. We incorporate orientation handoff into the DSR protocol by making the following changes: Each node maintains a neighbor-received power list containing the last received signal strength for packets originating from each neighbor. This list is updated whenever a packet is received.

Every node that is part of an active route checks its predecessor-link and next-link strengths for each route while receiving the data packet during normal communications (using omnidirectional antennas). If a node detects that either the previous or next link strength along an active route is predicted to fall below the orientation threshold, it initiates orientation handoff. If orientation with directional antennas cannot achieve strong link creation, no action is taken, the route is allowed to fail, and, consequently, standard DSR route repair procedure is followed.

Use of time-varying directional antenna patterns is envisioned to help in the establishment of directional links in multihop wireless networks that might otherwise not be possible with the use of omnidirectional antennas. We establish the links using directional antennas depending on the condition (signal strength) of the links within network between active nodes. Creation of directional link requires a prior knowledge of the location of the neighbor nodes or transmission/receiving direction. In this paper, we assume that wireless nodes can employ the techniques such as global positioning system (GPS) or AoA algorithms to determine the direction of communications by having position information of nodes and can use that information to orient their antennas.

While establishing the link, the transmitter and receiver may fall into one of the following scenarios: if the transmitter and the receiver are neighbors and the link between them is established by using an omnidirectional antenna, nodes are called near neighbors; or if two nodes are not near neighbors, they may be far neighbors; the link between them is established by using a directional antenna. In this scenario, the source and destination node may establish a directional link between far neighbors by orienting their receiver and transmitter antennas toward each other. In this case, the link is called an extended link and neighboring nodes are called far neighbors. The decision to establish an extended link between far neighbors for data transmission, instead of relying on multiple hops, depends on a number of factors such as delay in establishing the link, feasibility of a directed link, and the cost of the link in terms of its effect on aggregate throughput in the network.

The establishment of an extended link between far neighbors requires communications between the transmitter and receiver nodes through an immediate link (when they are near neighbors) to allow appropriate orientation of transmitter and receiver antennas. This signaling information must be exchanged during normal omnidirectional communications. The neighboring nodes can use an established omnidirectional link to send antenna orientation commands for establishment of an extended directional link. The establishment of extended links can result in a decrease in the end-to-end delay between end nodes (source and destination), since a directional link between those intermediate nodes would eliminate (or reduce) the costly operation of route rediscovery. Depending on communication traffic load and relative positions of other nodes, the extended link will have the varying effect of increasing the overall network throughput.

We keep updating the neighbor-received power list on all the nodes while using directional link. In some mobility patterns, it may be possible to go back and use omnidirectional antennas when received signal strength is more than orientation thresholds.

Performance Evaluation

Simulation Environment

Now, we illustrate a simulation scenario that uses the developed antenna models along with OPNET [1] to characterize network performance. The wireless communication channel is modeled by 13 pipeline stages, including antenna gains, propagation delay, signal-to-noise ratio, calculation of background noise and interference noise, transmission delay, etc. This powerful simulation environment enables designers to create realistic wireless scenarios. In this work, we have modified the MANET node model to make it to work with four individual antennas. We have used predefined and fixed beams and created an antenna pointer model. The antenna gain pattern specified in the antenna pattern editor is used to provide the gain values. The model includes four antennas to cover four directions (northeast, northwest, southeast, southwest). Antennas maintain this configuration with respect to the earth's meridian even if the node changes its orientation. Modern antennas can achieve it with the aid of magnetic needle that remains collinear to the earth's magnetic field. The antenna patterns have a directional gain of 10 dBi with 90 degree beamwidth.

We use the network of 50 nodes placed randomly over an area of 2,500 by 2,500 square meters. We have five random sources of constant bit rate (CBR), each of which generates 1,024-byte data packets to a randomly chosen destination at a rate of two to 50 packets per second. All five sources start data transmission with different times. All connections/communications end when simulation ends. For realizing realistic transmission range, the Hata Okumara path loss model [9, 10] is adopted. We use random waypoint mobility model to simulate different patterns of mobility (speed of nodes and pause time). The 802.11 standard protocol is used as MAC layer protocol for simulation. Simulations are run for 600 seconds, and all results are averaged over five seeds. We compare our results with original DSR over omnidirectional antennas. *Table 1* shows the set of parameters used for our simulations.

We also modified the MANET node model to implement our scheme and provide all necessary interfaces with the antenna models. The designed model is shown in *Figure 1*.

Simulation Results and Discussions

Aggregate throughput is the total number of bits transmitted from one node to other nodes in the network per unit of time. It is the total traffic sent and received in bits per second for entire network. We collected this statistic in our simulation scenario and compared with original DSR (with omnidirectional antenna nodes). Our scheme (Dir) performs better, since preventive link maintenance scheme allows a link to live longer than in the omnidirectional case. An establishment of directional link postpones the route rediscovery process and also reduces frequent disconnection. *Figure 2* shows that aggregate throughput is higher in our scheme using a directional antenna. Aggregate throughput (performance) of the entire network for our scheme is better for higher packet/data rate.

End-to-end delay (latency) of packets for the entire network is the time elapsed between the creation of a packet at its source and its destruction at its destination. Our scheme performs better when we have five flows communicating simultaneously in compare to smaller number of simultaneous flows. In other words, the average end-to-end delay per packet increases much more sharply for original DSR algorithm than our scheme as shown in *Figure 3*. It can be concluded that routing performance improves when we have many simultaneous connections. It is also because of the number of directional antennas (four) we use in our simulation.

When a route is not found to the destination or the next-hop reachability confirmation is not received (acknowledgment not received), the node drops the packets queued for the destination after the maximum number of attempted transmissions. This statistic represents the total number of appli-

TABLE 1

Simulation Parameters

Parameters	Value
Area	**2500 x 2500 sq. m**
Number of nodes	**50**
Directional gain	**10 dBi**
Orientation threshold	**3 dBi**
Packet size	**1,024 bytes**
CBR packet arrival interval	**1 ms to 10 ms**
Simulation time	**10 minutes**
Number of simultaneous connections	**Five, with a starting-time lag of 10 seconds**
Mobility model	**Random waypoint mobility model**

FIGURE 1

OPNET MANET Node Model

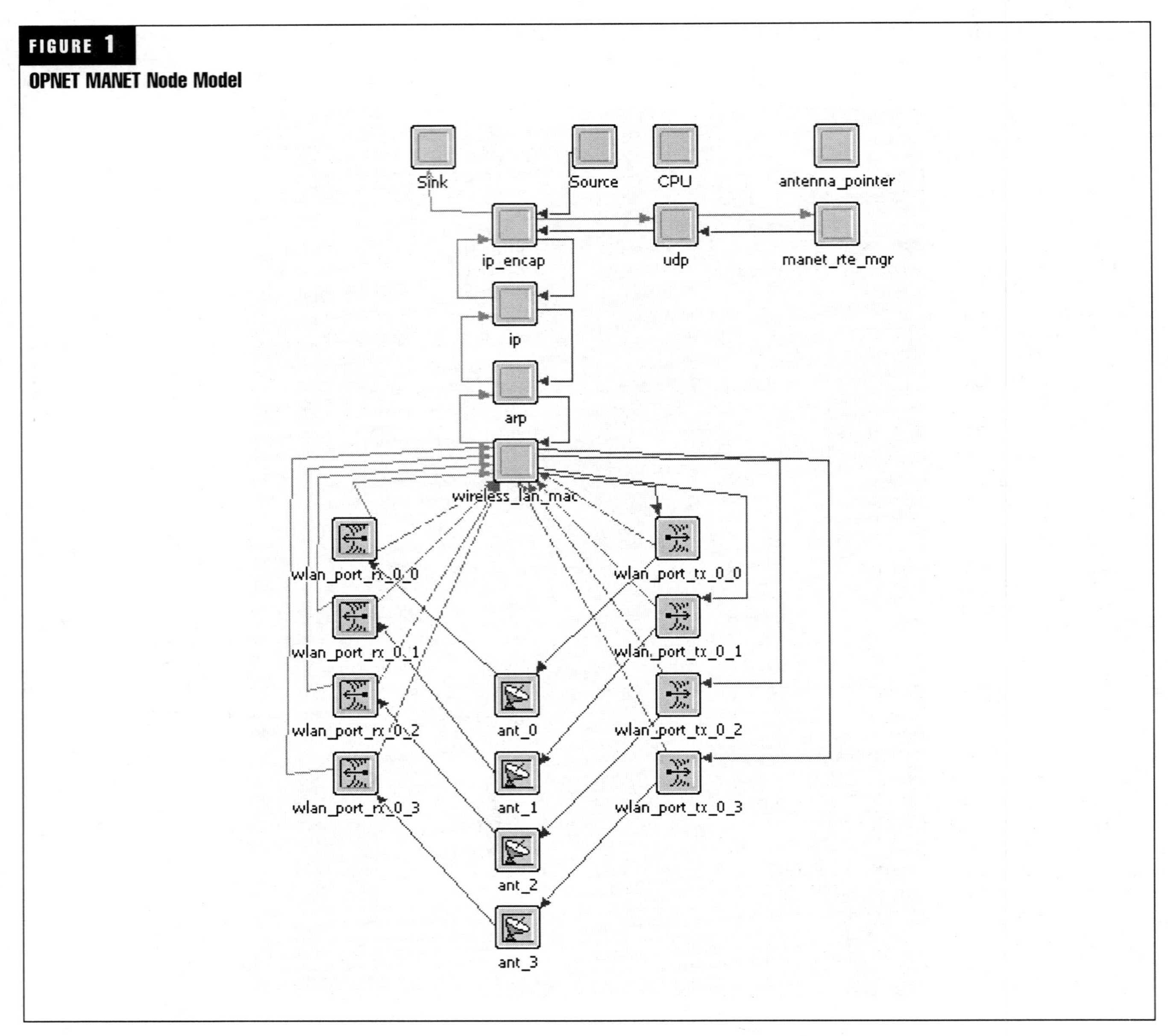

FIGURE 2

Aggregate Throughput

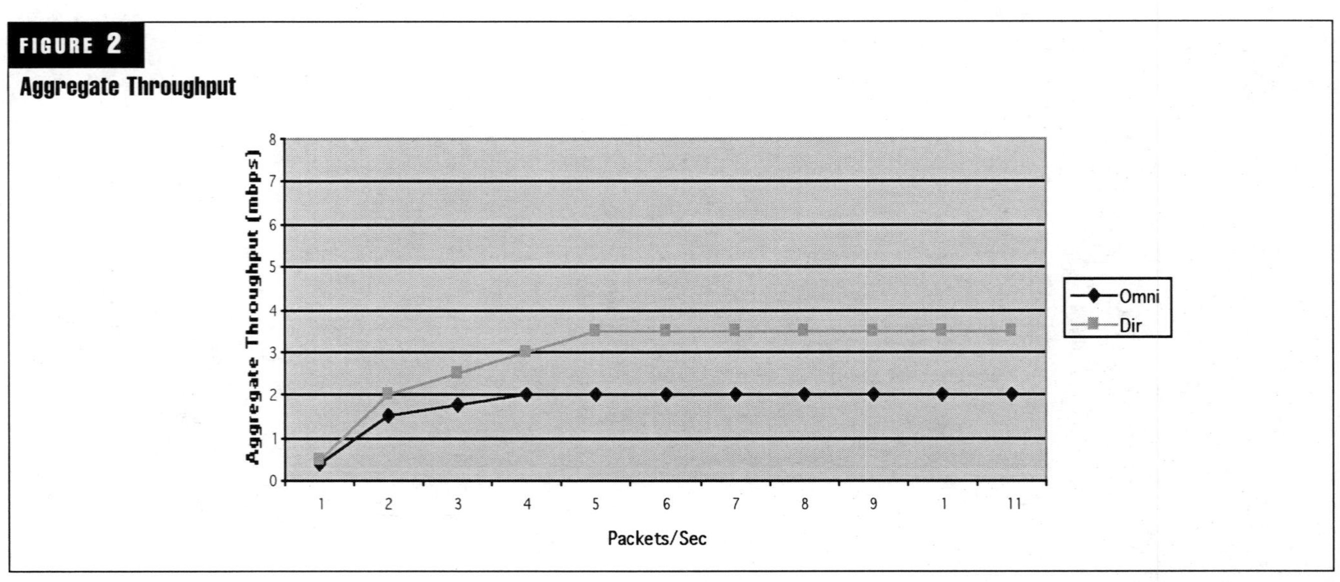

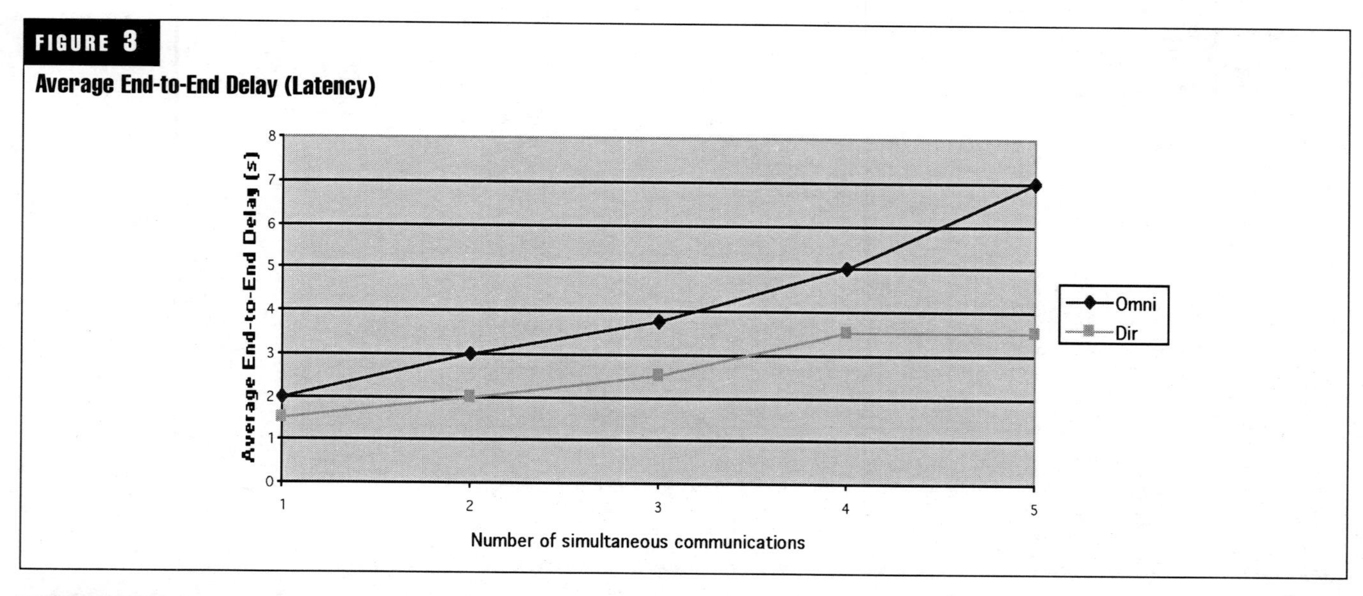

FIGURE 3

Average End-to-End Delay (Latency)

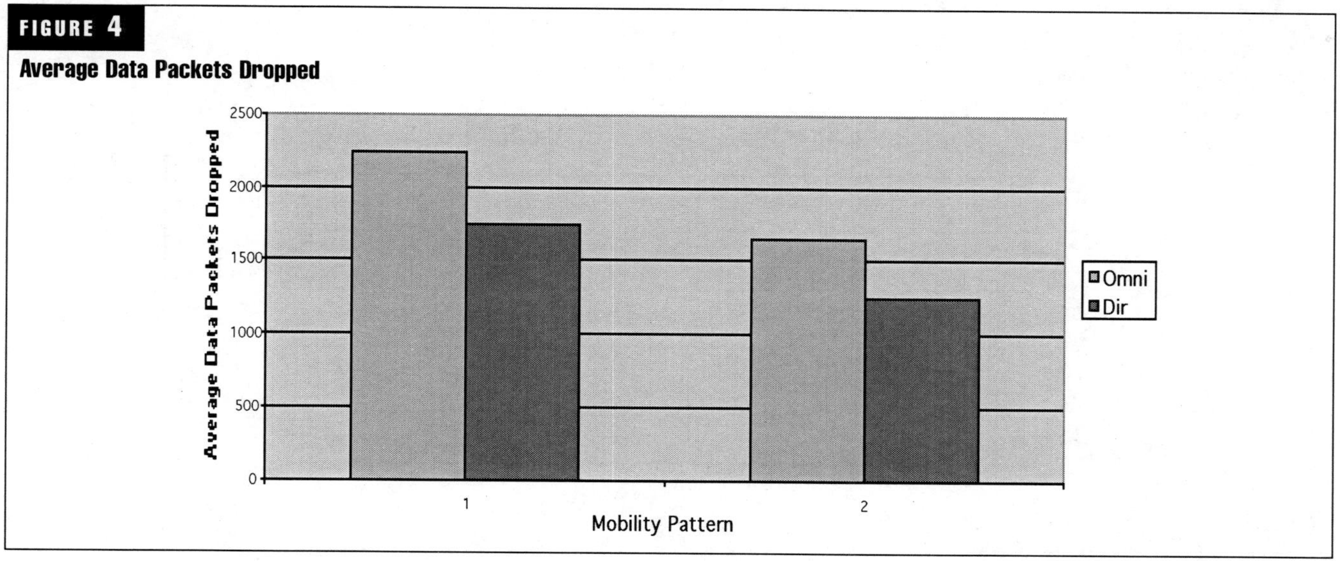

FIGURE 4

Average Data Packets Dropped

cation packets discarded by all nodes in the network. Average data packets dropped are far less in the simulation for our scheme than for original DSR. We have simulated two types of mobility patterns by choosing a different pause time (0 seconds pause time/continuous mobility and 20 seconds pause time) in the random waypoint mobility model in OPNET, as shown in *Figure 4*.

Packet delivery ratio is the ratio of packets received to packets generated. *Figure 5* shows this statistic. The result is intuitive that the packet delivery ratio is much higher for our scheme than for original DSR, since our scheme protects from link failure by creating directional links preventively. Packet delivery ratio drops drastically for our scheme for higher mobility, as we do not encourage using our scheme for a higher-mobility scenario. Still, its performance is better than original DSR with omnidirectional antennas.

Routing overhead per received packet is the ratio of the total number of routing control packets (including RREQ, RREP, and RERR) generated/forwarded to the number of data packets received correctly at the destination. *Figure 6* shows that routing overhead is higher in the case of original DSR. For a high-mobility scenario, original DSR produces a larger routing overhead, whereas our scheme has a lower routing overhead. It increases as the speed of the mobile node increases, but the increment is slight. In summary, our result is consistent with what is expected in different scenarios.

Conclusions

A directional antenna module is implemented in an OPNET simulator with two modes of operations: omnidirectional and directional. The antenna module has been incorporated in a wireless node model, and simulations are performed to characterize the performance improvement of DSR ad hoc routing protocol. The simulation model is developed to evaluate performance improvement when we use directional antennas to extend the life of a link that is about to break. Link breakage happens due to node movement and

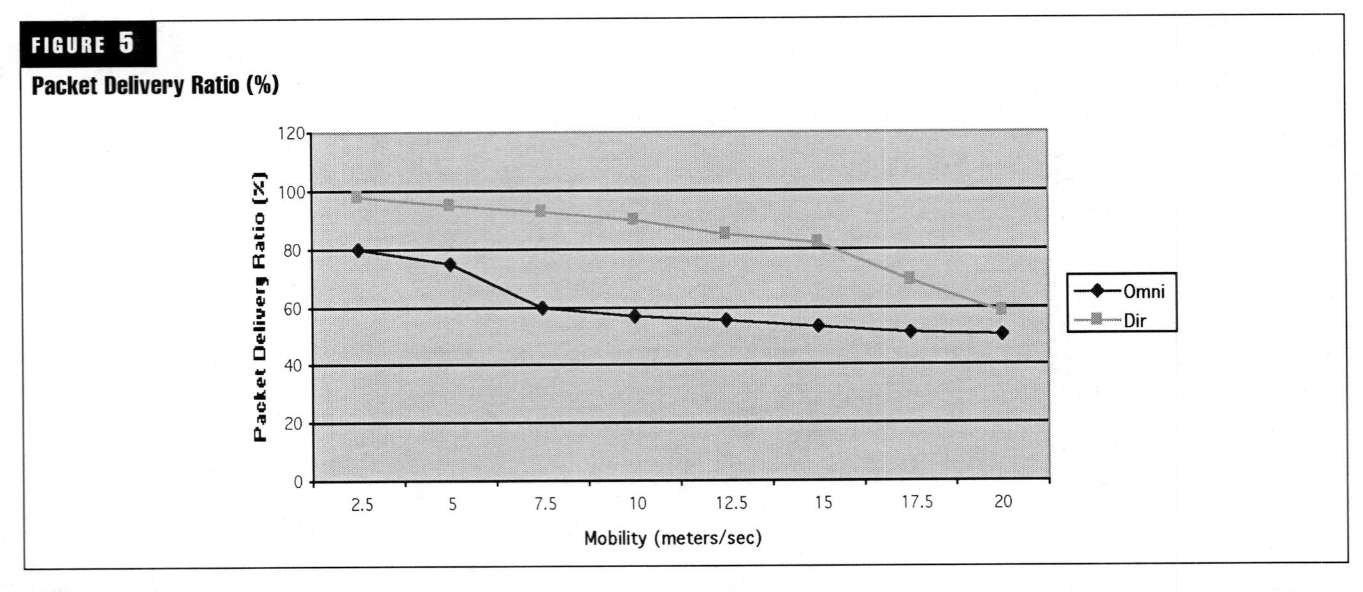

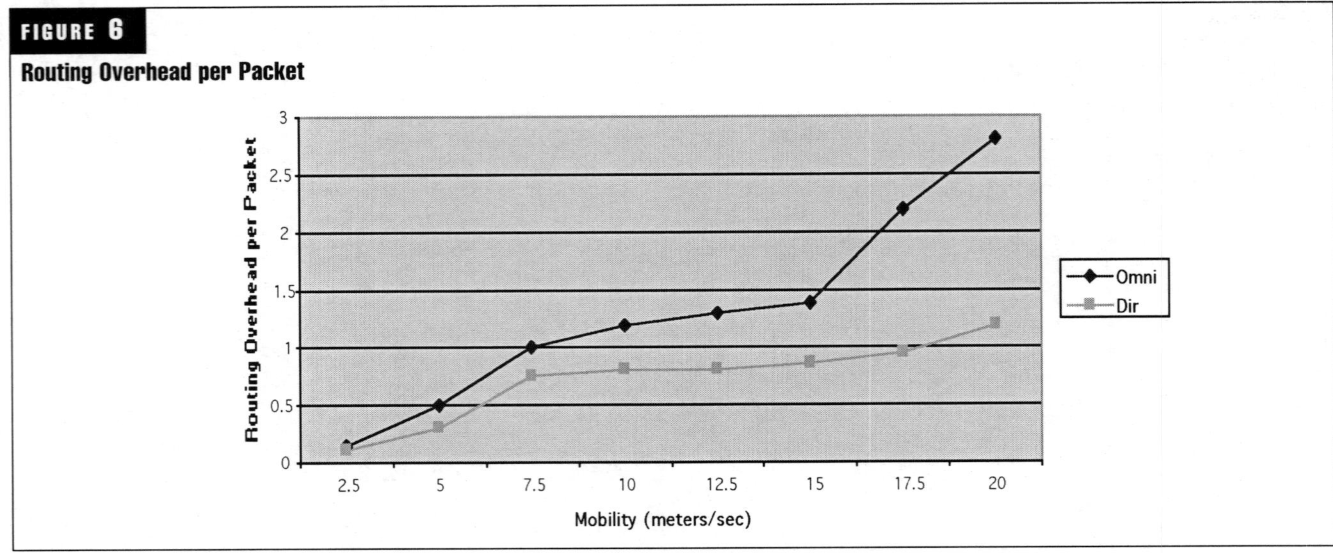

subsequent reduction in signal strength of receiving packets. A novel preventive link maintenance scheme using directional antennas is proposed to characterize the performance gain. It is achieved by orienting neighboring node antennas toward each other by using the position/location information of nodes. It helps ad hoc source nodes avoid or postpone the costly operation of route rediscovery in on-demand routing protocols. We compare the simulation results of our preventive link maintenance scheme (directional scheme) and omnidirectional scheme (original DSR algorithm) by collecting the statistics of aggregate throughput, end-to-end delay, number of data packets dropped, packet delivery ratio, and routing overhead. Using directional antennas has been found encouraging for on-demand routing protocols of MANET using our proposed preventive link maintenance scheme.

References

1. Online Documentation, "OPNET Modeler," www.opnet.com. Date visited: February 2006.
2. D. B. Johnson, D. A. Maltz, Y.-C. Hu and J. G. Jetcheva, "The dynamic source routing protocol for mobile ad hoc networks," *IETF Internet draft (November 2001), draft-ietf-manet-dsr-06.txt.*
3. H. Dajing, J. Shengming and R. Jianqiang, "A link availability prediction model for wireless ad hoc networks," *Proc. of the International Workshop on Wireless Networks and Mobile Computing*, Taipei, Taiwan, April 2000.
4. A. Nasipuri, S. Ye, J. You, and R. Hiromoto, "A MAC protocol for mobile ad hoc networks using directional antennas," *Proc. of the IEEE Wireless Communications and Networking Conference*, Chicago, Illinois, Vol. 3, September 23–28, 2000, pp. 1214–1219.
5. Z. Huang and C. Shen, "A comparison study of omnidirectional and directional MAC protocols for ad hoc networks," *In Proc. of IEEE Globecom '02*, 2002.
6. L. Bao, and J. J. Garcia-Luna-Aceves, "Transmission scheduling in ad hoc networks with directional antennas," *in ACM/SIGMOBILE MobiCom 2002*, 23–28 Sept. 2002.
7. M. Takai, J. Martin, R. Bagrodia, and A. Ren, "Directional virtual carrier sensing for directional antennas in mobile ad hoc networks," *Proc. of the ACM MobiHoc'02*, Lausanne, Switzerland, June 9–11, 2002, pp. 183–193.
8. R. Choudhury, X. Yang, R. Ramanathan and N. H. Vaidya, "Using directional antennas for medium access control in ad hoc networks," *Proc. of the MOBICOM*, Atlanta, Georgia, September 23–28, 2002, pp. 59–70.
9. Okumura et al., "Field strength and its variability in VHF and UHF band-mobile radio service," *Review of the Electrical Communications*

Laboratory, Vol. 16, No. 9–10, September–October 1968.

10. Hata, "Empirical formula for propagation loss in land mobile radio services," *IEEE Transactions on Vehicular Technology*, Vol. 29, No. 3, Aug 1980.
11. Abhilash P., Srinath Perur, and Sridhar Iyer, "Router Handoff: An Approach for Preemptive Route Repair in Mobile Ad Hoc Networks," *Proc. of High Performance Computing*, 2002.
12. Z. Huang, C. Shen, C. Srisathapornphat, C. Jaikaeo, "A busy-tone based directional MAC protocol for ad hoc networks," *Proc. IEEE MILCOM*, pp. 1,233–1,238, Oct. '02.
13. R. Ramanathan, "On the performance of ad hoc networks with beam-forming antennas," *Proc. of the ACM MobiHoc*, Long Beach, California, October 4–5, 2001, pp. 95–105.
14. S. Roy, D. Saha, S. Bandyopadhyay, T. Ueda, and S. Tanaka, "A network aware MAC and routing protocol for effective load balancing in ad hoc wireless networks with directional antenna," *In ACM Mobihoc '03*, June 2003.
15. S. Horisawa S. Bandyopadhyay, K. Hausike and S. Tawara, "An adaptive MAC and directional routing protocol for ad hoc wireless networks using ESPAR antenna," *in ACM/SIGMOBILE MobiHoc '01*, Oct. 2001.
16. M. Sanchez, T. Giles, and J. Zander, "CSMA/CA with beam forming antennas in multi-hop packet radio," *Proc. of the Swedish Workshop on Wireless Ad Hoc Networks*, March 5–6, 2001.
17. T. Goff, N. B. Abu Ghazaleh, D. Phatak, and R. Kahvecioglu, "Preemptive Routing in Ad Hoc Networks," *In Proc. of ACM SIGMOBILE*, Rome, pp.43–52, 2001.
18. D. Johnson, D. Maltz, "Dynamic source routing in ad hoc wireless networks," edited by T. Imielinski and H. Korth, Kluwer Academic Publisher, pp. 153–181, 1996.
19. C. Perkins, B. Bhagwat, "Highly dynamic destination sequence distance-vector (DSDV) routing for mobile computers," *in ACM SIGCOMM Symposium on Communication, Architectures, and Protocols*, 1994.
20. C. Perkins, E. Royer, *Ad-hoc On-demand Distance Vector (AODV) Routing (Internet-Drift)*, Aug. 1998.
21. J. Broch, D. Maltz, D. Johnson, Y. Hu, and J. Jetcheva, "A performance comparison of multi-hop wireless ad hoc network routing protocols," *Proc. Mobile Computing and Networking (MobiCom)*, pp. 85–97, 1998.
22. S. R. Das, R. Castaneda, J. Yan, "Comparative performance evaluation of routing protocols for mobile ad hoc networks," *7th International Conference on Computer Communication and Networks (IC3N '98)* pp. 153–161.
23. S. Murthy, J. J. Garcia-Luna-Aceves, "An efficient routing protocol for wireless networks," *ACM Mobile Networks and Applications Journal*, Special issue on Routing in Mobile Communication Networks, 1996.
24. Z. J. Haas, "A routing protocol for the reconfigurable wireless network," *in 1997 IEEE 6th International Conference on Universal Person Communications Record. Bridging the Way to the 21st Century, ICUPC '97*, vol. 2, pp 562–566, October 1997.
25. Y.-B. Ko and N. H. Vaidya, "Location-Aided Routing (LAR) in mobile ad hoc networks," *in Proceedings of the Fourth MobiCom'98*, pp. 66–75, October 1998.
26. R. Choudhury, N. H. Vaidya, "Performance of Ad Hoc Routing using Directional Antennas," Journal of Ad Hoc Networks, Elsevier Publishers, November 2004.

Whence and Whither Mobile Communications and Impacting Technologies

Augustine C. Odinma
Professor, Electrical and Computer Engineering
Lagos State University

Abstract

Many mobile network generations have been defined and standardized in the past two decades. Each next-generation network (NGN) has had a very short life of a few years, and many are concern that many of the standards are being defined concurrently. Second generation (2G), which is already out of vogue in major advanced countries, is still being deployed in Third World countries. Third generation (3G) was expected to solve all the problems of the previous generations, but the persistent demand for higher rates and the non-compatibility of 3G standards has shifted the enthusiasm to beyond 3G (B3G) or fourth generation (4G) networks. This paper takes a critical view at the various generations of network standards along with impacting technologies with the intention of focusing on the mindset of the telecom planners and designers and raising their research consciousness.

Introduction

The introduction of mobile NGNs has become an unending phenomenon, a phrase with real meaning that has a life of only a few years. This in part is due to unprecedented innovations in digital electronics and telecom technologies. On the other hand, the phrase could have an unstable meaning because, so far, each generation of standards has unfortunately failed to accomplish its proclaimed objectives. It is becoming increasingly difficult to keep track of NGN developments. In Third World countries, the network operators are still struggling to deploy 2G when it is already out of vogue in major advanced countries. This paper not only builds on earlier papers [1, 2, 3, 4], but also is expected to focus on the mindset of the telecom planners and examine the direction in which the NGNs are heading. It also reviews the technological advances impacting the next generation of mobile networks.

It began with the first generation (1G) networks, which were built to support analog mobile phones in 1984. The 1G networks were more or less a proprietary standard, with almost every nation or region having its own standard. Moreover, the network could only support voice traffic with very poor quality. In the European Union (EU) (1980), the need to roam among member countries and for a relatively reliable digital mobile phone network led to the development of the first 2G network standard, the Global System for Mobile Communications (GSM) network. The GSM network is based on time division multiple access (TDMA) air interface technology. A variation of TDMA mobile technology (IS–136) was later adopted in the United States by AT&T and a few other operators as personal communications service (PCS). The dominant 2G standard in the United States that followed GSM was code division multiple access (CDMA), which was introduced in 1992. The 2G technologies were optimized for voice at about the same time the Internet was becoming popular and the ability to move data among computers and networks was beginning to be a routine, so, the text messages and low bit rates, which were all that 2G networks could support, proved very inadequate. While the older generation of mobile users began to demand to be able to transfer relatively large amounts of data and images, the teenagers and younger generations have the need for music and video clips, video streaming, etc. which are also capacity-intensive. Coupled with this was also the fact that many mobile operators were facing unprecedented churn and turnover and were, as a result, searching for a new service differentiation to maintain their revenue. Customers' expectations and operators' desire to maintain revenue led to the approval of 3G wireless network standards by the International Telecommunication Union (ITU) in 1999. The 3G standard was touted as a panacea for all the customer expectations that could not be met in 2G. In terms of speed, the theoretical data rate that GSM mobile networks could support is 9.6 Kbps, while the CDMA networks could support up to 14.4 Kbps. The standard [5] that the ITU approved was called International Mobile Telecommunications–2000 (IMT–2000), and the following bandwidths were specified:

- 144 Kbps or higher for higher-mobility traffic
- 384 Kbps for pedestrian traffic
- 2 Mbps for indoor or in-building traffic

The short life span of each of the generations of wireless networks makes it difficult for operators to carry out full migration to the next generation. The operators are in business to make a profit and desire to protect prior investments. The need to recoup their investment, meaning that operators go for a step-wise or phased migration—for example, rather than moving to 3G, they first migrate to 2.5G and then to 3G. Although Japan's Docomo has been offering 3G since 2002, and many other operators have thereafter, but enthusiasm for 3G is waning. This is primarily due to the data rates observed from 3G networks. The two main evolution standards for 3G are wideband CDMA (WCDMA) and variations of CDMA2000, with CDMA2000 1x evolution data optimized (EVDO) expected to offer a higher rate. Tried-and-tested [6] 3G technologies of CDMA2000 1x EVDO cannot maintain a 1 Mbps speed on average, and the probable alternative of WCDMA promises an average throughput of around 1 to 2 Mbps. Furthermore, 3G services have proven so disappointing because instead of one standard worldwide, there are three incompatible systems in the United States alone [7]. The demand for higher rates and the non-compatibility of 3G standards has shifted the enthusiasm to 4G wireless networks. When 4G finally goes to market, it will support rates above 100 Mbps [8] and integrate all wireless networks, and it is expected to be application-independent. Particularly, it will support location-based services, mobile shopping, e-mail and multimedia data transfer, video conferencing, video streaming, etc.

This paper is therefore intended to retrace the various generations of mobile wireless networks. It will examine each generation and take comparative studies of each in relation to subsequent generations. Broadly, the discussion will cover 1G, 2G, 2.5G, 3G, 3.5G, 3.75G, 3.9G, and 4G. The 4G visions will be examined in detail.

Some NGNs Impacting Technologies

There are many technological innovations that have and are still forcing the change from one generation of mobile networks to another. We have singled out the following innovations:

Vocoders and Digital Signal Processing (DSP)

Advances in DSP have continued to impact technologies that influence the rapid change in wireless NGNs in many ways. One such technology is a voice coder (vocoder). Traditionally, an audible speech can be pulse code modulation (PCM)–coded and transported using 64 Kbps, but the fewer bits per second used, the more that scarce spectrums and transport facilities can be efficiently used. This led to the invention of linear predictive coders (LPCs) [9] in the '70s and code excitable linear predictors (CELPs) in the early '80s [10]. LPC is a low-bit coding technique that supports bandwidth of 2.4 to 4.8 Kbps. On the other hand, CELPs are a coding and compression technology for transmitting voice over packetized networks. Variations of CELPs use different codebooks, which contain sets of voice samples. CELPs use 80 PCM voice samples, which are 10 ms representations of a voice stream. The set of voice samples is compressed to remove silence and redundancy [11], and the compressed data set is compared to a set of index samples in the codebook. The index number is what is transmitted across the network. The index is sent across the network in a block of 160 bits every 10 ms, which gives 16 Kbps. Algebraic CELPs (ACELP) improves on CELP compression by a factor of 2:1 and, as such, gives a good quality voice at 8 Kbps. ACELPs use a codebook that comprises algebraic expressions of each set of samples, contrary to CELPs, which use a set of a series of numbers. ACELPs are standardized in ITU G.729 as conjugate structure–ACELPs (CS–ACELPs). Low-delay CELPs (LD–CELPs) compress voice at the CELP rate but at a much lower delay level. It uses five PCM samples, compared to 160 for CELPs. LD–CELPs have been standardized by ITU–T as G.728.

The coding and decoding of codes in the codebook is complex and requires enormous computational resources, which can be very expensive. However, with the advances in DSP technology, complex computation can be achieved in a cost-effective manner. Thus DSP is a powerful compression technique for CDMA and TDMA technologies. IS–136 (TDMA) ACELP algorithm [12] has been used to achieve 7.25 Kbps. Similarly, for IS–95 (CDMA), a 13 Kbps vocoder [13] has been described. Although the emphasis has shifted from voice and data to multimedia services, DSP has made possible the processing of complex computational requirements from one generation to another.

Packetization and Network Convergence

Telecom infrastructure has evolved dramatically. Until the last decade, information systems and telecom infrastructure were predominantly defined to meet the demands and characteristics of voice traffic. Advances in digital and packet technologies have systematically altered this perception [14]. Moreover, the availability of bandwidth to support highly skewed and sporadic traffic patterns for data and video has called for more efficient means of transporting these services. Voice is circuit-oriented and ubiquitous; it leaves little room for timely service differentiation, development, and deployment. Thus, operators concerned about future business turned to packet technology. The adoption of the Internet the world over has also made packet transport a strong contender to be the underlying technology behind all future communication systems. As a matter of fact, the ITU released three UMTS 3GPP [5] standards evangelizing packet technology as the backbone technology for the NGNs. Release 99, which came out in 1999, supports an ATM core technology. Release 4, which came out in 2000, supports a multiservice core, and Release 5 standards were published in 2001. Release 5 took the packetization notion beyond the core to "all–IP" across the board. Packet technology enables the use of a single type of network to carry all the types of services at the core. This convergence of the types of networks simplifies the NGN architecture, particularly in the core. The simplified architecture implies less infrastructure cost and allows operators enormous room for service differentiation. The quest for service differentiation drives the quest for a new NGN.

Adaptive Intelligent Antennas (AIAs)
Improving capacity in mobile systems has been a major active research area. The demands for high-speed and varied resource-intensive services have been major drivers for a new generation of mobile phone systems. AIAs are one of the technological innovations, and advances in DSP have made complex computation in AIAs affordable. AIAs primarily increase capacity, but they also offer [15] increased cell area coverage, improved call quality, improved mobile battery life, reduced base station amplifier power, improved robustness to imperfect cell location, and the ability to support higher data rates. The ability to support higher data rates makes AIAs a catalyst for new generations of mobile systems.

Whence the NGN

Advances in digital and communication technologies have created an insatiable desire for access to high-bandwidth applications, anywhere and at any time. This desire has led to rapid developments and, therefore, the ephemeral nature of generations of wireless networks. It is said that the mobile radio system [8], which is a prelude to current mobile networks, went into operation as far back as April 7, 1928. It is also said to be the world's first successful system, which was used by the Detroit Police Department. The system used amplitude modulation (AM) and as a result suffers from spike noise, which is prevalent in electromagnetic-oriented systems. However, in 1935, frequency modulation (FM), which proved to be more resistant to the noise problems than AM, was developed and tested by Edmond H. Armstrong. Aside from the noise problem, another major issue at the time was the large carrier bandwidth required. To transmit a voice grade signal of 3 KHz, a spectrum of 120 KHz was required. However, FM achieved this feat with 30 KHz, which makes it about 400 percent more efficient than AM.

World War II was also a major catalyst for developments that led to mobile communications. During the war, the FM–based mobile radio was the main thrust of communication. Shortly after the war, AT&T developed what was called the improved mobile telephone service (IMTS). Primarily, the system was introduced to cater to a metropolitan area and consisted of a broadcast system and a high-powered transmitter. IMTS evolved into a limited cellular system and was integrated with the fixed telephone system in the '50s. This led to development of paging systems that took advantage of this technology. Bell Labs, the innovative arm of AT&T at the time, continued its cellular research and testing, and in 1970, they were successful in convincing the Federal Communications Commission (FCC) to allocate spectrum space to cellular systems. The allocation of the cellular frequency gave AT&T the impetus to propose the first 1G network, known as advanced mobile phone system (AMPS), which was implemented in 1983 in the United States. This was followed by total access communications system (TACS), implemented in the United Kingdom in the 900 MHz frequency range. TACS is an offshoot of AMPS. In Japan, a variation of TACS, JTAC, was deployed in the 800 and 900 MHz range. At the same time, 1G analog cellular systems in Sweden, Norway, Denmark and Finland were deployed as Nordic Mobile Telephone (NMT) systems in the 450 and 900 MHz range. Germany was the main user of the 450 MHz. AMPS was the 1G system widely used in North America.

In the EU, the story was a bit different. Because of the different systems that existed in the different countries, mobility and roaming beyond those system areas were major problems. The problems were of major concern to EU member states when 1G was being deployed, and research into a unifying digital system was soon commissioned. This unifying 2G system was named Global System for Mobile Communications (GSM). Thus, GSM, the first 2G mobile technology, was deployed in 1991 by Radiolinja in Finland. The United States was not in a haste to evolve their 1G to 2G because the single AMPS standards adopted provided reasonable mobility and roaming among more than 40 countries. It was not until two years after the deployment of GSM that the U.S. firms and other countries began to deploy 2G. *Table 1* shows the four common types of 2G air interfaces. GSM uses a variation of TDMA air interface technology. The personal digital cellular (PDC) was the 2G system deployed in Japan.

In the United States, there are three 2G mobile standards: Telecommunications Industry Association and Electronic Industry Association (TIA/EIA, now known as TIA) interim standard (IS)–95, IS–41, and IS–136. IS–136 defines a 2G TDMA mobile technology and digital control channel (DCCH) for use in the United States. It supersedes IS–54, which defines dual-mode analog and digital standard for cellular phone service. The IS–136 enabled TDMA and AMPS to coexist in the network and share the same resources, thus allowing 1G AMPS to gracefully evolve to 2G TDMA. On the other hand, IS–95 is a 2G digital CDMA standard for use in the United States and South Korea. It is a direct sequence spread spectrum (DSSS) patented by QUALCOMM. In other words, any company using CDMA or a variation thereof pays a royalty to QUALCOMM. On June 3, 1997, CDMA development group (CDG) named the 2G CDMA CDMAOne.

There are other 2G systems around the world, but they are variations of GSM, predominantly, or CDMA. A simple discussion on GSM and CDMA is given in [17].

Table 1 gives theoretical data rates for each generation. 1G could move data at up to 9.6 Kbps using a V.32 modem, which was the highest-rated at the time of the deployment of AMPS in the early '80s. *Table 2* gives ITU–defined voice modem standards and associated data rates. 2G technologies do not need a modem to convert analog to digital, as in the case of 1G, which could not benefit from modem enhancement years later. The coding scheme of 2G base stations can only allow 9.6 Kbps for GSM/TDMA and 14.4 Kbps for CDMA technologies. In the '80s, computer users of LANs became used to moving large amounts of data and documents from one computer to another. As a result, rates of less than 14.4 Kbps eventually became insignificant, so while developments were still on for 2G, work began on 2G enhancements (2.5G) in the mid-'90s.

The 2.5G general packet radio service (GPRS) and enhanced data for GSM evolution (EDGE) are enhancements of 2G GSM. Similarly, IS–95B and CDMA2000 (IS–95C) are enhancements of CDMAOne. GPRS was deployed in 1999

TABLE 1

Comparing Wireless Generations

Generation	1G	2G	2.5G	3G	3.xG	4G
Talks begin	1970	1980	1997	1985	2001	2000
Deployment	1983	1991	1999	2003	2003/2005	2010?
Standards	AMPS, TACS	TDMA, GSM	GPRS, EDGE	WCDMA, UWC	CDMA EVDO	One standard
	NMT, etc.	CDMA, PDC	1xRTT, IS–95B	CDMA2000	WCDMA	
Services	Analog Voice	Digital voice messages	High-speed packet data	High-speed broadband	High-speed packet data	Complete IP multimedia
Multiplexing	FDM	TDMA, CDMA	TDMA, CDMA	CDMA	CDMA	OFDM
Core Network	PSTN Circuit	PSTN Circuit	PSTN Packet	Packet	Packet	Internet
Modulation	FDM	GMSK	8–PSK	Walsh	???	???
Data Rates	< 9.6 Kbps	< 14.4 Kbps	< 384 Kbps	< 2 Mbps	< 100Mbps	> 200 Mbps

to support GSM rates of up to 144 Kbps, while EDGE could support rates of up to 384 Kbps. IS–95B was designed to deliver 64 Kbps initially but ultimately supported 115 Kbps in 2.5G, whereas IS–95C is 3G. *Figure 1* is a compilation of the evolution path for 2G through 3G and beyond.

As we have already seen, 2G has very low bit rates and therefore could not support applications demanding very high bit rates. The enhancements to 2G could offer rates beyond 144 Kbps, but other shortfalls countered the enhancements. We have also seen that there were many 2G frequency ranges—450, 800, 900, 1,800, and 1,900 MHz—allotted by different countries, making roaming between networks complicated and limiting. Most of the 2G infrastructures are proprietary in nature, but infrastructure development trends are now leaning toward open-system architecture that supports third-party application development. Another major factor limiting 2G was technological developments in the core of the network and optical domain. In the '80s, switches and routers could only switch E1/T1 and E3/T3 rates, but in the '90s, routers and switches with gigabits per second capability became commonplace. Dense wavelength division multiplexing (DWDM) that could switch in the terabits per second (Tbps) was also developed in the past decade. These technological developments influenced new thinking among designers and led to new services development that 2G could not support seamlessly. These observed shortfalls in the '90s led to a demand for a new generation (3G) of mobile technology for the new millennium. These demands were given effects by ITU–Radiocommunications (ITU–R) in conjunction with other national standards organizations in countries and continents around the world. The ITU set up an initiative to coordinate the defining of specifications and standards for

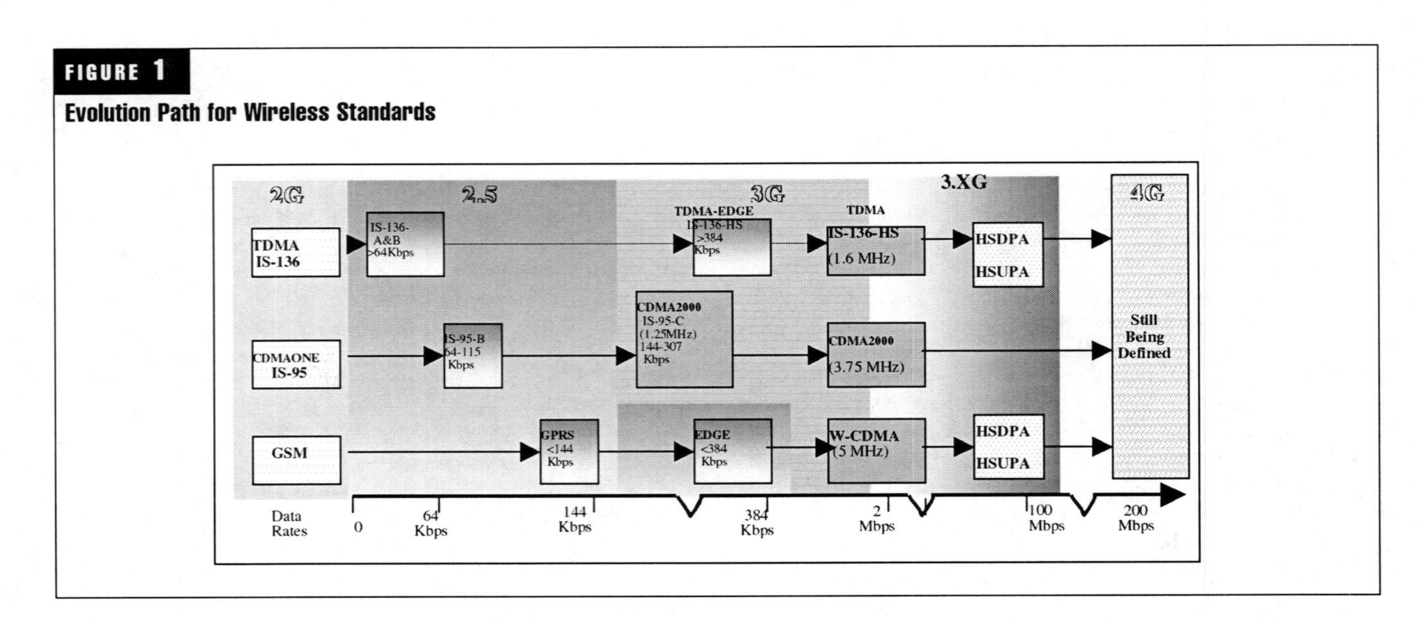

FIGURE 1

Evolution Path for Wireless Standards

3G. The initiative [5] was named International Mobile Telecommunications for the 21st Century (IMT–2000). The European Telecommunications Standards Institute (ETSI), a research and development projects funded by EU, set up special mobile group 2 (SMG2) [18] to define and submit specifications that would evolve GSM. WCDMA and TD–CDMA were proposed to ETSI in January 1998. In Japan, the Association of Radio Industries and Business (ARIB) submitted a WCDMA proposal and a CDMA2000 proposal to evolve their 2G. China Wireless Telecommunications Standards (CWTS) also proposed a variation of the WCDMA (time division synchronous code division multiple access [TDSCDMA]) in 1998 and became a member of 3GPP in 2001. The Telecommunications Technology Association (TTA) of Korea proposed WCDMA and CDMA2000 versions to evolve its TDMA and CDMA 2G technologies to 3G. In North America, the following four proposals were submitted:

- A CDMA2000 proposal (TIA–45–5) submitted via Samsung, Motorola, Lucent, Nortel, and QUALCOMM
- A TDMA (UWC–136) evolution proposal (TIA–45.3) submitted by AT&T, South Western Bell, and BellSouth
- A WIM evolution proposal (TIA–46.1) submitted by Golden Bridge Technologies
- A WCDMA evolution proposal (ATIS TIP1) for GSM networks

The foregoing discussions show that many 3G proposals were submitted to ITU. Indeed, in response to the ITU circular letter of April 1997, 10 terrestrial proposals and five satellite proposals were submitted.

Eight of the terrestrial proposals were variants of WCDMA and CDMA2000. The remaining two dealt with the migration of TDMA to WTDMA: universal wireless communications–136 (UWC–136) and pedestrian digital enhanced cordless technology (DECT). This led to three main 3G technology proposals—WCDMA, CDMA2000, and UWC–136—to enable backward compatibility with 2G and hence protect investment in 2G. Two partnership project groups were set up by ITU to help build consensus among the proponents of the proposals and work out specifications for 3G. The two 3G partnership groups were as follows:

- 3GPP had four other technical specification groups (TSGs) that focus on WCDMA [5].
- 3GPP2 had six TSGs that attempt to evolve the North American proposals to CDMA2000 [19].

Japan NTT was the first to implement 3G, in the last quarter of 2002. There are many other 3G vendors, but the enthusiasm for 3G is dying a natural death.

Whither the NGN

In the past couple of years, several efforts have been expended to increase the data rates of 3G networks to between 1.8 and 14.4 Mbps. This has led the following companies to deploy various 3.5G networks:

- KDDI deployed EVDO in December 2003 in South Korea.
- Cingular deployed high-speed downlink packet access (HSDPA) in December 2005 in the United States.
- O2 deployed HSDPA in November 2005 in the United Kingdom.

HSDPA is expected to boost network data transmission by up to four times and would allow many wireless users per site. HSDPA is expected to make efficient use of existing new resources and help deliver high quality to end users while reducing operating cost [27]. Another enhancement to 3G is 3.75G, which is high-speed uplink packet access (HSUPA). Like HSDPA, HSUPA is UMTS enhancement to WCDMA 3G technology. HSUPA is a data access protocol with an extremely high upload speed of up to 5.76 Kbps. It is still being defined and expected in UMTS Release 6. Austria's T-Mobile and Italy's Telecom Italia plan to introduce HSUPA in 2007. Plans are already in place to converge the HSDPA, HSUPA, and WiMAX in a 3.9G standard. 3.9G, which will be finalized by mid-2007, will also be known as UMTS long-term evolution. It is expected to support up to 100 Mbps in the downlink and 50 Mbps in the uplink [28].

One of the main reasons for the initial clamoring for 3G was to overcome the incompatibility of the various 2G networks, though 3G now has three incompatible standards as well. Moreover, most of the capability set [16] promised by ITU for 3G, including various multimedia services and 2 Mbps, was far-fetched. The 3GPP and 3GPP2 efforts to eliminate previous incompatibilities associated with 2G and make 3G become a truly global system were not successful. The 3G system would have higher-quality voice channels as well as broadband data capabilities of up to 2 Mbps. Unfortunately, the two groups could not reconcile their differences [7], which led to multiple mobile standards. These and other reasons based on advances in electronics and optical technology have now shifted the thinking to a new-generation initiative, now called 4G. The IEEE sometimes refers it to beyond 3G (B3G). The 4G Forum, which now meets annually, has the task of defining the migration path for 4G. The annual 4G Forum started in 2003 under the leadership of Samsung Electronics.

The 4G vision is depicted in *Figure 2*. It will support open and programmable architecture, thus enabling third-party developments. It will use multiple inputs multiple outputs (MIMO) orthogonal frequency division multiplexing (OFDM). It will support various data rates, including 144

TABLE 2

Voice Modem Standards

Standard	Year	Data rates (Kbps)
V.22	1980	0.6 or 1.2
V.22 bis	1984	1.2 or 2.4
V.32	1984	4.8 or 9.6
V.32 bis	1991	V.32, 9.6, 12 or 14.4
V.34	1994	Up to 28.8
V.90	1997	Up to 56

FIGURE 2

The 4G Vision

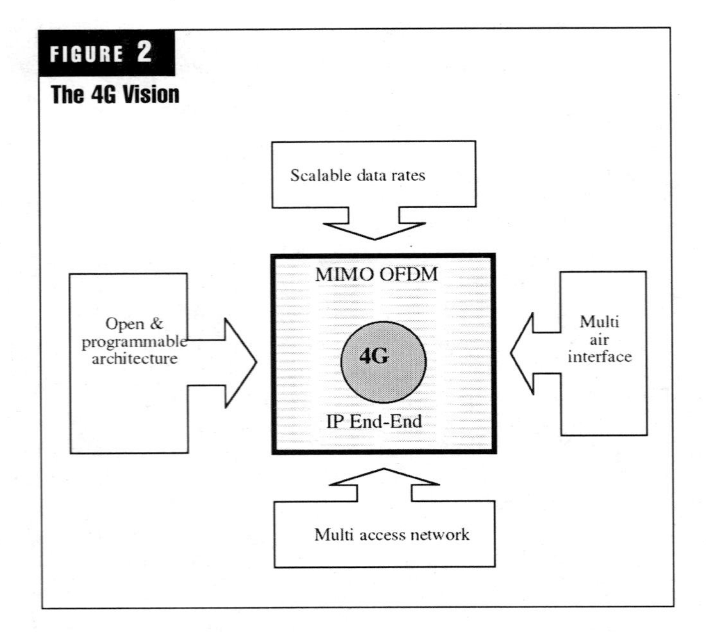

Kbps, 384 Kbps, 2 Mbps, and 10 Mbps, to meet existing technologies. The core of the network will be IP. The emerging 4G will be entirely packet-switched networks. It is expected to support IP from end to end. It will support rates of up to 100 Mbps to carry various forms of multimedia services at a lower cost. All the network elements must be digital, and the network will include water-proof-network security.

4G networks are expected to support global mobility and service portability. The network, when operational, would support location-based services, mobile shopping, e-mail, multimedia data transfer, and video streaming at reasonable prices. The plan is to introduce 4G services around 2010. Recently, Japan NTT Docomo announced plans to introduce 4G services in 2006, four years earlier than previously planned [26]. 4G is expected to converge several wireless systems, including wireless local loop (WLL), fixed wireless access (FWA), wireless LAN (WLAN), personal area network (PAN), WiMAX, and Universal Mobile Telecommunication Service Terrestrial Radio Access Network (UTRAN).

B3G and 4G are very active research areas. A detailed discussion on the vision for 4G was given in [20, 21, 22]. Yang et al. [23] considers in detail the combination of MIMO–OFDM with multiple antennas and concludes that this multiplexing technique is a very suitable and efficient method for high-data-rate wireless transmission. The application requirements and challenges of B3G/4G were explored in [24], and some solutions are proffered for the emerging system. A major aspect of the 4G vision is its envisaged open wireless architecture. 4G would support [25] varied access methods, multi-standard air interface, scalable data rates, scalable architecture, etc. 4G is expected to be 2 to 5 GHz technology.

Observation and Recommendations

The research for 4G poses great challenge for developing nations. Telecommunications business will be dictating future business for a long time to come, but the Third World nations play little or no active role in the definition of infrastructures of tomorrow. This paper is intended to stimulate the consciousness of the telecom world, particularly in the developing nations. While the world was already embracing 3G, operators in most developing nations, including Nigeria, were beginning to deploy 2G. Furthermore, the 2G switches deployed by most operators were circuit-oriented rather than packet technology, almost four years after legacy switches were no longer in use in developed nations. This is a challenge to the relevant engineering institutions in the developing nations, the telecom companies, and the government to create a forum to enlighten and keep stakeholders abreast of current technologies. We saw earlier that companies in the industry made most of the proposals and developments in advanced countries. Take TIA in the United States, which was responsible for submitting three 3G proposals to ITU, for example. TIA is a consortium of telecom companies with the staff of each company playing an active role in the definition and specification of NGNs. Similarly, the ETSI is an EU–funded standards body with members from the companies in the industry and educational institutions. The Association of Telecom Companies (ATCOM) in the developing nations should emulate what TIA is doing in the United States. ATCOM could get its employees involved in some level of research, no matter how small it is. The 4G discussions are currently going on, but companies in developing nations have no participation in it. Consequently, that means that poor investment decisions will probably be made for years to come out of ignorance. The governments of developing nations should be encouraged to set up research bodies just like the EU is doing with ETSI and other government-funded research projects in most countries in such an important area as telecommunications. ATCOM and other engineering institutions should work with the leadership of academic institutions to raise research consciousness among lecturers.

Conclusion

Telecommunications technology is evolving at a rapid pace due to unprecedented advances in electronics and optical technology and demands for higher data rates. The change is also a catalyst for new thinking in the industry and demand for new high-bandwidth services. In two decades, the mobile generations have evolved from 1G to 3G and beyond, when each generation infrastructure was supposed to have had a life of about 20 years or more. 2G failed because of its lack of support for high bit rates and incompatibility of networks in different countries. On the other hand, 3G was expected to ameliorate the ills of 2G, but the many proposals aimed at evolving 2G to 3G led to many incompatible 3G standards as well. Moreover, the various 3G standards have not met their promised capability set of delivering up to 2 Mbps and supporting multiple variations of multimedia services. This has propelled the quest for a new generation of mobile network, known as 4G or B3G. This new initiative is currently being defined, but it is projected to support any service anywhere and at any time.

The emerging 4G will be made up of entirely packet-switched networks. It is expected to support all–IP from end to end. It would support rates of up to 100 Mbps to carry various forms of multimedia services at a lower cost. All the network elements must be digital, and the network will have water-proof-security. The 4G would support global mobility and service portability. The network, when operational, will support location-based services, mobile shopping, e-mail, multimedia data transfer, and video streaming at reasonable prices. The 4G is also expected to support varied access interfaces such as WLAN, WLL, various radio access networks (RANs), and base station transceivers. 4G is the future.

Telecom is the driving force of most businesses in the world and will continue to be for many years to come. It is therefore a pivotal field that any forward-looking nation, particularly developing nations, must follow closely. It is therefore recommended that the telecom associations of developing nations should engage in some level of research to keep abreast of current thinking and trends in the industry. Also, the governments of developing nations should follow the practices of many industrialized nations and set up bodies that would support research, particularly in telecommunications, so the industry may be well-informed.

References

[1] Odinma, A, "Advances in Media Convergence for Information Systems," NICOMM2002, Abuja, November 11–13, 2002.

[2] Odinma, A., "Circuit and Packet Implications for Telecom Operators," NSE, Electr. Div. National Conference, Lagos, October 6–7, 2004.

[3] Odinma, A., "Recommendations for Pragmatic Migration Strategies to Next Generation Network for Nigerian Telecom Operators," NSE, Electr. Div. National Conference, Lagos, October 6–7, 2004.

[4] Odinma, A., "The Convergence of Telecom Networks and Migration Strategies for Operators," NSE, Electr. Div., National Conference, Lagos, October 6–7, 2004.

[5] www.3gpp.org.

[6] Tae-gwu, Kim, "4G Forum Offers Glimpse of New Telecom Tech," Staff Reporter, The Korea Times, July 2004.

[7] US Network Magazine, Vol. 1, No. 1, December 2002.

[8] Black, Uyless, "Mobile & Wireless Networks," Prentice Hall, N.J., 1996.

[9] Atal, B. S., Hanauer, S. L., "Speech Analysis and Synthesis by Linear Prediction of the Speech Wave," Journal of Acoust. Soc. Am., Vol. 50, No. 2, August 1971, 637–655.

[10] Atal, B. S., Schroeder, M.R., "Stochastic Coding of Speech Signals at Very Low Bit Rates," Proc. Int. Conf. Commun. Vol. 3, Amsterdam, May 14–17, 1984, 637–655.

[11] Horak, Ray, "Communication Systems and Networks," Hungry Minds, 2000.

[12] TIA/EIA, "TDMA Cellular/PCS—Radio Interface Enhanced Full Rate Voice Codec," TIA/EIA IS–641, May 1996.

[13] Atal, B. S., Cuperman, V. and Gersho, "Speech and Audio Coding for Wireless and Network Applications," Kluwer Academic Press, Boston, 1993, 85–92.

[14] Odinma, A. C., "Telecom Networks Convergent Architecture and Migration Strategies for Operators," NSE Tech. Transaction, Vol. 40, No. 4, October–December 2005, 1–12.

[15] Zysman, G., Tarallo, J., Howard, R., Freidenfields, J, Valenzuela, R. and Mankiewich, P., "Technology Evolution for Mobile and Personal Communications," Bell Labs Tech. Journal, Vol. 5, No. 1, January–March 2000.

[16] ITU–T Draft Recommendation Q1701, www.itu.org.

[17] Odinma, A., "The GSM Hype," *BusinessDay*, September 16, 2002; *ThisDay*, February 20, 2003.

[18] www.etsi.org.

[19] www.3gpp2.org.

[20] Evans, B. G. and Baughan, K, "Visions of 4G," Electronics and Communication Engineering Journal, December 2002.

[21] Huomo, H., Nokia, "Fourth Generation Mobile," presented at ACTS Mobile Summit99, Sorrento, Italy, June 1999.

[22] Pereira, J. M., "Fourth Generation: Now, It Is Personal," Proceedings of the 11th IEEE International Symposium on Personal, Indoor and Mobile Radio Communications, London, United Kingdom, September 2000.

[23] Yang, H., "A Road to Future Broadband Wireless Access: MIMO–OFDM–Based Air Interface, IEEE Communications Magazine, Vol. 43, No. 1, January 2005.

[24] Yu, X., "Toward Beyond 3G: The Future Project in China," IEEE Communications Magazine, Vol. 43, No. 1, January 2005.

[25] Lu, W., "4G Mobile Research in Asia," IEEE Communications Magazine, March 2003.

[26] www.eurotechnology.com/4G.

[27] Eazel, W. W., "Super-Fast HSDPA Mobile Broadband Coming Soon," www.vnunet.com, April 2, 2006.

[28] Thomson, I., "Nokia Insists WiMAX will not kill 3G," www.vnunet.com, June 13, 2005.

The Evolution of WiMAX Certification

Monica Paolini
President
Senza Fili Consulting

WiMAX certification has become reality almost a year ago, when the first WiMAX Forum Certified products were announced in January 2006. This was a key benchmark for the entire WiMAX community. After much hype and anticipation, the market has been able to assess WiMAX performance in real networks instead of making educated guesses from abstract specifications. Since then, twenty eight products have been certified as of October 2006, and certification for Mobile WiMAX is planned to start at the beginning of 2007.

These first WiMAX products, however, mark only the beginning of a certification process that will ultimately include numerous released and waves of testing. Each release or wave will include new certification profiles and/or new functionality to support new frequencies and different access modes (fixed, nomadic, portable, and mobile). The changes in the program are driven mostly by technological advances and product availability. They are crucial for ensuring that certified products have the functionality the market requires and can support new applications and services. Understanding how the certification process evolves is necessary if we are to have accurate expectations of certified products.

Certification is often perceived as a binary attribute: a product is either certified or uncertified. The reality is more complex. A product may be certified for only some of the functionality it supports. For instance, Fixed WiMAX products certified in the first wave will not be certified for quality of service (QoS). As a result, two certified products may work together in their basic configuration, but certified interoperability would not extend to QoS. This is a substantial limitation for a service provider that wants to offer QoS–based services such as voice over Internet protocol (VoIP) and is planning to use base stations and subscriber units from different vendors.

Product certification is an inherently complex process, especially when it involves interoperability among vendors, as is the case for WiMAX. The Wi-Fi Alliance, for instance, has been very successful in guaranteeing interoperability for certified products, but this has required a constant expansion of the number of profiles and the functionality that is tested and has taken over five years to get to where we are today. Some of the additions are certified as optional add-ons, but in some cases they soon become an integral part of the mandatory testing.

The WiMAX Forum is following a similar path. It is defining different system and certification profiles for classes of products that interoperate with each other, and setting subsequent certification releases and waves, each including additional functionality.

Before Certification: System Profiles and Certification Profiles

Not all WiMAX products will interoperate with each other[1]. A subscriber unit that operates in the 3.5 GHz band, for instance, will not be able to establish a connection with a 5.8 GHz base station. However, both products are based on the same standards—Institute of Electrical and Electronics Engineers (IEEE) 802.16 and European Telecommunications Standards Institute (ETSI) high-performance radio metropolitan-area network (HiperMAN)—and meet the same requirements. The WiMAX Forum uses the following two types of profiles to address the need for different classes of products that use the same technology:

- *System profiles*: System profiles set a basic level of common requirements that all WiMAX systems have to meet. To date, three system profiles have been defined:
  - Fixed WiMAX, optimized for fixed and nomadic access, is based on the 802.16-2004 version of the IEEE 802.16 standard and uses Orthogonal Frequency Division Multiplex (OFDM) multiplexing.
  - Mobile WiMAX, optimized for mobile and portable access, but supporting also fixed and nomadic access, is based on the IEEE 801.16e-2005 amendment and uses Orthogonal Frequency Division Multiple Access (OFDMA) multiplexing.
  - Evolutionary WiMAX is based on IEEE 801.16e-2005 and uses Orthogonal Frequency Division Multiplex (OFDM) multiplexing. It represents an evolution over Fixed WiMAX by adding portable access.

- *Certification profiles*: For each system profile, there are multiple certification profiles. For the Fixed WiMAX system profile, five certification profiles have been defined so far (see *Table 1*). No certification profiles have been announced yet for the Mobile WiMAX system profile, but the first ones will be in the 2.3-2.4 GHz, 2.5-2.7 GHz and 3.4-2.6 GHz bands. Certification profiles are defined by a system profile, the relevant spec-

TABLE 1

WiMAX Forum certification profiles for Fixed WiMAX

Spectrum band	Duplexing	Channel width
3.4-3.6 GHz	TDD	3.5 MHz
3.4-3.6 GHz	FDD	3.5 MHz
3.4-3.6 GHz*	TDD	7 MHz
3.4-3.6 GHz*	FDD	7 MHz
5.7-5.8 GHz*	TDD	10 MHz

(*) Eligible certification profiles. No product has been certified in this band as of October 2006.
Source: WiMAX Forum

trum band, what kind of duplexing is used (time division duplexing [TDD] or frequency division duplexing [FDD]), and the channel width.

These profiles are defined ahead of certification, based on spectrum availability in different countries and on market demand and vendor interest. Vendor interest is clearly a prerequisite for interoperability, as a minimum of three vendors are needed to get the interoperability testing under way.

The Certification Process: Compliance and Interoperability Testing

Ultimately, products need to pass the following two stages in the testing process to gain certification:

- Conformance testing to ensure that the product complies with the test specifications set forth in the system profile.
- Interoperability testing to ensure that subscriber units and base stations from different vendors operate in the same network.

While conformance testing is relatively straightforward, interoperability testing can cause delays during the initial phase of testing for a profile, since different vendors are often required to make changes to their products but may not be able to do so within the anticipated time frame due to disagreements over the changes or the difficulty of implementing them on a given vendor's system.

The Changing Scope of Certification

Technologies change with time, and their success is tied to their ability to support new applications and, more generally, to meet the growing requirements of their users. Certification programs need to adapt to these changes, but this requires a delicate balancing act between supporting innovation and maintaining compliance and interoperability across products certified at different stages.

WiMAX is no exception. The certification program has been structured using a framework of releases and waves, with each successive release and wave adding new profiles and functionality along the way. The products certified under the initial release and wave undergo an early and more limited set of tests. This is a wise choice as it focuses on the air interface protocol of WiMAX and sets the entire program on a solid footing. However, this also means that interoperability between a product certified during the first wave and one certified during the second wave will be limited to the set of features tested during the initial wave. Software or hardware upgrades will be required if a network operator wants to use the newly introduced features. Service providers and individual users will need to pay attention to the type of certification a product has received to make sure it includes the features they care about.

The certification releases and waves announced to date by the WiMAX Forum are shown in *Table 2*. The first Fixed WiMAX products submitted for certification in Release 1.0 Wave 1 were announced in January 2006 and early deployments have started later in the year. Certification of Fixed WiMAX products will continue in the second wave during the first half of 2006, and it will include QoS, and advanced security as optional modules. QoS is needed to support VoIP and, more generally, to prioritize access based on users or applications (e.g., for subscribers who pay higher fees or for real-time applications).

Certification for Mobile WiMAX is planned to start in early 2007, with the first certified products expected during the first half of 2007. The first products to be certified will include PCMCIA cards, laptop and PDA modules, indoor modems and base stations. In 2008, we expect to see the first certified mobile devices.

Different Vendor Strategies

Not all vendors have followed the same approach to certification. This reflects their market strategy, product timeline, and overall resource availability.

TABLE 2

WiMAX certification waves

	Certified products	Functionality supported
802.16-2004 WiMAX		
First wave Air protocol	4Q2005	Air protocol interoperability
Second wave Outdoor services	1H2006	Outdoor CPE in fixed deployments with QoS, security, advanced radio features*
Third wave Indoor services	2H2006	Indoor CPE and PCMCIA cards in fixed and nomadic networks*
802.16e WiMAX		
Third wave Portable services	1Q2007	Handoffs, simple mobility*
Fourth wave Mobile services	2007	Full mobility*

** = Expected. Test specifications have not been finalized yet.*
Source: WiMAX Forum, vendors, Senza Fili Consulting.

Some vendors (Airspan, Aperto, Proxim, and Redline) have entered Fixed WiMAX products for certification in the first wave and are committed to have the first WiMAX–certified products. Their initial involvement is not only motivated by a first-to-market strategy, but also by a desire to play a more active role in the certification process from the beginning.

Other vendors have chosen to join the certification process at a later stage. Certification requires a substantial effort, and they do not want to shift the focus away from enhancing their existing products for current customers deploying commercial networks to seek certification during the first wave since it will include only basic air interface testing.

The focus on portability and mobility has led several vendors to skip Fixed WiMAX certification altogether and focus on Mobile WiMAX. Alcatel, Motorola, Navini, and Nortel have all decided to develop exclusively Mobile WiMAX products, which they expect to see deployed in fixed, portable, and mobile networks. In several cases, these vendors believe that Fixed WiMAX does not offer sufficient advantages over their proprietary products to justify the development of a new product.

The Choice for Operators

Where does all this complexity leave those operators that are trying to decide what solutions to deploy and when to do so? They will certainly have some homework to do if they want to deploy gear from different vendors.

Knowledge of the requirements for system and certification profiles and of the functionality added during subsequent waves is needed to assess how products will interoperate in a WiMAX network. The level of interoperability among products may not initially cover features they consider essential or desirable (e.g., QoS or AES for Fixed WiMAX). After multiple certification released and waves, the degree of interoperability may vary depending on the wave during which the products involved were certified. Operators will need to look beyond the certification stamp and understand what features are covered by certification and whether they match their requirements.

Note

1. This is of course also the case with other wireless technologies that operate in different frequencies (cellular, Wi-Fi and several proprietary BWA products) or in different modes in the same frequency (b and g in Wi-Fi)

Prime Time for Mobile Television

Extending the Entertainment Concept by Bringing Together the Best of Both Worlds

Rob van den Dam
EMEA Leader, Telecom Sector
IBM Institute for Business Value

Jeanette Carlsson
Global Communications Sector Leader
IBM Global Business Services

Executive Summary

Not long ago, the thought of watching television on a small screen seemed far-fetched. However, since 2005, interest in mobile television has grown rapidly. Broadcasters and content providers increasingly deploy mobile TV as a vehicle for distributing their content on a larger scale and a source of new revenue. Telecom operators aim to increase average revenue per unit (ARPU) and reduce churn, and they view mobile TV as offering enormous potential to achieve both of these objectives. From a consumer perspective, market research indicates strong latent demand for mobile TV.

Many operators are already offering mobile TV services on a commercial basis over their wireless networks. Early deployments suggest that more people are watching television on their mobiles than originally anticipated. Meanwhile, subscriber numbers and volumes of video traffic are still low enough to avoid network problems. However, with increasing mobile TV adoption, networks will become increasingly congested. To overcome the capacity problems of wireless networks, many operators worldwide are carrying out trials with complementary broadcasting networks. In a broadcasting network, a mobile picks up the signal directly from the ether. In Southeast Asia, some operators have already launched commercial services on this basis.

One problem, however, is the lack of consensus on a universal technology standard. Nokia has adopted the Digital Video Broadcast – Handheld (DVB–H) standard, while Samsung is pushing Digital Multimedia Broadcasting (DMB). QUALCOMM has developed its own proprietary standard, MediaFLO. In Japan, the Integrated Services Digital Broadcasting – Terrestrial (ISDB–T) standard will be used for mobile TV. The Pacific region is driving the DMB technology, while in the United States, both MediaFLO and DVB–H will be the main standards. In nearly all European countries, mobile operators are piloting the DVB–H standard. Complicating this issue further is the fact that none of these standards is compatible with another.

The business model for mobile TV is also not yet clear. Early industry examples suggest that partnerships among operators, broadcasters, and other media parties—including content providers—will become the most commonly adopted model. Whatever the specific business model, cross-industry collaboration is critical to bringing the best of the mobile and TV worlds together, supported by suitable revenue-sharing arrangements. It enables the parties to profit from new revenue streams, from access fees, additional traffic, subscription services, premium messaging, and download sales to "m-commerce" (buying and selling of goods and services through wireless handheld devices) and advertising. The revenue sharing model, however, is yet to evolve.

For mobile operators, mobile TV offers an opportunity for differentiation to help drive new revenue growth, improve network utilization, and reduce subscriber churn. Mobile TV is seen as having good revenue potential because of its mass appeal across the customer base, unlike many other content services such as games, which appeal only to selected customer groups such as teenagers. The combination of a mobile and a complementary broadcasting network enables interactivity between user and content, seen as a key driver of further mobile TV adoption and usage. Without the reverse link that the operator provides with their mobile network, much of the technological and economic benefits of the broadcast model are negated.

For the consumer, mobile TV has the potential to deliver new and revolutionary content that will extend the entertainment concept beyond new ways of watching TV. Mobile TV enables consumers to personalize their viewing experiences with the content to suit their tastes, wherever and whenever they want—a first step toward personal TV in a multidevice (TV, computer, and mobile) and networked environment. Southeast Asian countries are leading the way in mobile TV, with the rest of the world likely to follow, as TV is an easy concept to sell to consumers. However, before widespread mobile TV can become a reality, telecom providers need to resolve key issues, including the following:

- *Content:* Availability of the "right" content is key. Consumers want access to specific content to suit their tastes. "Must-see" content, interactivity, and personalization are key adoption criteria.

- *Pricing*: The success of mobile TV depends on consumers' willingness to pay. Bundled, tiered, flat-rate pricing is more likely to be accepted than a la carte (metered) pricing. Additional fees can be based on pay-per-view, clip-casting and interactivity-driven purchases.

- *User-friendliness*: Key issues such as form factor, power consumption, added functionality, and TV picture quality have yet to be resolved. Handsets must be attractive and affordable.

- *Business model*: All parties involved have to agree on their roles in the mobile TV value chain, including the adoption of a revenue-sharing model that is acceptable to all.

- *Digital rights management (DRM)*: Content providers see mobile TV as a new channel to market their content and want to enforce content rights and payments.

- *Technology standards and spectrum allocation*: The coexistence of several incompatible standards and problems with harmonization of the frequency bands, particularly in Europe and Asia, are key hurdles to be overcome.

Planning for a Future in Mobile TV

Media Companies Seek to Exploit Their Content on a Larger Scale

Mobile TV combines two popular consumer products of our time—the mobile phone and the TV. As *Figure 1* shows, TV is more popular than ever and mobile penetration is very high in most major markets.

With mobile market penetration now greater than 80 percent in many countries, mobile TV has the potential to appeal to the mass market. For this reason, broadcasters and content providers are increasingly interested in mobile TV. It provides them access to an untapped market of more than 2 billion cell phone subscribers worldwide.[1] Extending content delivery to the mobile device offers a new distribution channel on a much larger and more profitable scale. In addition, the mobile phone enables them to enrich their services with interactivity and personalization. For instance, viewers may chat, vote, or respond directly to TV programs. Other options include downloading special backstage shoots, and interactively reserving cinema/theater tickets on the basis of previews of the latest movies watched on the mobile. Mobile TV has indeed enormous revenue potential for media companies.

The increasing interest of the media world in mobile TV is illustrated by recent market trends: MTV, Warner Music, and CNN have formed alliances with operators to provide content specially tailored for cell phones.[2] The Walt Disney Company has announced its new mobile TV station, Disney Mobile.[3] As one of the first, Fox Entertainment has started developing one-minute episodes for mobile TV.[4] An example of these so-called "mobisodes" is the "24 Conspiracy" series. More and more media parties such as ABC and Endemol have followed Fox in producing mobisodes. In the United Kingdom, Endemol has also introduced new TV channels such as the Extreme Reality Channel and Comedy Channel especially for mobile phone users.[5]

Operators Seek to Increase ARPU and Reduce Churn

Mobile operators are increasingly active in mobile TV. With mobile/wireless Internet access ramp-up driving demand for multimedia content and other data services "on the move," many operators are pinning their hopes on a more profitable future for mobile TV than, for instance, mobile phones with built-in cameras. Despite the hype, camera phones have not created value for telephone companies (telcos) to date, as they have not driven additional traffic. In

FIGURE 1

TV Consumption and Mobile Penetration

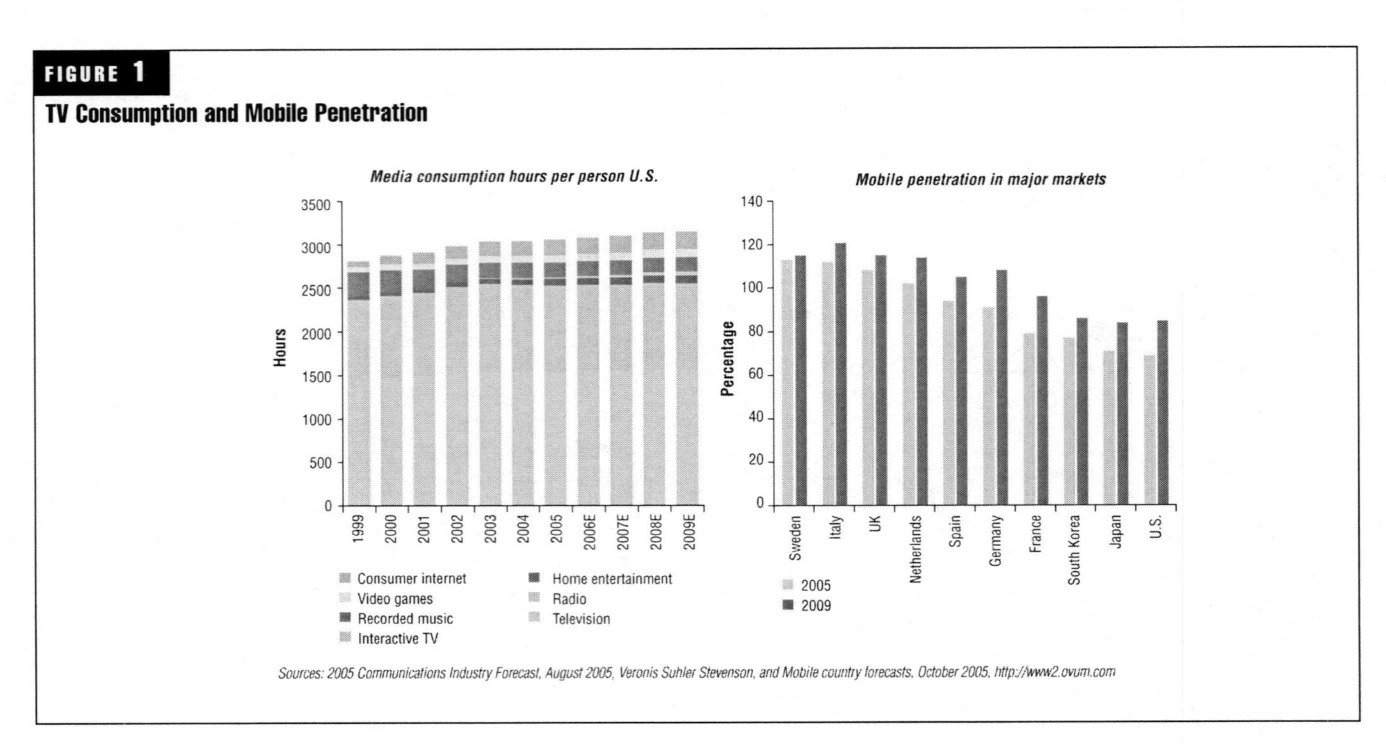

Sources: 2005 Communications Industry Forecast, August 2005, Veronis Suhler Stevenson, and Mobile country forecasts, October 2005, http://www2.ovum.com

fact, industry forecasts suggest that the interest in mobile TV might be greater than all other prospective mobile multimedia services, including games and music.[6]

Aside from its potential to drive additional data traffic, mobile TV may also make important contributions to telco customer retention and loyalty strategies through customer lock-in. However, successful execution hinges on operators offering high-quality content tailored to individual customer needs, which makes collaboration with broadcasters and media producers a critical success factor. In recognition of this, some operators are starting to make acquisitions in the media world. For example, SK Telecom is the main shareholder in the Korean media company TUMedia, and NTT DoCoMo recently acquired shares in Fuji Television.[7] In Europe, H3G Italia has acquired Canal 7, an Italian TV channel.[8]

Delivering Mobile TV over Cellular Networks

Today's Mobile TV More Popular than Anticipated

Many operators are already offering mobile TV services on a commercial basis. They have observed that more people are watching television on their mobiles than originally anticipated. In the United States, Verizon Wireless provides (non–real time) mobile TV services under the brand V CAST over its evolution data optimized (EVDO) network.[9] Sprint offers Sprint TV (packaged video clips) and Sprint TV Live (real-time TV) over its new EVDO network and its PCS network.[10] And Cingular Wireless has the MobiTV service over its Enhanced Data for Global Evolution (EDGE) wireless network.[11] Both Sprint and Cingular get all their content from MobiTV, a company that specializes in taking television feeds and sending them over cellular networks.[12]

In Europe, mobile TV is provided by operators such as O2, Orange, T-Mobile, and Vodafone. Orange in France offers more than 50 TV channels on its UMTS network.[13] Vodafone offers mobile TV on its Universal Mobile Telecommunications System (UMTS) Vodafone Live service in 12 countries. Most recently, Vodafone launched global mobile TV channels across its markets as an extension of domestic TV programming.[14] What these commercial services have in common is that signals are transmitted over wireless networks. This involves video streaming, where the supplier sends the television signal to every user over the network in separate streams. This is known as unicasting, which can be real time or non–real time (see *Figure 2*).

Video streaming has the advantage of working across different mobile standards such as general packet radio service (GPRS)/EDGE and third generation (3G). Another advantage is that, in principle, an unlimited number of television channels can be offered. Finally, the two-way directional transmissions—upstream and downstream—allow a high level of interaction with the mobile user, which some players view as key to driving adoption and usage. Overall, this gives operators many opportunities to generate additional revenues. Indeed, many operators see television on the mobile phone as one of the important drivers for the success of 3G.

Popularity of Mobile TV over Cellular Networks Results in Congestion

Unicast transmissions, however, consume more network resources as usage grows. At present, 3G networks have plenty of spare capacity. However, as mobile TV commoditizes and the number of television viewers grows, unicast networks will become more heavily loaded and approach capacity constraints. That will certainly be the case with the current low 3G network speeds of 150 to 220 Kbps, but even with steady increases in transmission speeds over the coming years to the 3G maximum speed of around 384 Kbps, this will be far from sufficient. With the evolution toward high-speed downlink packet access (HSDPA) technology, faster transmissions can be realized. But even the theoretical

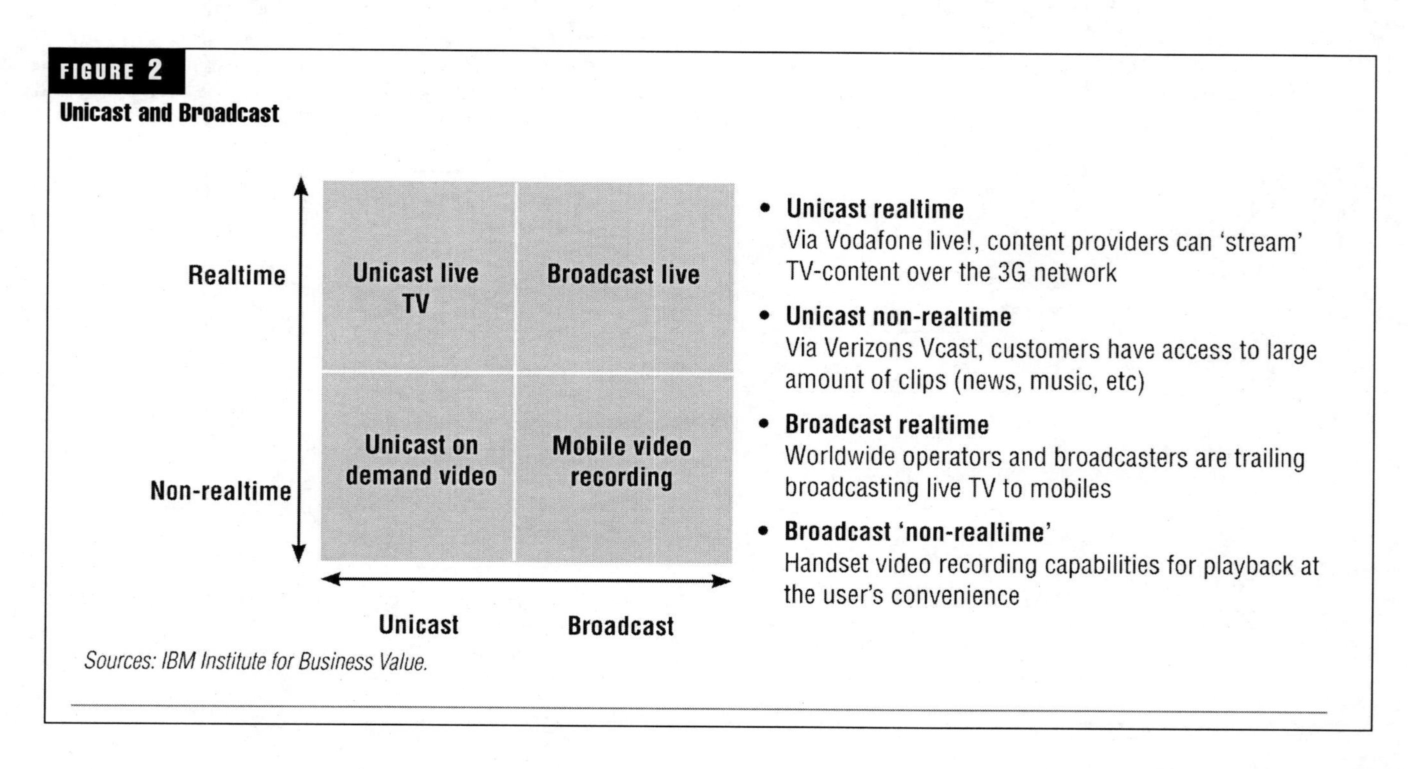

speed of 14.4 Mbps will be insufficient in areas with many mobile television viewers.

Another technology alternative to improve video streaming over mobile networks is multimedia broadcast and multimedia services (MBMS) technology, which is being developed as part of 3G partnership project (3GPP) Release 6.[15] MBMS enables efficient broadcasting over UMTS but is only suitable for a small quantity of video streaming over only one to three television channels. Moreover, commercial products based on this standard will be available in 2007 at the earliest.[16]

The reality is that 3G networks were never intended for real-time television. It is an inefficient use of limited network capacity to use the mobile network to transmit a multicast service, and also inefficient from a financial point of view. According to Markus Lindquest, Nokia director of rich media services, a three-minute video clip delivered to 100,000 users over current 3G networks could take one and a half to two days to deliver.[17] Over a broadcast network, the same transmission could take only a few seconds. In a broadcast network, a mobile picks up the signal directly from the ether. Across the globe, therefore, trails are carried out with complementary broadcasting networks, to overcome the capacity problems of mobile networks.

TV Broadcasting over a Mobile

War of the Standards

The big challenge with broadcasting mobile TV is power consumption. Mass market rollout will depend on the ability to provide TV pictures without significantly reducing battery life. This has been one of the main reasons for developing new digital television standards that are compatible with mobile phones. One key problem, however, is that the involved parties cannot agree on a universal standard (see *Figure 3*).

The DMB standard is the TV version of the European Digital Audio Broadcasting (DAB) standard. The Asia-Pacific region is driving the development of this technology, backed by Samsung in particular. This standard can be divided into two headline technologies: satellite (S–DMB) and terrestrial (T–DMB). In 2005, SK Telecom launched the world's first commercial television based on the S–DMB standard with seven channels for a $13 per month subscription fee.[19] In Japan, satellite mobile TV using DMB is provided by the Mobile Broadcasting Corporation (MBCO). The DAB/DMB standard is also in trial stages in various countries in Europe, including the United Kingdom[20] and the Bayern region of Germany.[21]

The dominant development in Japan is the construction of a terrestrial broadcasting network based on the ISDB–T standard. ISDB–T is the same standard that is used in Japan to deliver digital TV (DTV) services to the home. ISDB–T will be employed in Japan only. Commercial mobile TV services started on April 1, 2006.[22]

QUALCOMM in the United States has developed a proprietary broadcasting standard, MediaFLO, based on its forward link only (FLO) technology.[23] FLO is a proprietary standard that operates in the 700 MHz frequency spectrum only. QUALCOMM has already obtained exclusive nationwide rights to use this frequency band. QUALCOMM's MediaFLO can offer up to 20 channels at 30 frames per second. MediaFLO will mainly be used in the United States. Japanese operator KDDI has announced plans to go head to head with leading rival NTT DoCoMo with the revelation that it will set up a mobile TV joint venture with QUALCOMM.[24]

According to Informa Telecoms & Media, the DVB–H standard will be the dominant standard by 2008, reaching 74 million users worldwide by 2010, equal to almost 60 percent of all broadcast mobile TV users.[25] DVB–H is the mobile version of DVB–Terrestrial (DVB–T), with reduced power consumption by as much as a factor of 10, as a result of a time-slicing scheme (content is delivered in bursts, allowing the receiver to be "on" only around 10 percent of the time). It can theoretically offer 30 TV channels and has a frame rate of 25 frames per second, comparable with standard television. It is pushed by Nokia and has already been approved

FIGURE 3

Standards for Mobile TV

	MBMS[A]	BCMCS[B]	DVB-H	ISDB-T	MediaFLO	DMB
Standard	Open	Open	Open	Open	Proprietary	Open
Regions	Europe, Asia	U.S., Asia	U.S., Asia, Europe	Japan	U.S.	Korea
Technology	W-CDMA; EDGE/GPRS	CDMA2000; EV-DO	OFDM[C]	OFDM	OFDM	OFDM
Availability service	2006/2007	2006/2007	2006	Available	2006/2007	Available
Availability handsets	2006/2007	2006/2007	2005/2006	Available	2006/2007	Available

Notes: A) Multimedia Broadcast and Multimedia Services, allowing broadcasting over UMTS; B) BroadCast and MultCast Service (MBMS-variant for CDMA2000); C) Orthogonal Frequency Division Multiplexing interface technology.
Source: IBM Institute for Business Value.

by the European Telecommunications Standards Institute (ETSI). A key advantage of DVB–H is that it can be used over several frequency bands. It needs a separate network of terrestrial transmitters, but it is likely for the transmitters to be situated at the same sites as the GSM/UMTS transmitter masts. Nokia has recently formed an alliance, called the Mobile DTV Alliance, with Intel, Modeo, Motorola, and Texas Instruments.[26] Recently Microsoft has joined the alliance.[27]

Research Firms Predict a Bright Future for Mobile TV

Despite incompatible standards, there is much optimism that mobile TV will live up to its promise based on this new digital broadcasting technology. As already shown, this is backed by a wide range of research that forecasts the market will grow to between $5 billion and $27 billion worldwide by 2010. For instance, Datamonitor has created a conservative forecast for the take-up of mobile TV and is expecting 69 million global subscribers in 2009, generating revenues of around $5.5 billion (see *Figure 4*).[28]

Frost & Sullivan predict a similar trend; they expect mobile TV revenues to hit $8.1 billion by 2011.[29] Other researchers forecast even more mobile TV users by 2009/2010. ABI Research, for instance, forecast a market worth some $27 billion by 2010, though the services will be spread much more widely than in today's mobile ecosystem.[30] Research by Informa Telecoms & Media predicts that by 2010, almost 125 million people around the world will possess a handset on which television pictures can be viewed.[31] Finally, Juniper Research expects that by 2010, 65 million people around the world will be subscribing to mobile TV services.[32]

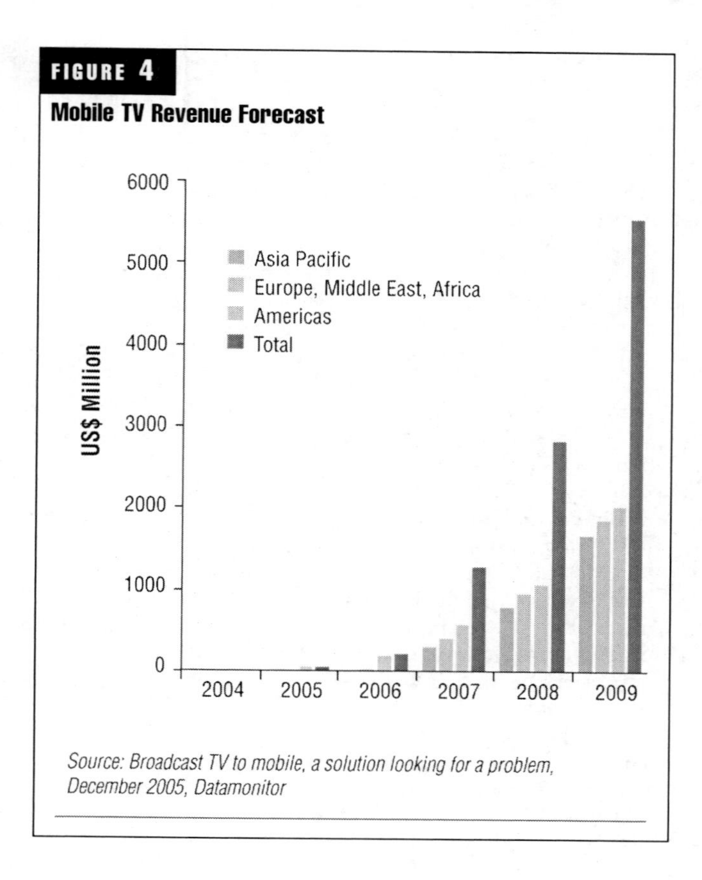

FIGURE 4

Mobile TV Revenue Forecast

Source: Broadcast TV to mobile, a solution looking for a problem, December 2005, Datamonitor

Preparing for Liftoff

The United States' Finale between QUALCOMM and Modeo

Mobile TV trials are now widespread across the United States, Europe, and Asia. In the United States, spectrum has been allocated and sold to QUALCOMM (700 MHz range) and Crown Castle (1,670 to 1,675 MHz range). Both parties are working on national deployment of their broadcasting networks in approximately 30 markets across the United States. QUALCOMM'S MediaFLO network is operated by MediaFLO USA. The DVB–H network is operated by Modeo, formerly known as Crown Castle Mobile Media. Both broadcast network operators will wholesale their services to mobile operators, which in turn will offer the service to their subscribers. QUALCOMM plans to launch its network in October 2006. Verizon Wireless, the largest U.S. CDMA carrier, will be the first mobile operator to use the MediaFLO network.[33] Modeo, which has piloted its network in Pittsburgh, plans to launch the service in mid-2006 in selected major U.S. markets, including New York.[34] Cingular Wireless, the largest U.S. Global System for Mobile Communications (GSM) carrier, could be one of the main mobile operators using the DVB–H network.

European Countries Are Preparing for a Future in DVB–H

In nearly all European countries, mobile operators are piloting the DVB–H standard to evaluate technology, business models, consumer preferences, and new revenue opportunities. In Italy, media group Mediaset and Telecom Italia Mobile (TIM) have reached an agreement that looks set to become the basis for the world's first commercial launch of digital terrestrial TV on mobile phones using DVB–H technology.[35] However, 3 Italia, the Italian subsidiary of Hutchison Whampoa, is also in the race and planning to launch its DVB–H mobile TV service in June.[36] Italy is expected to be a fertile ground for the new technology as Italians number among the world's keenest watchers of television and owners of mobile phones. In Berlin, Germany, a DVB–H transmitter is permanently in the air to test new services. Germany wants to get ready to broadcast the 2006 football world championships on mobile devices.

There is trial activity in many other European countries, including France (Orange, Bouygues), Netherlands (KPN), Switzerland (Swisscom), Spain (Telefonica Moviles), and the Czech Republic (T-Mobile). In the United Kingdom, in the Oxford region, 360 users are taking part in a DVB–H trial by O2, Nokia, and Arqiva. In Finland, Nokia, Elisa, TeliaSonera, Digita, MTV, YLE, and Nelonen have carried out a mobile TV pilot with 500 users in the Helsinki region. Finland has awarded Europe's first operating license for mobile TV broadcast to French media group TDF's Digita unit, which plans to open commercial service in the Nordic country in the second half of 2006.[37] In general, the lack of spectrum in European countries and the slow process of allocating frequency bands may seriously delay the development of mobile TV services there. In addition, there is the question of whether the allocation can be harmonized across the various countries.

Bringing the Best of Two Worlds Together

Business Model Innovation Is the Strategic Differentiator

It is not yet clear how the business model for mobile TV will evolve. Some operators may decide to buy themselves into the media world. SK Telecom, NTT DoCoMo, and 3 Italia have made initial moves in that direction.

On the other hand, some media companies may aim to become mobile virtual network operators (MVNOs) themselves. Walt Disney may be one such example, creating a direct channel for its huge catalogs of movies, TV, games, and images, and also—through the MVNO—its own branded mobile phones. BSkyB, a pay-TV satellite broadcaster in the Unite Kingdom that already provides mobile TV services via Vodafone's 3G network, recently bought broadband provider Easynet, which could be a first step in that direction.[39] It would also be a logical response to the recent merger of Virgin Mobile with cable company NTL, which created a quadruple-play mobile, fixed line, broadband, and TV service under the Virgin brand.

But in general, innovative business models in which operators, broadcasters, and media companies work closely together have the greatest potential for success. The 2006 IBM CEO study reveals that business model innovation and collaboration are the key components for differentiating organizations in a highly competitive environment. In fact, an important finding of the study is that innovative collaboration separates winners and losers.

Therefore, partnerships among operators, broadcasters, content providers, and other media companies is the logical way to bring the worlds of mobile and TV together (see **Figure 5**).

Operators have a large subscriber base and means for secure personal interactivity but seek to increase ARPU and reduce churn. Mobile TV complements 2.5/3G services with a one-to-many broadcasting capability that enables the operators to develop their customer relationships further. Operators have little experience in quality content that only the media world can provide. Working together with broadcasters will enable them to offer unique programs that provide opportunities for brand differentiation.

FIGURE 5

Business Model and Roles: Multiple Opportunities for Business Partners

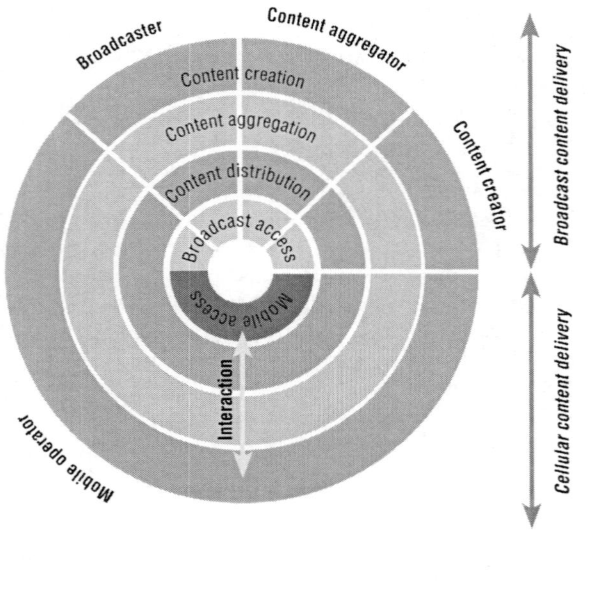

Broadcasters have attractive content and an efficient broadcast channel but seek to distribute expensive content on a large scale and enrich their services with interactivity and personalization. Mobile TV provides a new platform for taking broadcast TV into virtually any location outside the home. It expands the broadcast business into the increasingly mobile lifestyle of viewers. It creates additional, incremental audiences to receive broadcast entertainment and information during new prime times: the commuting periods and lunch breaks. Mobile TV is a real opportunity for broadcasters to build a unique mobile audience for their programs and brands.

From a business model perspective, content providers see mobile TV as an additional distribution channel to extend their reach. Content payment and revenue sharing agreements, as well as DRM, are critical to getting content providers on board. DRM involves the delivery of secured and accounted-for digital content over IP networks specifying usage rights and prohibiting unauthorized usage or access. DRM covers key content-related concerns such as the number of times a particular piece of content may be viewed, whether content may be shared with other devices, and content expiration dates, as well as enabling the previewing of content by a customer prior to purchase. Enabling conditional access (making sure only those paying for the service get it) is not a trivial matter, with a number of issues still pending resolution.

Integrating Cellular Networks and Broadcasting Networks for Interactivity

The combination of a mobile and an overlay broadcasting network enables interactivity between user and content, widely considered key to further growth and essential for mobile operators to generate revenues from their mobile networks. This can be executed via a parallel model; both the broadcaster and operator will be able to send signals to the mobile user, with a cellular backhaul (the return channel). Alternatively, an integrated model may be used, in which the operator is responsible for sending both cellular and broadcasting signals to its subscribers, and gets its content from broadcasters, content providers, content aggregators, or other media parties (see *Figure 6*).

What the revenue sharing model will look like is not yet clear. More than likely, it will be based on a combination of subscription fees, network use, advertising revenue, and download sales. But it is clear that cooperation among operators, broadcasters, media producers, advertisers, and retailers can benefit all. Telcos will profit from service subscriptions, additional traffic, premium messaging (SMS/MMS revenues from viewer voting and user polls), download sales (on-demand video clips), and m-commerce (for example, purchasing, data, and Web-based services) via their networks' return channels. For the media players, it offers new revenue streams from access fees (for basic and enhanced packages of channels and content), pay-TV subscription services (premium channels, pay-per-view), and

FIGURE 6

Parallel Model versus Integrated Model

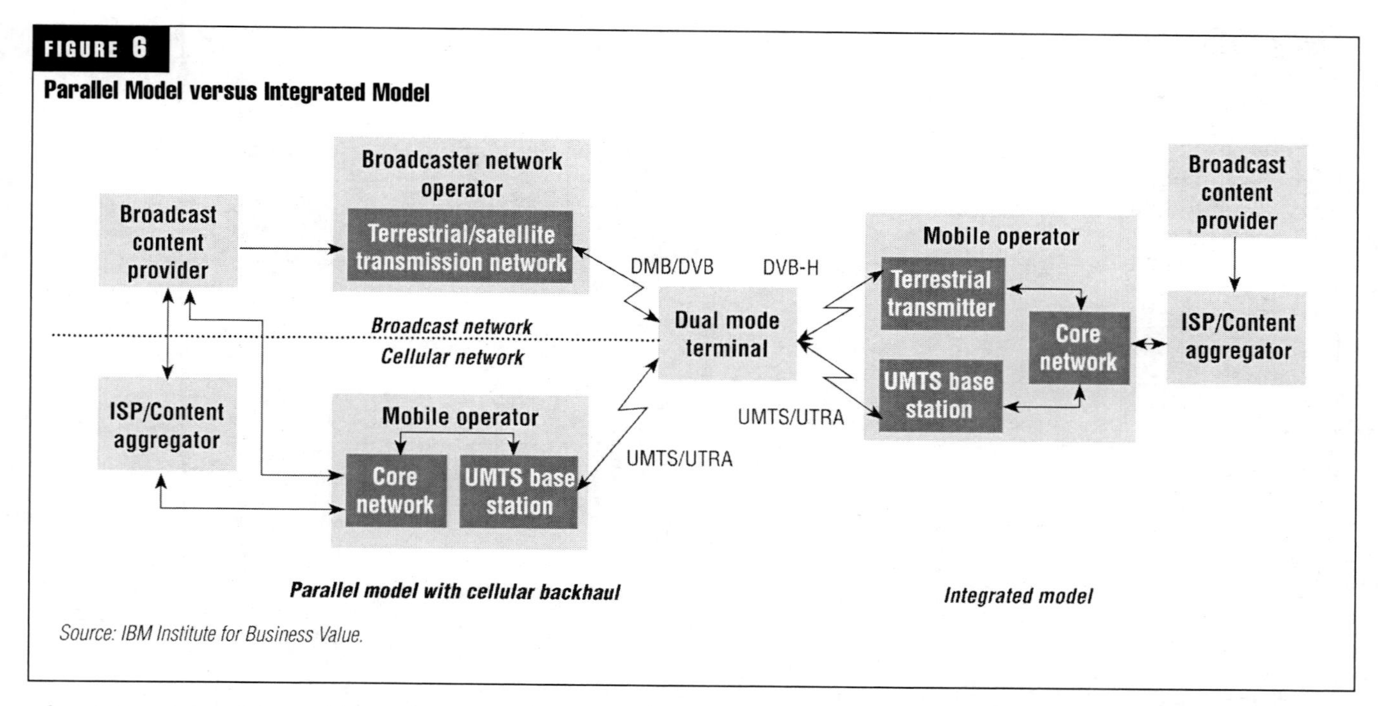

Source: IBM Institute for Business Value.

advertising. Advertising will be a source of income only from the moment that mass adoption has actually occurred.

Mobile TV Offers an Opportunity for Differentiation to Help Reduce Churn

Mobile TV is seen as having good revenue potential for mobile operators because of its mass appeal, unlike other content services such as games, which only appeal to selected user groups, for example, teenagers. Overall, mobile TV provides the following opportunities for mobile operators:

- A new service that is easy to position and with obvious usage scenarios, and hence the potential to increase ARPU
- A complementary service to current 2.5/3G services with one-to-many broadcasting capability, enabling new service bundling and deepening of customer relationships (lock-in)
- Scope for interactivity via network return channels to provide additional revenue potential through services such as voting, chatting, purchasing, and data and Web-based services using the mobile
- Scope to improve network utilization through additional traffic via the return channel (such as on-demand clips)
- Deepening of business partner relationships (and revenue streams) through need for billing and e-commerce system upgrades
- Opportunity to offer unique program channels that allow brand differentiation

Mobile TV offers a high-quality, interactive mobile TV experience that could help to improve customer loyalty. Operators should focus on differentiation around exclusive content, high-profile brand names, and more personalized programming. Exclusive deals with content providers to provide customers with previews and additional interactive services, such as ring tones and games, are key examples. Operators can also introduce pay-per-view options for exclusive content such as movies or pre-screenings of popular TV series. An example of such an exclusive joint venture is that of KPN and Endemol. Together, they form one of the largest European entertainment companies.[41]

Operators play several important roles in making mobile TV commercially feasible. As stated, they provide the reverse link needed to enable interactivity. Secondly, mobile operators already have in place the technology to recognize authorized users for the services. Thirdly, in most major mobile markets today, operators already serve as primary distribution channels for handsets.

Consumer Adoption Determines Success or Failure of Mobile TV

Recent User Trials Reveal Strong Demand for Mobile TV

Recent surveys and trials and current commercial mobile TV services show that the demand for mobile TV is higher than originally anticipated. For instance, in a Siemens survey of 5,300 mobile users in eight countries (United States, Canada, Brazil, Germany, Italy, Russia, China, and South Korea), 59 percent of respondents expressed an interest in mobile TV.[42] Also, the results from recent DVB–H trials[43] and BT Movio pilot tests[44] have demonstrated apparent consumer demand for mobile TV.

Interim results from the pilot in Oxford revealed that 83 percent of the participants were satisfied with the mobile TV service and 76 percent said they would take up the service within 12 months of commercial launch. In France, 68 percent said they would pay for mobile TV services, while 55 percent in Spain were willing to do so. Nearly 75 percent of Spanish participants would recommend the service to friends and family.

FIGURE 7

Results of Pilot Mobile TV Services in Four Countries

	Finland	UK	Spain	France
Positive response to mobile TV	58 percent believe mobile TV services would be popular	83 percent are satisfied with the service	75 percent would recommend the service	73 percent were satisfied with the service
Willingness to pay for mobile TV	41 percent	76 percent	55 percent	68 percent
Acceptable monthly fee for mobile TV	€10	-	€5	€7
Average daily viewing	5 to 30 minutes of mobile TV per day on average	23 minutes per session with 1 to 2 sessions per day	16 minutes	20 minutes
Peak viewing times	-	Mornings, lunchtime and early evenings	While commuting and between 7pm and 8pm	Morning (9-10), midday (1-2) and evening (8-10)
Popular content	Local programs available through Finnish national TV and sporting events	News, soaps, music, documentaries and sports	News, series and music	News, music, entertainment, sports, documentaries and films

Above all, the success of mobile TV depends on consumer adoption. Mobile TV provides for the customer a new and desirable service that is easy to understand and has obvious usage occasions. But consumers will only be willing to pay for mobile TV if they are satisfied with the content, price, and user friendliness. The aforementioned surveys and trials have revealed important indications for these important issues.

Consumers Want Content at the Right Time

Key to success is the availability of the right content at the right time. The trials have identified the following particular situations in which mobile users like to watch mobile TV:[46]

- *Spare time*: Watching during spare time, for instance, while waiting for or traveling on buses, trains, and airplanes.
- *Must see*: When people do not want to miss their favorite programs, even though they are not at home.
- *Catch up*: In situations where people want to follow breaking news. During the bombing incidents on the London underground in 2005, for instance, people found out that they could follow the news on their mobiles all the time.
- *Quick escape*: Orange has observed that 36 percent of its mobile TV users are watching its 3G mobile TV during lunch or other breaks.[47]
- *Background TV/radio*: Many people use TV or radio as background; for instance, at work.

The results of a trial in the Helsinki area in Finland are illustrated in *Figure 8*.

FIGURE 8

Mobile TV Use at Particular Moments in Finland

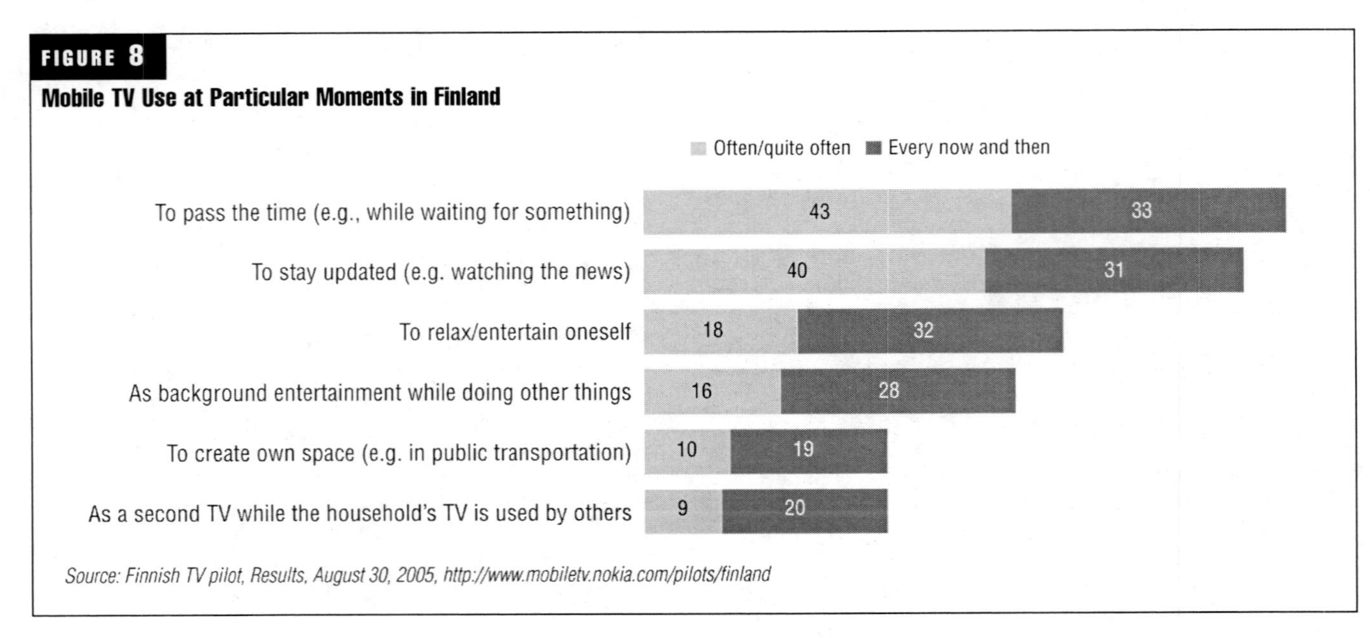

Source: Finnish TV pilot, Results, August 30, 2005, http://www.mobiletv.nokia.com/pilots/finland

New prime times for broadcasters and advertisers also emerged from the pilots. The Oxford results reveal a lunchtime viewing peak higher than the normal TV pattern, suggesting that viewers are enjoying their favorite TV content while on their lunch breaks. In France, participants watched TV for 20 minutes on average per day, with early morning, lunchtime, and mid-evening representing the periods of highest use. The Spanish pilot also reveals mobile TV viewing spread throughout the day, with early evening representing peak viewing.

In the Finnish trial, the average user watched up to 30 minutes of mobile TV, with entertainment, sports, and news content as the most popular.[48] Interim results of the O2 trial in Oxford showed that users watched, on average, 23 minutes per session, with one or two sessions per day. Overall, trial participants viewed an average of about three hours per week, with one group of enthusiasts viewing for more than five hours per week.[49]

Consumers Want the Right Content

Mobile TV is a crossroads of two powerful social trends: greater mobility and new forms of accessing media content. Program content will, of course, be key to the success of mobile TV and video services. Mobile TV is not just traditional television on a small screen. It is not a substitute for traditional television; instead, it will complement rather than replace it. The introduction of mobile TV will result in television without frontiers.

Mobile TV has the potential to deliver new, revolutionary content that will extend the entertainment concept rather than extending the ways in which traditional TV can be viewed. Indeed, if content providers embrace the opportunities mobile TV delivers, the long-term implications for traditional TV programming will be significant, as consumer demands shift and evolve. Mobile TV provides the highest degree of personalization and widest selection of desirable programming. Consumers may expect more choice and individual treatment.

The TV viewer on the move is likely to enjoy programming that is coming in shorter fragments. Breaking news, highlights of sport events, mobisodes, and music clips lend themselves well to this. Also, the "Big Brother" approach has proven to be a successful mode, where the consumers had the opportunity to watch behind the screen, which was not available to the general TV public. Creating made-for-mobile extensions of popular TV programs, usually in the form of clips, has appeal. And there is a potential for unique, made-for-mobile content, as Endemol is focusing on. The greatest gains are to be sought in the addition of interactivity to TV programs such as voting, user polls, additional information on programs, chatting, merchandising, ticketing, and on-demand downloads (see *Figure 9*).

Broadcast TV for mobile can be a powerful new service that further enables users to personalize their mobile handsets so that they can always have the content they want. In fact, it is a trend toward "personal TV," and the addition of TV content digitally broadcast straight to users' mobiles is a huge part of that vision. Mobile TV will appeal to some demographic groups more than others. But unlike services such as games, mobile TV is attractive in its potential to appeal to a wide mass-market demographic. Mobile TV is likely to appeal beyond teenagers, the typical target group and early adopters of new content services. For example, people in the 30-to-40-year-old range have shown interest, according to recent surveys conducted by Siemens and Nokia.

Customers Want the Right Price

Getting the pricing right will be important. First we have the fees for the mobile services itself. There are the following two options:

- *Bundled, tiered, flat-rate pricing*: The most likely scenario involving a fee structure similar to that of traditional TV, this would likely involve an "all you can eat" fee structure revolving around tiered package plans with different fees for basic programming and incremental additions for premium programming.

- *Metered pricing*: This is based on one of several options, including total minutes of use, usage in proportion to network use, or by individual program subscriptions. A metered plan is not the most likely option, as it could hinder the widespread adoption of this new service.

Additional services and fees can be built upon the following flat-rate and metered structures:

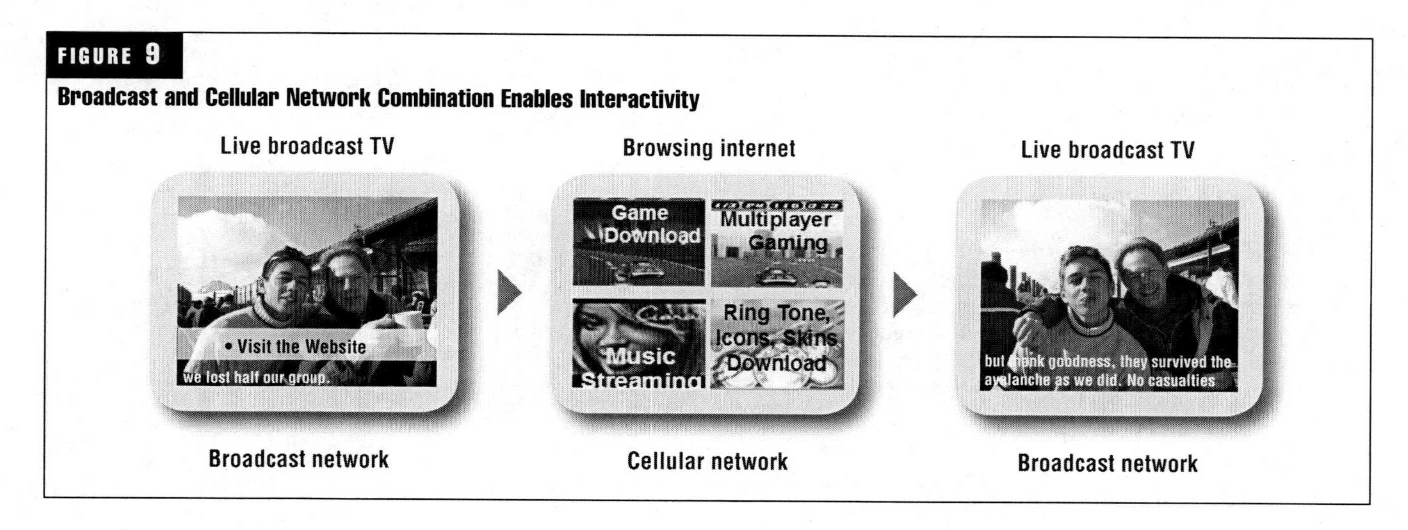

FIGURE 9
Broadcast and Cellular Network Combination Enables Interactivity

- Pay-per-view services for real-time events
- Clip-casting programs such as a gardening channel
- Real-time applets such as sports tickers or financial exchange tickers
- Regularly delivered game/application of the week type of service

Also, there is the pricing for interactive-driven purchases of ring tones and wallpaper; items such as songs and videos; and content-associated real-world items such as DVDs, movie tickets, and toys.

Furthermore, the prices for the handsets are important. Initial broadcast-capable phones will be around $500, beyond the means of the average subscriber. To overcome this problem, operators may decide to subsidize the handsets.

Customers Want Easy-to-Use Devices

Most handset manufacturers are optimistic about the success of mobile TV. DMB handsets, which support commercial mobile TV services, are already available in Southeast Asia, granting Korean handset manufacturers Samsung and LG first-mover advantages. Nokia has developed the Nokia N92, the first mobile phone with a built-in DVB–H television receiver. And Sharp is one of the device manufacturers investing in MediaFLO. Almost all handset suppliers have developed mobile TV devices or prototypes capable of supporting one or more standards.

However, before widespread mobile TV becomes a reality, a number of issues such as the following have to be resolved for the handsets, particularly with respect to user friendliness:

- *Power consumption*: Significant reductions in battery life will not be acceptable to consumers. The new digital mobile broadcasting standards can reduce power consumption by as much as 10 times; in addition, advances in semiconductor-process technology will further improve power consumption.

- *Form factor*: Customers are used to small devices and will not easily accept something larger. A tuner, a demodulator/decoder, and additional semiconductor chips must be integrated within acceptable "real estate" limits.

- *Good TV picture quality*: The picture quality must be comparable to that of normal television. The new standards make a speed of 20 to 30 frames per second possible, almost equal to the standard television frame rate. The average screen resolution is also good enough to view the pictures.

- *Added functionality*: Added functionality must be focused on making the handset easy-to-use for mobile TV and should include: easy navigation and fast channel change, personalized list of TV stations, program reminders, and recording capability. Extra features should not be added at the expense of losing existing functionality.

Recent handset launches, in particular in the South Korean market, indicate that manufacturers are on track to overcome these challenges. They are supported by component and semiconductor suppliers that are also increasingly active in this area. Philips was the first in Europe to demonstrate a complete TV-on-cellular chipset, which consists of a TV tuner, decoder, and peripheral component, and is now doing the same in the United States.[51] Thomson recently announced the acquisition of Thales Broadband and Multimedia to provide cellular operators with a complete set of mobile TV solutions.[52]

A Bright Future for Mobile TV

Mobile TV is a key example of the digital convergence among networks, devices, and content. Digital convergence will have a profound impact on consumer lifestyles, and as such, it carries new business opportunities for a plethora of players in the telecom and media worlds. To make mobile TV successful, a combination of mass-market content, low prices, and mainstream phones is crucial. User adoption will be greatest if the content has the widest possible appeal at the times of day when users are most likely to be away from standard television.

Mobile TV must have a new programming paradigm, different from normal television, in particular based on personalization and interactive experiences. The programming must have its own must-see element. And it must be delivered without comprising the TV experience. Channel changes must not be slower than on standard TV, and the phones must include understandable electronic program guides and picture quality that does not deteriorate in bright light. The service must be priced at a modest premium—after all, most of the content that underlies the basic service is not unique or inaccessible through other channels. Finally, adoption hinges on the availability of suitable, attractive devices priced to attract the mass market.

The Asian countries have already taken a lead in the mobile TV field. It is likely that the rest of the world will follow, as TV, unlike mobile Web browsing, is an easy concept to sell to consumers. Trials in Europe and the United States have shown positive results, though issues around business models and multi-player revenue-sharing arrangements are yet to be resolved. A number of commercial deployments are expected in 2006, but significant subscriber growth is not likely to gain momentum until 2007. Nevertheless, with many introductions in 2006 and the football World Cup as a key promotion, 2006 looks set to be the year mobile TV takes off.

References

1. Cellular Online. February 7, 2006. www.cellular.co.za/news_2006/feb/020206-over_21-billion.htm.
2. Press release. "Warner Music Group and MTV Networks sign landmark agreement for use of music videos in original mobile content and programming." Warner Music Group, September 26, 2005. www.wmg.com/news/story.jsp?article=27520025.
3. Sharma, Dinesh C. "Mickey Mouse goes wireless." CNET News, July 6, 2005. http://news.com.com/Mickey+Mouse+goes+wireless/2100-1039_3-5776327.html.
4. Press release. "'24: Conspiracy' From Fox Entertainment Group Available Exclusively in the U.S. to Verizon Wireless V CAST Customers." February 1, 2005. www.prnewswire.com/cgi-bin/stories.pl?ACCT=109&STORY=/www/story/02-01-2005/0002941793&EDATE.
5. "Endemol to introduce new Mobile TV channels." Telecomworldwire,

October 20 2005. www.findarticles.com/p/articles/mi_m0ECZ/is_2005_Oct_20/ai_n15737425.
6. Kharif, Olga. "The Coming Mobile-Video Deluge." BusinessWeek online, October 11, 2005. www.businessweek.com/technology/content/oct2005/tc20051011_9768_tc024.htm.
7. Whenham, T.O. "NTT DoCoMo buys part of Fuji Television Network." MobileMag, December 21, 2005. www.mobilemag.com/content/100/344/C5711.
8. Tilak, John. "H3G Italia to begin DVB–H broadcasting in 2006." November 28, 2005. www.dmeurope.com/default.asp?ArticleID=11582.
9. "V CAST." Verizon Wireless, accessed on March 24, 2006. http://getitnow.vzwshop.com/index.aspx?id=vcast.
10. "Sprint TV." Accessed on March 24, 2006. http://www1.sprintpcs.com/explore/ueContent.jsp?scTopic=multimedia100.
11. "MobiTV Channels." Accessed on March 24, 2006. www.mobitv.com/channels/cingular.html.
12. "MobiTV." Accessed on March 24, 2006. www.mobitv.com.
13. Faultline. "French to enjoy 50 mobile TV channels." The Register, September 15, 2005. www.theregister.co.uk/2005/09/15/french_mobile_channels.
14. Tanner, Ben. "Vodafone launches global Mobile TV." Dmeurope, December 6, 2005. www.dmeurope.com/default.asp?ArticleID=11820.
15. Jacobsen, Alan. "MBMS – The newest mobile multimedia standard." Converge! Network Digest, October 21, 2005. www.convergedigest.com/bp-ttp/bp1.asp?ID=266&ctgy.
16. Kharif, Olga. "The Coming Mobile-Video Deluge." BusinessWeek online, October 11, 2005. www.businessweek.com/technology/content/oct2005/tc20051011_9768_tc024.htm.
17. Wieland, Ken. "3G or not 3G." Telecommunications International, October 2005.
18. Reardon, Marguerite. "Streaming video too fat for 3G phone networks." CNET News, October 4, 2005. http://news.cnet.co.uk/software/0,39029694,39192910,00.htm.
19. "Korea's satellite mobile TV customers reach 20,000." Slashphone, May 16, 2005. www.slashphone.com/74/1914.html.
20. Kirk, Jeremy. "BT to launch digital Mobile TV in summer." Webwereld, January 13, 2006. www.webwereld.nl/articles/39306.
21. Bayern digital radio news. February 6, 2006. www.bayerndigitalradio.de/download/bdr_pressemeldungen/pm_060102_lfk.pdf.
22. "Japan launches digital TV on cellphones." Physorg.com, April 1, 2006.
23. "Sprint: Go with the FLO?" Dailywireless.org, September 27, 2005. http://dailywireless.org/modules.php?name=News&file=article&sid=4747&src=rss10.
24. "Japan's KDDI, QUALCOMM in Mobile TV business." Dev Shed, December 22, 2005. www.devshed.com/showblog/20253/Japans-KDDI-QUALCOMM-in-mobile-TV-business.
25. Press release. "Broadcast Mobile TV set to boom." Electronicstalk, May 2, 2005. www.electronicstalk.com/news/acg/acg104.html.
26. RCR. "DVB–H alliance formed." Wireless Industry News, January 24, 2006. www.wirelessindustrynews.org/news-jan-2006/0352-012406-win-news.html.
27. Press release. "Microsoft announces participation in Mobile TV Alliance to Help Accelerate DVB–H Deployment in North America." February 28, 2006. www.microsoft.com/presspass/press/2006/feb06/02-28DVBHDeploymentPR.mspx.
28. "Broadcast TV to mobile, a solution looking for a problem?" Datamonitor, December 26, 2005.
29 "Mobile TV Revenue to Hit 6.8 Billion Euro by 2011." European Tech Wire, October 7, 2005. www.europeantechwire.com/etw/2005/10/report_mobile_t.html.
30. "At CeBIT: Mobile TV proves strong draw, hype over Origami grows." Jamaica Observer, March 9, 2006. www.jamaicaobserver.com/lifestyle/html/20060308T210000-0500_100239_OBS_AT_CEBIT__MOBILE_TV_PROVES_STRONG_DRAW__HYPE_OVER_ORIGAMI_GROWS_.asp.
31. "124.8 million broadcast Mobile TV users worldwide by 2010." Informa Telecoms & Media. May 10, 2005. www.informamedia.com/itmgcontent/tcoms/news/articles/20017302411.html.
32. Global Mobile TV subscriptions to reach 65 million by 2010. Juniper Research. MobileTech News, September 6, 2005. www.mobiletech-news.com/info/2005/09/06/120854.html.
33. Press release. "QUALCOMM and Verizon Wireless Announce Plans for Nationwide Commercial Launch of MediaFLO's Mobile Real-time TV Services." Verizon Wireless, December 1, 2005. http://news.vzw.com/news/2005/12/pr2005-12-01.html.
34. Press release. "Crown Castle Mobile Media becomes Modeo." Modeo, January 4, 2006. www.modeo.com/press_04.asp.
35. "Mediaset, TIM Strike Deal for TV over Cellphones." PC Magazine, October 12, 2005. www.findarticles.com/p/articles/mi_zdpcm/is_200510/ai_n15667494.
36. Lovelace Consulting. "3 Italia aims for 500,000 DVB–H subscribers." Industry Consumer News, February 21, 2006, www.dtg.org.uk/news/news.php?class=countries&subclass=0&id=1491.
37. "Finland gives mobile TV license to France's Digita." Reuters, March 21, 2006. http://go.reuters.co.uk/newsArticle.jhtml?type=internetNews&storyID=11622998§ion=news&src=rss/uk/internetNews.
38. Thomson, Iain. "Britain gets first mobile TV in Europe." Vnunet.com News: Mobile Communications, February 14, 2006. www.vnunet.com/vnunet/news/2150278/britain-gets-first-mobile-tv.
39. "BSkyB unveils Mobile TV service." October 31, 2005. www.findarticles.com/p/search?tb=art&qt=Finland+%2F+Communication+systems.
40. "Expanding the Innovation Horizon: The IBM Global CEO Study 2006." IBM Global Business Services, March 2006.
41. Boumans, Jak. "Endemol and KPN in content joint venture, at last." Blog Spot: Buzialulane, February 8, 2006. http://buziaulane.blogspot.com/2006/02/endemol-and-kpn-in-content-joint.html.
42. "Mobile phone users want e-mail, music and TV." NewsFactor Network, March 8, 2006. www.newsfactor.com/story.xhtml?story_id=42004.
43. "Pilots reveal strong demand for DVB-H Mobile TV services Says Nokia." IT News Online, March 9, 2006. www.itnewsonline.com/showstory.php?storyid=3082&scatid=7&contid=4.
44. "BT Movio pilot tests demand for broadcast mobile TV." Informitv news, January 12, 2006. http://informitv.com/articles/2006/01/12/btmoviopilot.
45. "Pilots reveal strong demand for DVB–H Mobile TV services says Nokia." IT News Online, March 9, 2006. www.itnewsonline.com/showstory.php?storyid=3082&scatid=7&contid=4.
46. "TV goes mobile." November 2, 2005. www.mobiletv.nokia.com/news/events/files/nmc_2005_richard_sharp.pdf
47. "Pilots reveal strong demand for DVB–H Mobile TV services says Nokia." IT News Online, March 9, 2006. www.itnewsonline.com/showstory.php?storyid=3082&scatid=7&contid=4
48. Yapp, Edwin. "Mobile TV, anyone?" TechCentral: Corporate IT, September 20, 2005. http://star-techcentral.com/tech/story.asp?file=/2005/9/20/corpit/12066894&sec=corpit.
49. "O2 Study Shows Strong Demand for Mobile TV." MobileTech News, January 17, 2006. www.mobiletechnews.com/info/2006/01/17/220855.html.
50. "BT hails success of Mobile TV trial." Digital TV Group, January 13, 2006. www.dtg.org.uk/news/news.php?id=1405.
51. Press release. "Philips introduces TV-on-mobile solutions for the U.S. market." Philips, December 12, 2005. www.semiconductors.philips.com/news/content/file_1207.html.
52. Press release. "Thomson Expands its IPTV, Mobile TV and Digital Terrestrial TV Solutions through Potential Acquisition of Thales Broadcast and Multimedia." December 12, 2005. www.thomson.net/EN/Home/Press/Press+Details.htm?PressReleaseID=f283df2f-a0dd-4c7f-98b5-3235345a070b.

Technologies and Networks

WLAN Mesh Architectures and IEEE 802.11s

Derek Cheung
Member
The Institution of Engineering and Technology

Abstract

Mesh wireless local-area networks (WLANs) using 802.11 mesh access points (APs) to relay traffic among each other to increase the radio frequency (RF) coverage of the networks have become very popular for implementing wireless municipal networks in the past few years, including Philadelphia's infamous city-wide WLAN mesh network.

Depending on the wireless coverage and the expected traffic usages of the municipality, it is not uncommon to have a WLAN mesh network employ a few hundreds or even tens of thousands of mesh APs to cover areas in hundreds or even thousands of square kilometers.

Many network equipment vendors and start-ups offer 802.11 WLAN mesh products using proprietary technology and protocols. In order to increase the acceptance of the WLAN mesh technology, IEEE has formed the 802.11s workgroup to solve the mesh equipment interoperability issue by defining an architecture and protocols that support both broadcast/multicast and unicast delivery using radio-aware metrics over self-configuring multi-hop topologies (i.e., a mesh network). With the agreement between the two competing mesh proposals from Wi-Mesh and SEEMesh in the 802.11s workgroup in early 2006, a draft 802.11s proposal is expected to be available in 2007. This draft proposal is expected to become an Institute of Electrical and Electronics Engineering (IEEE) standard by 2008.

This paper introduces the operations of contemporary WLAN mesh networks and highlights the areas that are being worked on or standardized by the IEEE 802.11s workgroup. Important WLAN mesh equipment selection criteria and network engineering parameters that are applicable for current and future 802.11s–compliant WLAN mesh systems and networks are discussed.

Introduction

The WLAN mesh technology offers low-cost and high-bandwidth wireless access and is being adopted by many municipalities around the world for implementing wireless municipal networks. Traffic-light video streaming, electrical smart-meter telemetry, and public outdoor Internet access are some of the common network applications running over WLAN mesh networks. *Figure 1* shows a typical WLAN mesh network setup.

Both the mesh APs and the mesh portal provide wireless access and backhaul traffic-relaying functions. However, the mesh portal has a wired connection (e.g., Ethernet) to the backhaul network, and it is responsible for aggregating traffic for multiple mesh APs (usually three to 10 mesh APs) to form a "mesh community." In the above diagram, the mesh AP2 uses its wireless backhaul link to forward its access traffic from the WLAN devices such as wireless laptops or traffic-light video cameras to the Intranet or Internet via the mesh AP1 and the mesh portal.

The backhaul wireless links among the mesh APs are usually WiMAX– or 802.11a-based. When 802.11a (5 GHz) technology is used for implementing the wireless backhaul links among the mesh APs, the mesh APs will provide only 802.11b/g (2.4 GHz) wireless access service for the WLAN devices so that sufficient frequency separation is available between the access and backhaul radios inside the mesh AP to minimize self-interference.

Many WLAN mesh products utilize 802.11a OFDM modulation for implementing the wireless backhaul links. It is found that OFDM performs reasonably well for non–line of sight (NLOS) communication among the mesh APs to allow more flexible mesh AP deployment. It is likely that mesh equipment vendors will take advantage of the newer modulation scheme scalable orthogonal frequency division multiplexing access (SOFDMA) with sub-channelization offered in the IEEE 802.16e standard to offer improved NLOS communication among the mesh APs when the cost of the 802.16e radio is reduced.

The typical separation distance between two mesh APs is about a few hundred meters in metropolitan mesh network deployment. In a rural mesh network deployment where access traffic density is lower, the mesh APs can be spaced farther apart so that fewer mesh APs are required to cover a given rural area. Most mesh APs support the use of different external backhaul antennas for different deployment scenar-

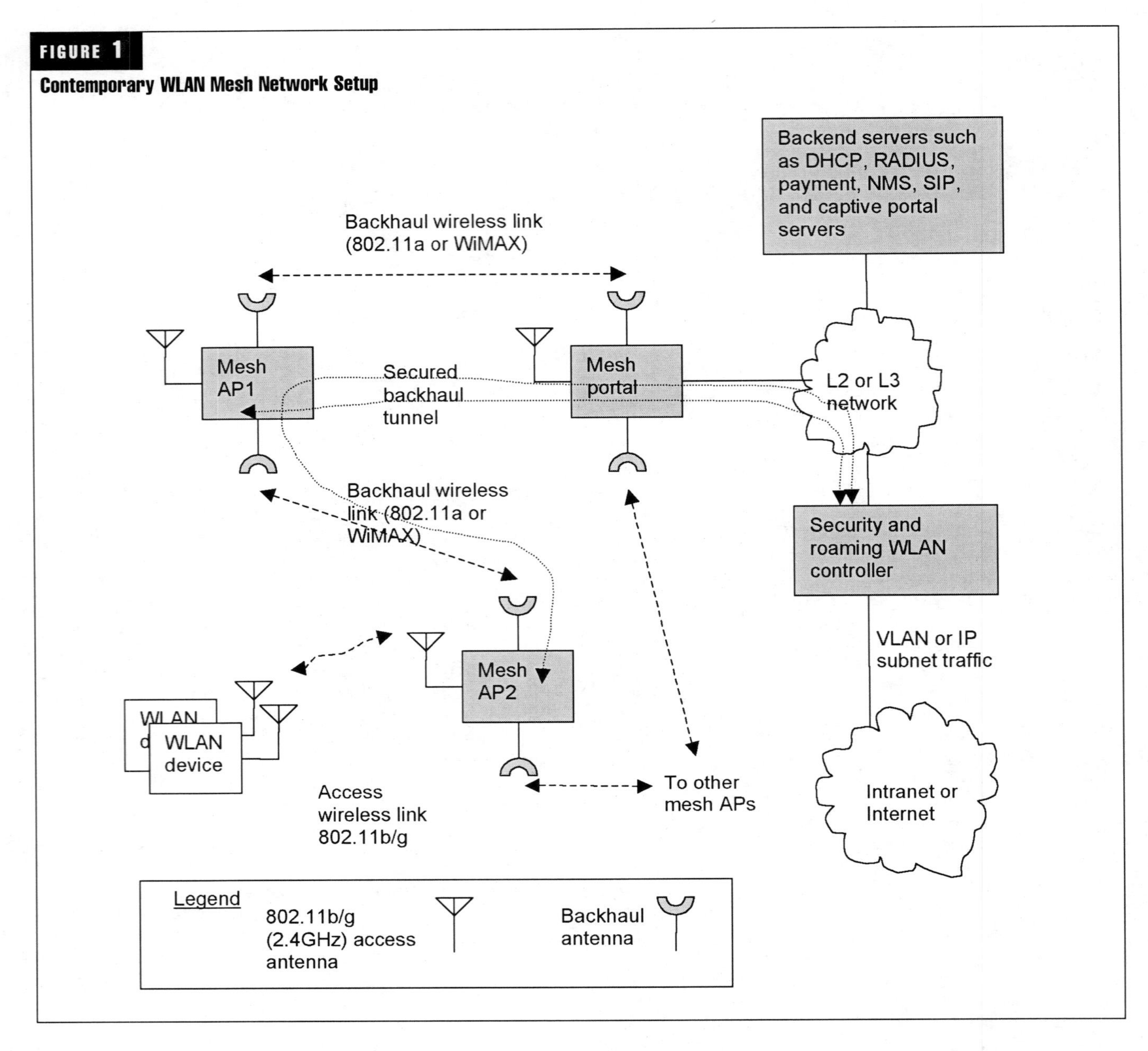

FIGURE 1

Contemporary WLAN Mesh Network Setup

ios. For example, with a typical mesh AP's backhaul radio outputting at 26 dbm (400 mW), a higher gain (e.g., 8 dbi) backhaul antenna can easily allow two mesh APs to space over 1 km in rural mesh network deployment while meeting the regulator's requirements (e.g., effective isotropic radio power [EIRP] of less than 36 dbm or 4 W in higher 5 GHz user network interface [UNI] band in Canada).

Most of the current WLAN mesh solutions employ centralized WLAN controller(s) for securely terminating mesh networks' traffic to and from the mesh APs. The WLAN controllers also offer subscriber mobility by allowing subscribers to roam among the mesh APs without session drop or user re-authentication. Depending on its design, one WLAN controller can usually support a few hundred mesh APs. Usually, the WLAN controllers can be cascaded together to support a larger WLAN mesh network that requires thousands of mesh APs.

Since the high-level design and architecture of WLAN mesh networks are very similar among mesh equipment vendors, the vendors attempt to differentiate their products in the following areas:

- *Mesh routing protocols*: Traffic forwarding optimization and resilience
- *Antenna design (MIMO, beamforming)*: Wireless distance and coverage
- *Backhaul radio design*: Multimedia traffic handling capability

Besides the mesh APs, mesh portals, and WLAN controllers, a WLAN mesh network needs the following backend servers for proper operations:

- *Dynamic host configuration protocol (DHCP) server*: Subscribers and network devices Internet protocol (IP) address assignment

- *Captive portal server*: Web user authentication for public Internet access
- *Remote authentication dial-in user service (RADIUS) server*: User and network device authentication
- *Payment server*: Network billing and payment such as on-line credit card authentication
- *Session initiation protocol (SIP) server*: Voice over mesh/IP operations
- *Network-management system (NMS)*: WLAN mesh network management
- *Firewall*: Real-time virus scanning and network attack

The above are commonly used backend servers for a WLAN mesh network. Other servers such as content cache servers may require depending on the requirements of the WLAN mesh network.

Almost all WLAN mesh equipment vendors claim that their products can be used to build a scalable, secured, cost-effective, and resilient wireless network that requires minimum configuration effort such as the following:

- Topology discovery and channel allocation to minimize configuration effort
- Path selection and forwarding for optimal and alternative traffic paths under network failure
- Access traffic security and subscriber authentication for secured access traffic protection
- Backhaul traffic security and mesh device authentication for secured backhaul traffic protection
- Traffic management for multimedia wireless access traffic
- Network management for ease of WLAN mesh network operations and maintenance

Coincidentally, these are also the areas and functions that the 802.11s workgroup will address and standardize for mesh device interoperability. The following sections describe the operations of each of the above functions in a contemporary WLAN mesh network. The information is useful for understanding the upcoming 802.11s draft and standard when they are available in 2007 or 2008.

Mesh Topology Discovery and Channel Allocation

When a mesh AP is powered up, it will scan its backhaul frequency channels to try to form backhaul wireless links with its adjacent mesh APs. The ultimate goal is to locate its corresponding mesh port for access traffic forwarding to the intranet or Internet.

Many WLAN mesh products utilize the unlicensed 5 GHz band for implementing the wireless backhaul links. Since the unlicensed 5 GHz band is not harmonized in the world, different 5 GHz UNI bands are being used in different countries. For example, in Canada, the unlicensed upper UNI band from 5.725GHz to 5.825GHz is usually used by WLAN mesh vendors to implement the wireless backhaul links. There are altogether six non-overlapping channels within this spectrum under the 802.11a modulation.

In general, a mesh AP will only form two to three wireless backhaul links with its adjacent mesh APs for performance reason. The discovery protocols exchanged by the mesh APs use many parameters to determine which backhaul wireless links should be formed among the mesh APs, including the following:

- Relative signal strength index (RSSI)
- Loading on the adjacent mesh APs
- Signal-to-noise ratio (SNR)
- Interference on the backhaul channel frequency
- Network redundancy and topology optimization
- Packet errors

Depending on the number of backhaul radios (physical or virtual) and the type of backhaul antennas (omnidirectional or directional) used at the mesh APs, the mesh network topology discovery and channel allocation protocols and processes can be fairly complex. Most mesh equipment vendors use proprietary network discovery protocols to achieve minimum or no-effort WLAN mesh network configuration and expansion.

802.11b/g communication modes are normally supported by a mesh AP for wireless access. There are 11 channels defined in the 802.11b/g 2.4 GHz spectrum, but only three of them (channels 1, 6, and 11) are non-overlapping in term of frequency bandwidth.

Some mesh equipment vendors offer automatic access-channel allocation with dynamic transmission power–level adjustment to ensure that the mesh APs provide adequate access link RF coverage for a given geographic area with minimum inter-mesh AP interference. The continuous RF coverage ensures subscribers can roam among the mesh APs without session drop or re-authentication. Normally, most mesh vendors can support device roaming at speeds up to 60 km/h. Some mesh equipment vendors claim to support fast device handoff to support roaming speeds of more than 100 km/h.

The goals for access link channel allocation and dynamic power level adjustment are as follows:

- Maximum access channel reuse
- Minimum inter-mesh AP interference
- Continuous access channel coverage for device roaming

Figure 2 illustrates an example for access link RF coverage to achieve the above requirements.

Control protocols such as control and provisioning of wireless access points (CAPWAP) are usually used to achieve automatic access-link channel allocation and dynamic power level adjustment.

Path Selection and Forwarding

A mesh AP usually forms more than one wireless backhaul link with its adjacent mesh APs for access traffic forwarding. This is done for network redundancy and load-sharing purposes. Most of the contemporary WLAN mesh products use Layer 3 (L3) IP routing for access traffic forwarding over multiple wireless backhaul links. This is also being adopted by the 802.11s workgroup instead of the L2 bridging mechanism.

Similar to IP routers in that they exchange interior gateway protocol (IGP) routing protocols such as open shortest path

FIGURE 2
WLAN Mesh Access Link Channel Assignment and Dynamic Power Adjustment

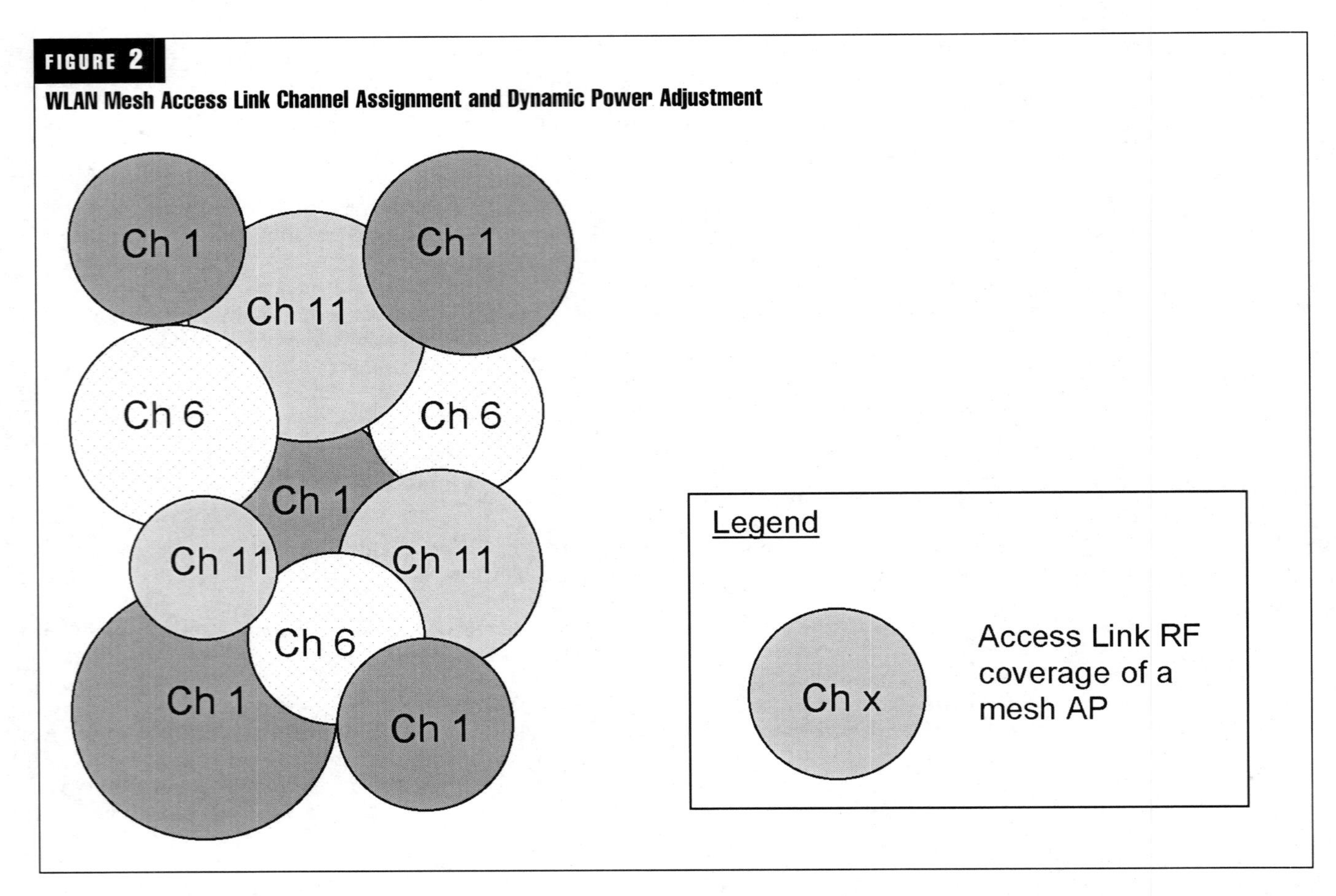

first (OSPF) or intermediate system to intermediate system (ISIS) among each other to learn the optimal network topology under normal and failure conditions, mesh APs also exchange routing protocols among themselves over the control channel of the wireless backhaul links. Through the routing protocol exchange among the mesh APs, the mesh APs form the optimal network topology for relaying access traffic via the wireless backhaul links. When a network outage occurs due to equipment failure or RF interference, the mesh network routing protocol will update the forwarding table of the mesh APs to forward access traffic via alternate paths over the mesh network.

WLAN mesh networks' routing protocols differ from the standard IP's IGP protocols due to the different characteristics of a WLAN mesh network such as link latency and reliability and RF interference. There are more than 70 competing mesh routing schemes available, including the following:

- Ad hoc on-demand distance vector
- Dynamic source routing
- Optimized link state routing protocol
- Temporally ordered routing algorithm

A few WLAN mesh equipment vendors champion their products around their proprietary mesh network routing protocols and claim that they can always find the optimal path for traffic forwarding.

When access traffic arrives at the WLAN controller, the WLAN controller either tags the access packets with the designated virtual local-area network (VLAN) identifier (ID) or route the packets according to the IP subnet information. Usually, different user groups are assigned to different service set IDs (SSIDs) and their traffic is then segregated and forwarded based on either VLAN IDs or IP subnets at the egress ports of the WLAN controller. There is not much difference between the VLAN and IP subnet forwarding designs in term of broadcast and multicast traffic containment.

One of the key deliverables from the 802.11s workgroup is to standardize the mesh AP routing protocol over a WLAN mesh network for device interoperability.

Access Traffic Security and Subscriber Authentication

The existing WLAN access security and authentication mechanisms offered by the IEEE and the Wi-Fi Alliance are normally supported by a WLAN mesh network so that a user can use the same WLAN hardware and software for secured WLAN mesh access.

WLAN security is a big alphabet soup with many terminologies such as TKIP (temporal key integrity protocol), WPA2 (Wi-Fi protected access 2), EAP–TTLS (extensible authentication protocol–tunneled transport layered security), EAP–MD5 (EAP–message digest 5), and AES (advanced encryption standard).

In general, the access traffic security of a WLAN mesh network is as good as any modern WLAN or wired enterprise LAN as long as the WLAN mesh network and devices are properly set up.

Backhaul Traffic Security and Mesh Device Authentication
As mentioned in the mesh network topology discovery section, when a mesh AP is powered up, it will attempt to form wireless backhaul links with its adjacent mesh APs. In order to ensure that it will not form adjacent links with rogue mesh APs to allow hackers from gaining illegal access to the WLAN mesh network, standard X.509 digital certificates with user-defined passwords are usually employed at the mesh APs and WLAN controller for proper device authentication.

Once a mesh AP forms adjacent wireless backhaul links with its neighboring mesh APs, it normally establishes a secured traffic tunnel from the mesh AP to the WLAN controller to securely transport access traffic over the WLAN mesh network. Through digital key exchange, only the mesh AP and the WLAN controller have the right pair of keys for encrypting and decrypting the traffic between them. This secured traffic tunnel design has been adopted by many WLAN mesh vendors. When evaluating existing WLAN mesh products, one should ask the equipment vendor how mesh APs' traffic is transported over the WLAN mesh network and if there is a hardware accelerator inside the WLAN controller to expedite the encryption and decryption processes, which are very central processing unit (CPU)–extensive.

Traffic Management
Best-effort data over a large-scale WLAN mesh network involving thousands of mesh APs has been implemented and proven in many large WLAN mesh network deployments. The next challenge for the mesh equipment vendors is to support a high volume of multimedia traffic such as voice and video over a WLAN mesh network. This is becoming important as many handset vendors are releasing their second-generation dual-radio handsets that support both cellular (code division multiple access [CDMA] or Global System for Mobile Communications [GSM]) and WLAN access. These second-generation dual-radio WLAN handsets have the same appealing form factor as the latest second generation (2G)/3G handsets. The ability for a subscriber to roam between a 2G/3G cellular network (e.g., cover a larger geographical area at a higher connection cost) and a WLAN mesh network (e.g., a smaller coverage area than cellular but with higher data rates at lower connection costs) is very appealing to users.

It is very likely that the 802.11s workgroup will adopt the IEEE 802.11e (Wi-Fi multimedia [WMM]) standard for traffic management for WLAN mesh networks. This enables the mesh APs to perform some sorts of traffic prioritization to meet the latency requirements of multimedia traffic. Note that WMM is simply a class of service (CoS) instead of a resource-reserved quality of service (QoS) mechanism due to the limitation of the original 802.11 protocol. However, in light of the widely successful 802.11 protocol, changing a new WLAN multiple access (MAC) protocol to offer better QoS is not possible in the near-term; WMM is a reasonable alternative for supporting multimedia traffic over a WLAN mesh network.

Network Management
A WLAN mesh network can comprise hundreds or thousands of mesh APs, each with multiple wireless backhaul links. Managing a WLAN mesh network of this size and complexity demands a powerful and intelligent network management system.

One way to promote mesh device interoperability is to offer standardized mesh devices and simple network-management protocol (SNMP) management information bases (MIBs) so that operators can import various standard mesh network MIBs into their network management system.

While the 802.11s workgroups will address many mesh device interoperability issues, they will also deliberately leave out some WLAN mesh architecture designs for equipment vendors to differentiate their products. Two noticeable issues that are left out are the design of the backhaul radio (e.g., single versus multiple radios) and interference mitigation in the unlicensed bands.

WLAN Mesh Network Design and Engineering

The design and engineering of a WLAN mesh network is very similar to that of any other wireless network that involves site survey, equipment selection and evaluation, RF propagation simulation, and user traffic requirement gathering. However, a WLAN mesh network has some additional attributes, including the following:

- *Unlicensed spectrum and interference*: Many 3G carriers spend billions of dollars to secure the licensed 3G frequency spectrum in their countries for 3G cellular services. The fact that a WLAN mesh network uses the unlicensed 2.4 GHz and 5 GHz frequency spectrums is a double-edged sword. It makes deploying WLAN mesh networks fast, and they have a much lower capital cost. However, it also makes the network susceptible to innocent interference in these unlicensed frequency spectrums from devices such as microwave ovens and wireless home phones.

- *802.11 MAC layer does not support QoS, even with WMM*: Network engineering, planning, and management are more important in WLAN mesh networks to ensure that no mesh devices are overloaded during normal and peak operations to ensure adequate multimedia traffic support.

- Some hold the untrue assumption that a WLAN mesh network is self-configured and easy to scale, with minimum engineering and management efforts.

The following summarizes some of the network engineering issues that are specific to WLAN mesh networks. They are applicable for the current as well as the future 802.11s–compliant WLAN mesh networks:

- Understanding the current and/or projected traffic pattern, mobility requirements, traffic types (e.g., best-effort data, voice over mesh), wired and wireless (e.g., 3G cellular) interworking requirements, and resilience strategy for discussion with the equipment vendors or network integrator.

- Using a directional instead of omnidirectional backhaul antenna to minimize interference in the backhaul frequency spectrum whenever possible.

- Selecting a WLAN mesh system that uses a different backhaul channel for each wireless backhaul link for better interference mitigation (see the Appendix).

- Fully utilizing the number of non-overlapping backhaul frequency channels among the mesh APs when designing the WLAN mesh network.

- Performing a proper site survey to identify the locations for mesh AP deployment and adjust the transmission power of the mesh APs to minimize interference among them if the WLAN mesh system does not offer automatic AP power adjustment.

- Selecting the spectrum that has less interference if the WLAN mesh system and the local regulatory allow a choice of operating frequency. For example, the 4.9 GHz frequency spectrum used for U.S. homeland security or the 3.5 GHz for 802.16–2004 fixed point-to-point wireless link has less interference than the 5.8 GHz unlicensed frequency spectrum.

- Minimizing or eliminating access traffic support at the mesh portal. This is because the mesh portal is responsible for aggregating traffic from multiple mesh APs within a mesh community, and it is usually the performance bottleneck of the system. When a single CPU is used inside a mesh AP (see the Appendix), the less work for the mesh portal to process access traffic, the more CPU cycles it can spare for forwarding mesh APs' traffic.

- Selecting mesh APs with multiple physical backhaul radios for better throughput and latency performance—if the WLAN mesh network is to support a high volume of multimedia traffic such as voice and video—though it is generally more expensive.

- Selecting a WLAN mesh system that has built-in access and backhaul WLAN intrusion and interference detection and containment functions.

- Limiting the mesh hop count to ensure that latency along the mesh APs and the mesh portal falls within the requirements of the multimedia traffic. For example, the one-way latency for voice over a WLAN mesh network should be less than 100 ms.

- Understanding that a mesh AP with multiple physical backhaul radios can support multimedia traffic better than a mesh AP with only a single and/or multiple virtual backhaul radios. The industrial trend is to have multiple physical backhaul radios implemented inside a mesh AP.

- If the WLAN mesh system utilizes centralized WLAN controller(s) for securely terminating mesh traffic, checking with the vendors on the scalability of the WLAN controller(s) to make sure that multiple WLAN controllers can be cascaded together to support a large WLAN mesh network. Also, ensuring that the WLAN controller has dedicated hardware for secured traffic encryption and decryption.

Designing, installing, and managing a WLAN mesh network for a municipality with hundreds or even thousands of mesh APs is not a small task. The mesh equipment vendors' promise that WLAN mesh networks are self-configured without much manual intervention is far from the truth. Incumbent carriers with extensive experience and infrastructure in wired and wireless networks are in a better position to perform WLAN mesh network design, engineering, installation, and management.

Conclusion

With the upcoming standardization of the WLAN mesh technology via the IEEE 802.11s workgroup, one can expect that WLAN mesh networks will be more acceptable by both carriers and users.

Many municipalities around the world plan to operate their own WLAN mesh networks, with WLAN mesh equipment vendors' promises of easy network expansion, lower cost, and minimum management efforts. However, it has been found that designing, installing, and managing a WLAN mesh network is very similar to designing, installing, and managing any existing carrier wireless network and, in many ways, is not as simple. Incumbent carriers with significant wired and wireless network experience and infrastructure are in a better position to design, engineer, install, and manage a WLAN mesh network. This is especially true when the WLAN mesh network needs to interwork with the incumbent carrier's 2G/3G cellular network for device mobility.

Appendix

Since WLAN mesh networks are normally operating in the unlicensed frequency spectrum, interference mitigation is an important design aspect of a WLAN mesh system. For the contemporary WLAN mesh products, the following are two backhaul frequency reuse designs that can affect how well the products mitigate interference:

One common backhaul channel for a mesh community—Recall that a mesh portal is responsible for aggregating access traffic from multiple mesh APs to form a mesh community. In this design, all wireless backhaul links within a mesh community comprising the mesh portal and the mesh APs use the same backhaul frequency channel. Normally, omnidirectional backhaul antenna is used in this design.

The advantage of this design is that the there is no switching latency between the wireless backhaul links when the mesh AP has only one physical backhaul radio. This is good for implementing fast handoff and supporting multimedia traffic. The downside is that when there is RF interference on the backhaul frequency, all the mesh APs and the mesh portal will have to change to a new backhaul channel and is likely to cause network downtime to the mobile subscribers.

Different backhaul frequency for each wireless backhaul link—In this design, a mesh AP selects different backhaul channels for each of its backhaul links with its adjacent mesh APs. The mesh APs intelligently utilize all the available backhaul channels for implementing the wireless backhaul links among them. This design is very good for inter-

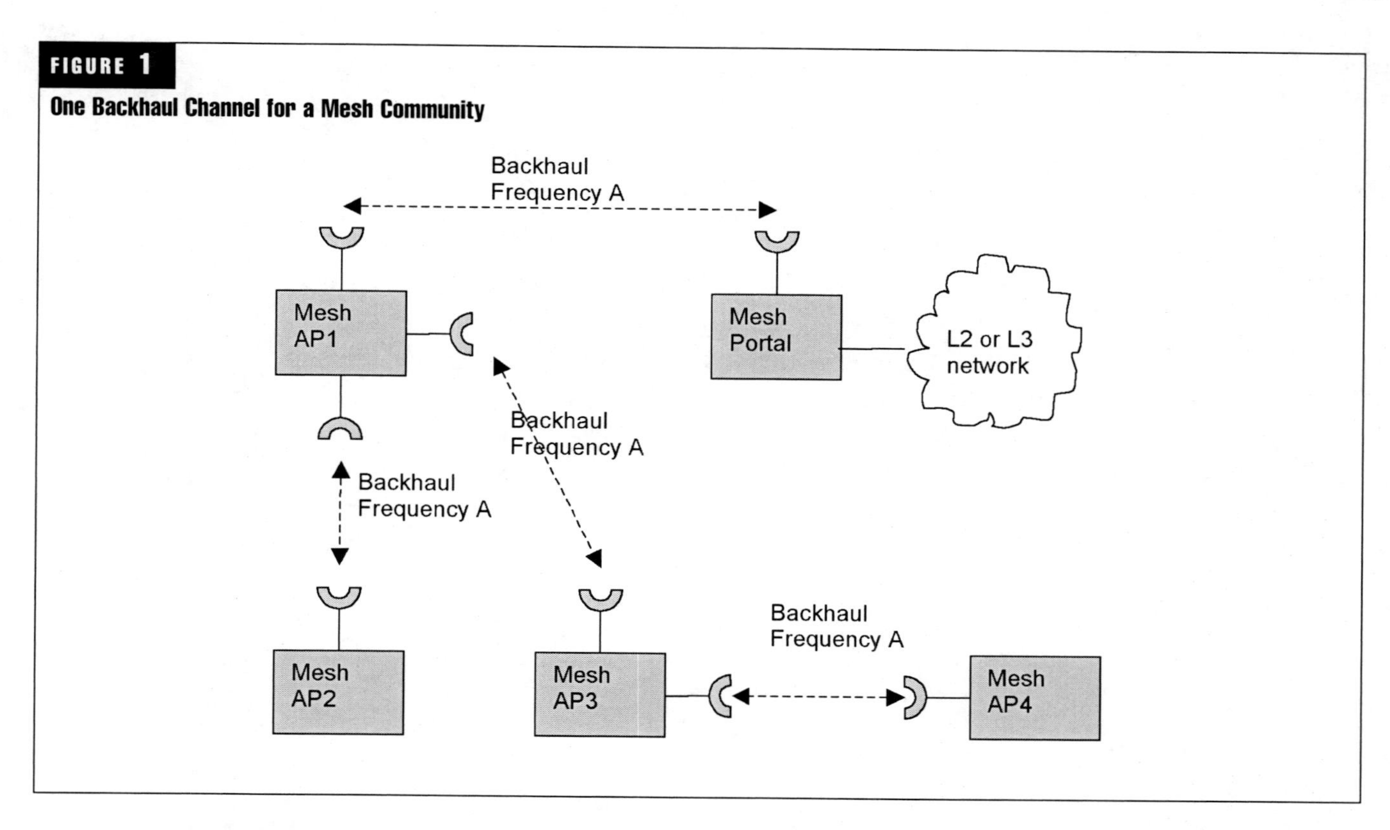

FIGURE 1

One Backhaul Channel for a Mesh Community

ference migration because only one mesh AP link need to change its operating frequency under RF interference. However, if a single physical backhaul radio is being used in the mesh AP, the design will incur additional switching latency among the wireless backhaul links due to frequency changes of the backhaul radio. This may affect its support of multimedia traffic.

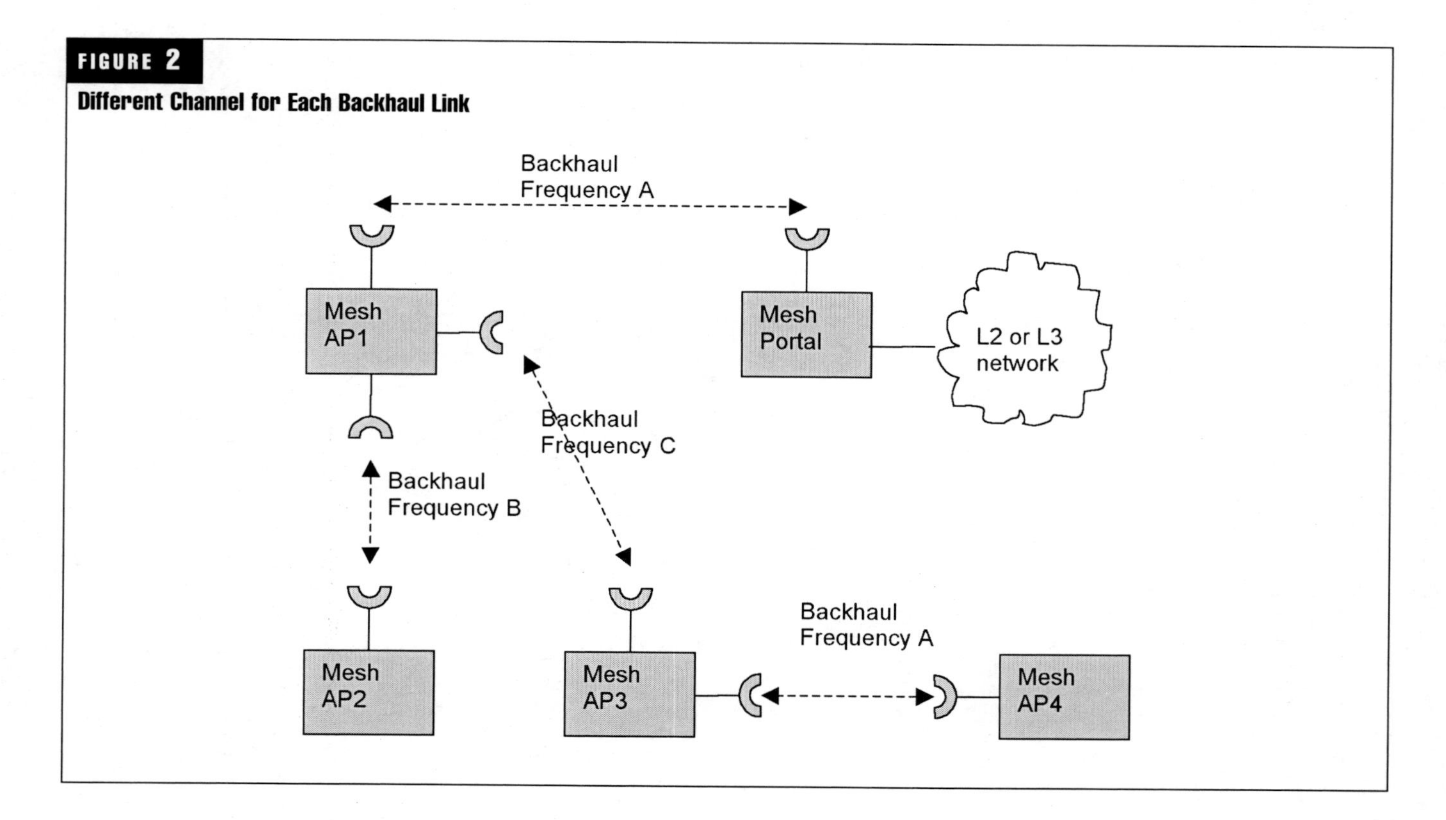

FIGURE 2

Different Channel for Each Backhaul Link

Wireless Billboard Channels: Vehicle and Infrastructural Support for Advertisement, Discovery, and Association of UCWW Services

Paul Flynn

Postgraduate Researcher, Telecommunications Research Centre

University of Limerick

Dr. Ivan Ganchev

Lecturer and Deputy Director, Telecommunications Research Centre

University of Limerick

Máirtín O'Droma

Director, Telecommunications Research Centre

University of Limerick

Abstract

This paper describes wireless billboard channels (WBCs) used for advertisement, discovery, and association (ADA) of wireless and mobile services in future fourth-generation wireless world (4GWW) and next-generation wireless networks, and in ubiquitous consumer wireless world (UCWW) business environments. A WBC service discovery model is proposed along with a format and ASN.1 encoding for service description (SD). Categorization of services is provided and service templates elaborated. Steps are taken to determine the best format of the data on the WBC channel and the best technology to allow mobile terminals to access the channel. Critical review and comparison of possible WBC carrier technologies is given.

Introduction

A 4GWW is seen within the European Union as a world encompassing all existing and future mobile and fixed wireless networks, both terrestrial and satellite [1]. It is envisioned that the mobile terminals (MT) of the future will be reconfigurable to be capable of using many access technologies. This gives rise to the always best connected and served (ABC&S) paradigm, by which it will no longer be enough for a terminal to be able to be connected anywhere and at any time—it will have to be connected in the best possible way to be provided with the best possible mobile service. Mobile users (MUs) will have profiles that specify what parameters (e.g., cost-to-performance ratio) determine what "best" means in any particular scenario.

The current subscriber-based business model that exists in the wireless world is not seen as ideal to support ABC&S. In this model, the MU is a subscriber to a user home access network provider (UHANP). Only service providers (SPs) that have a business agreement with the UHANP can provide services to the MU. The MU may roam on foreign networks that have a business agreement with the UHANP. The MU is billed for all used services by the UHANP.

Proposed in [2] is what is considered to be a superior model and the foundation for ubiquitous consumer wireless world (UCWW) [43]. In this model, the MU is not a subscriber, but a consumer. There is no UHANP. The MU is not confined to any particular ANP and may use any available service through any available access network (AN). This reflects the view for the next-generation networks (NGN) as "all–IP" networks where all services are offered independent of the AN they are accessed through [3, 30].

Taking into consideration the potentially huge range of services available to the MU in the putative 4GWW, the demand for an efficient and easy mechanism for service advertisement, discovery, and association (ADA) adapted to the terminal capabilities, user preferences, and user location, is clearly foreseen [4]. The approach we are proposing to facilitate ADA is to use wireless billboard channels (WBCs). A WBC is a channel that will provide the terminal with information about the services available to it. The terminal can compare this information with the terminal capabilities, the current user profile's preferences, the user's location, and other information such as the time of day, to select the best services to use to achieve a particular goal (e.g., make a phone call). As well as advertising the service, the WBC will also provide information to help with the process of discovering and associating with that service (e.g., access networks' physical layer information). Unlike the bidirectional channel attributes of the not-too-dissimilar proposal [5], offering point-to-point services, our proposal is that this channel would be point to multipoint, simplex, unidirectional and narrowband, and would be used solely for service advertisement purposes (including information on discovery and association procedures) and not for actual association execution or for software downloading for terminal reconfigurability functions.

WBCs should have the following characteristics:

- *Simplex and broadcast service*: Simplex here applies not just to the unidirectional physical nature of the channel, but also to the unidirectional nature of the service. This attribute has the additional benefit of easing their physical deployment and operation. If the channel is duplex, as in [5], then in a way it simply becomes another wireless AN infrastructure for two-way wireless communications, and thus in the end no different in its general attributes from any other wireless AN infrastructure. And if this is the case, then bandwidth-spectrum allocation becomes a much more significant issue, as it has been for existing cellular spectrum allocations, for instance.

- *Limited bandwidth*: Given the proposed usage (point-to-multipoint unidirectional service of advertisements), bandwidth requirements will be relatively narrow. This has the added advantage of enhancing their likely success, e.g., of global agreements on spectrum allocations for them. With limited spectrum available, this would improve a WBC's chances of becoming a worldwide standard.

- *Maximum coverage area*: the channels should ideally be available anywhere and at any time. No matter where it is, a terminal should have the ability to discover what services are available to it from local to regional and international service providers. Terminal mobility should not affect the ability to receive information on the channel.

- *Different versions for different areas*: The number and types of WBCs could correspond to the local regional/national/international interests of advertisers and users. In practice, there would be growth—perhaps a start-up situation would be one national WBC channel, advertising all the services that are relevant on a local, regional, or interregional significance. And then separate regional WBC channels could advertise the services available in that region.

- *Non–ANP service*: WBCs will need to be regulated and be fully independent and physically separate from ANPs and their ANs. This is necessary to ensure fair competition and equity of access to WBC advertisement space (i.e., equally open to all ANPs). For this, it is better that they be operated by non–ANP SPs.

This paper details the steps taken to come up with the best design for WBCs. Section 2 explains the methods by which terminals discover ANs' communications services (ANCSs) and application services (ASs), and how WBCs will draw from and improve on these. Section 3 gives examples of attribute lists describing ANCS and AS services. Section 4 shows the proposed structure of WBCs, including schemes for data scheduling and data indexing. Section 5 covers possible broadcast technologies to carry WBC channels. Section 6 concludes the paper.

WBC Service Discovery Model

The purpose of a WBC is to allow services to be discovered by terminals. There are already a number of protocols that deal with SD. Three of the most widely used protocols, namely Jini [6], SLP [7], and Salutation [8], were used as a basis for the development of WBC service advertisement and discovery. Short descriptions of these protocols are given in Appendix A. Though very different, all these use the same basic service model, as follows:

- There are central registries of SDs.
- The SDs include a list of service attributes.
- Service providers can register SDs with the registries.
- Anyone wishing to discover available services queries the registries based on desired service properties (type, attributes).
- The registries respond with one or more SDs that match the query.

This service model needs to be modified and/or adjusted to be applied in a WBC. WBC service providers have a central registry of services similarly to the above model. However, a WBC should be a simplex, broadcast channel. Because there is no "back channel," it is not possible to query the registry. There is no way to know what SDs are required, so the only choice is to broadcast every one, repeatedly, in turn. The WBC service discovery model, therefore, is as follows:

- There is a central registry of SDs.
- The SDs include a list of service attributes.
- Service providers can register SDs with the registry. Since there is no back channel, this can be done separately to the WBC, via a Web portal.
- The registry broadcasts all its SDs in turn, repeatedly, on a WBC.

WBC SD Format

A WBC SD was elaborated on the basis of the SDs of the three already mentioned service discovery protocols (see Appendix B). A WBC SD includes the following fields:

- *Service type*: Similar to the service type in SLP and the functional unit ID in Salutation, this field identifies the type of service. The purpose of the service type is to group together services that perform the same function so that SDs on the WBC can be easily found. Each service type has a template, which SDs follow. These templates can be managed by the WBC provider and published for all service providers to follow.

- *Scope list*: As in SLP, the scope list identifies to which scopes the service belongs. Scopes are a way to group services together. As an example of how this can be useful, consider the following scenario: A service provider offers a number of news alert services to MUs who pay a monthly subscription. Those MUs can then be assigned the same scope. The MUs, when discovering services using the WBC, will see the scope of these services and will know they are allowed to use them. Non-subscribing MUs will see that these services have a scope that they have not been assigned and can ignore any such SDs.

- *Attribute list*: This field specifies the list of attributes and their values similarly to the attribute list in SLP and the set of attribute records in Salutation. There is no need to include the "compare function" subfield from Salutation, as this is only used in queries. The attributes are to carry information for service advertisement and association.

WBC SD Encoding

One of the desired properties for a WBC is that it should use as little bandwidth as possible. In Jini, SDs are Java objects, and passing the descriptions involves serializing the objects. Serialized Java objects are a very inefficient way of encoding data. In SLP, the SDs are encoded as text, which, while more efficient than Java objects, is still not as efficient as we would like. The Salutation SD, which uses abstract syntax notation (ASN.1) [9], is the most efficient approach in this case. ASN.1 data can be encoded in a number of ways, depending on which encoding rules are followed. The packed encoding rules (PERs) [10] give the most efficient encoding (close to optimal).

To save bandwidth, the service type field should be an integer, as in Salutation, rather than a string, as in SLP. Scopes should also be integers. Attributes should be like Salutation attributes, i.e., of any ASN.1 type. There is only a need to include attribute values. The attribute ID field, used in Salutation, can be dropped, since the PER encodes which attribute is which. Service templates can be specified in ASN.1. Any MU that has these service templates will be able to decode the PER–encoded SDs and know which values belong to which attributes. The following is an ASN.1 encoding sample of WBC service description:

```
        ServiceDescription          ::= SEQUENCE
{
        serviceType                 INTEGER
        scopeList                   SEQUENCE OF
                                    INTEGER
        attrList                    ServiceTemplate
}
```

Service Types and Templates

We use the term service in this paper to mean both ANCSs and ASs. Many service types are under both headings. Each service type needs a service template, which is an ASN.1 specification of the attributes used to define a service of that type.

Access Networks' Communication Services (ANCSs)

A service type is used to group together services that perform a similar function. There are different types of AN: multi-access wireless networks and single-access wireless networks. There are also many types of access technology—UMTS, GSM/GPRS/EDGE, CDMA2000, Wi-Fi, WiMAX, and Bluetooth—and there will be more in the future. It makes sense to group ANCSs by their access technology so that there is one service type for each technology. To reduce complexity, a multi-access wireless network can be advertised as several single-access wireless networks, one for each access technology.

The attribute list of ANCS needs to contain information for ADA procedure. Advertisement information consists of attributes (e.g., cost and performance) that allow a terminal to decide whether it wants to use a particular ANCS. Discovery and association information consists of attributes that allow a terminal, which has decided to use a particular ANCS, to discover and associate with the corresponding AN. This information will generally be specific to a particular access point (AP) of the AN. The best way to accommodate this is to create a separate service type for an access point. This means that for each ANCS, there are two service types: one for an AN and the other for an AP.

Each service type needs a service template, which is an ASN.1 specification of the attributes used to define a service of that type. First it must be decided what attributes are necessary to describe each AN and AP.

AN Attributes

A mobile terminal (MT) trying to discover what ANs are available will want to know several details about each AN so that it can choose which one to use. We consider here a universal mobile telecommunications system (UMTS) AN as an example. The attributes defining this service type are shown in *Table 1*.

The following, specified in ASN.1, forms the UMTS AN service template:

```
        UMTSAccessNetwork       ::= SEQUENCE
{
Description                 IA5String
networkId                   PLMN-Identity
cost                        UMTSNetworkCost
performance                 UMTSNetworkPerformance
}
```

AP Attributes

Having selected an AN, a terminal must now discover and associate with an AP of that AN. To see which attributes would be useful in the procedure of discovering and associating with an AP, we looked at the steps taken by a terminal performing this procedure normally (i.e., without a WBC) to see where additional information would have been helpful.

The following example shows the steps taken by a UMTS terminal to discover and associate with a UMTS cell (AP) of a particular PLMN (AN) [11, 12]:

- Search all RF channels in UMTS terrestrial radio air (UTRA) interface bands to see which are carriers.

TABLE 1

Attributes for the UMTS AN Service Type

Attribute	Meaning	MT/MU use
Description	A string containing a short description of the access network.	The MT can display this to allow the MU to choose the best access network.
Public land mobile network (PLMN) identity	An identifier for the access network that is broadcast in every cell of the network.	The MT can recognize which cells belong to this access network when searching for cells.
Cost	Contains information relating to the charging structure of a UMTS network, such as what quality of service options it offers and how much it charges for each.	Based on cost-to-performance ratio, the MU/MT decides whether to use this access network for a particular application service delivery.
Performance	Contains third-party measurements of the performance of the access network.	

- For each found carrier, determine the downlink scrambling code and frame synchronization of the cell.
- Read the system information on the cell's broadcast channel (BC) and check if the cell is suitable. A suitable cell is a non-barred cell of the correct PLMN that does not belong to a forbidden location area (LA), which passes the cell selection criteria. The cell selection criteria is a formula involving the received signal quality, the terminal's maximum transmit power, and the cell selection parameters received in the system information.
- If the cell is suitable, attempt to register with the cell. If registration is successful, cell search is finished. If the cell is not suitable or if registration is unsuccessful, continue cell search, repeating steps 1 through 4.

The following are situations that can speed along the process of finding a suitable cell:

- If the terminal knows beforehand which channels are carriers, it will not have to search all the channels.
- If the terminal knows the synchronization code and scrambling code beforehand, achieving frame synchronization is simplified.
- If the terminal is given the relevant system information beforehand, it knows everything it needs to check if the cell is suitable except for the received signal quality. It only needs to take measurements for cells that pass all the other criteria.

From that, we can compile the following sample list of attributes [13] for the UMTS AP service type:

The ASN.1 service template for the UMTS AP service type is as follows:

```
UMTSAccessPoint          ::= SEQUENCE
{
    networkId      PLMN-Identity,
    lac            LocationAreaCode,
    cellID         CellIdentity,
    cellAccRes     CellAccessRestriction,
    freqInfo       FrequencyInfo,
    cellInfo       CellInfo,
    cellSelInfo    CellSelectInfo,
    coverage       EllipsoidPointUncertCircle
}
```

Similarly, when performing the same analysis for APs of other access technologies, we can form ASN.1 templates for their service types. The following are examples of IEEE 802.11 [14] and Bluetooth [15] AP service-type templates:

```
IEEE80211AccessPoint     ::= SEQUENCE
{
    networkId      SSID,
    supRates       SupportedRates,
    capInfo        CapabilityInformation,
    bssid          BSSID,
    beacInt        BeaconInterval,
    timeStamp      TimeStamp,
    security       SecurityInformation
    phyInfo        PhySpecificInformation,
    coverage       EllipsoidPointUncertCircle
}
```

```
BluetoothAccessPoint     ::= SEQUENCE
{
    netwId         BluetoothNetworkID,
    bdaddr         BluetoothDeviceAddress,
    scanRep        ScanRepetition,
    clk            BluetoothDeviceClock,
    scanMode       PageScanMode,
    version        BluetoothVersion,
    class          ClassOfDevice,
    supFeatures    SupportedFeatures,
    coverage       EllipsoidPointUncertCircle
}
```

TABLE 2

Attributes for the UMTS AP Service Type

Attribute	Meaning	MT/MU use
PLMN identity	Made up of the mobile country code (MCC) and mobile network code (MNC), this identifies the PLMN to which the cell belongs.	To check if the cell belongs to the correct PLMN.
Location area code	Identifies the LA, within the PLMN, to which the cell belongs.	To make sure the cell does not belong to a forbidden LA.
Cell access restriction	Indicates whether the cell is barred or barred to certain access classes.	To check that the cell is not barred (or barred) to the access class to which the terminal belongs.
Frequency information	Identifies the frequency channel that the cell is on. In the case of frequency division duplex (FDD), it identifies both the uplink and downlink channels.	To know which frequency or frequencies to look for the cell on.
Cell information	Contains various information about a cell, including scrambling and midamble (for time division duplex [TDD]) codes.	To simplify the procedure of synchronizing with the cell.
Cell selection information	Contains parameters for cell selection, including the maximum transmitting power level a MT may use to access the cell, as well as the minimum received signal level and quality at the MT for which it is permitted to access the cell.	To check if the cell fulfils the cell selection criteria.
Cell identity	A unique identifier for the cell, within the PLMN.	To allow an MT reading a description to know if this is a cell it already knows about. Also, after selecting a cell based on its description, to check if the selected cell is the one that was described.
Coverage area	The geographic area in which the MT can expect to find an acceptable strength signal from the cell.	To determine the likelihood of being able to achieve the required signal quality from the cell at the MT's position. The MT can prioritize which cells to take measurements for based on this likelihood.

When designing the ASN.1 specification, it is important to accommodate for future service types and updates. For instance, there should be information as to where software defined radio (SDR) downloads can be acquired so a terminal can access ANs that use a technology unknown to it.

Application Services (ASs)

The vision for next-generation networks (NGNs) is for all–IP networks where MTs will receive application services from server providers over IP. There are many AS types available to MTs, and in an NGN, that number will be greater. A service type should include all services that perform a similar function. For example, a voice-call service should have one service type rather than many service types depending on the voice over IP (VoIP) protocol being used. These, instead, should be included as attributes of the voice-call AS.

In addition, service types can themselves be put into broader categories—for example, communication services, information services, location-based services, etc. This allows similar service types to be grouped together on a WBC channel. The International Telecommunication Union—Telecommunication Standardization Sector (ITU–T) NGN Focus Group [30] Working Group 1 is concerned with NGN services. Using its NGN Release 1 Service Classification [16] as a guide to what services will be available in NGN, possible AS categories and types advertised to MUs/MTs over a WBC are as follows:

- *Communication services*: Voice call, video call, push-to-talk, conference call, multimedia interactive communication service, incoming call, voice mail, call diversion

- *Messaging services*: SMS, MMS, e-mail, fax, instant messaging, chat rooms, bulletin boards
- *Information services*: News, weather, sports, finance, directory
- *Entertainment services*: Games (download, play on-line, on-line multiplayer), music (download, streaming), video clips (download, streaming), other downloads (ring tones, logos, icons, screen savers, etc.), television (streaming), radio (streaming), adult entertainment, competitions, gambling, dating, voting/surveys (perhaps related to TV programs, e.g., reality TV)
- *Education services*: M-Learning (tutorials, labs, tests, examinations)
- *M-commerce services*: M-pay facilities, m-shopping portals, m-banking and share trading, bookings and ticketing, m-market-type services (eBay)
- *Location-based services*: Nearest restaurant, cinema, etc.; local weather; city guide; walking/driving directions; car parking; traffic information; public transport services/timetables; location-based advertising; location-based dating
- *Other services*: On-line data storage, organizational services (calendar, address book, etc.), Web browsing, file transfer, virtual private network (VPN)

Each service type needs a service template to specify the attributes that are included in an SD. Similarly to the AN service types, the purpose of these attributes is for ADA. Making decisions between ASs will be semiautomatic, as for ANs. The terminal reads the SDs from the WBC and provides relevant information to the MU. For instance, for a given AS type, an MT will show all available AS providers (sorted by preference based on attributes such as cost, quality, and supported features specified in the AS profile), and the MU will choose from these the best AS provider for the desired AS.

After the MU chooses the best AS, the MT will need to know how to associate with the AS. The first thing needed is for the client-side software to be installed on the terminal (if not already installed or preloaded). For this, the terminal needs an attribute to tell it how to download the software. There should then be an attribute specifying the software itself and its version so that the terminal can see if it needs to be updated. Rather than having an attribute that indicates where to get the software or software update, it could be better to have one software-download service (which would also be advertised, as any other service on a WBC) that would allow downloading of all additional software needed to use ASs advertised on a WBC (this could also be used for SDR downloads for ANs' communications service types). After the software is installed, there are some attributes specific to a particular AS (e.g., IP address and port numbers) that also need to be known.

Considering the voice-call AS as an example, the list of its attributes for advertisement and association is given in *Table 3*.

Specified in ASN.1, the voice-call AS definition is as follows:

```
        VoiceCallService  ::= SEQUENCE
{
        cost              VoiceCallCost
        performance       VoiceCallPerformance
        protocols         VoiceCallProtocols
        features          VoiceCallFeatures
        software          SoftwareIdentifier
        versions          SoftwareVersion
        softParams        VoiceCallSoftwareParameters
}
```

It is important to design the ASN.1 specification so that it can also accommodate future service types and updates.

WBC Structure

Data scheduling

An SD contains the values of all the attributes specified in the service template. These are encoded into binary data using the PER of ASN.1. Depending on the service type and the values of certain attributes, SDs are not all going to be the same size. A WBC broadcasts SDs continuously, so terminals listening to the channel can receive the service descriptions they need. But how can a WBC know which SD to broadcast at any given time? The simplest solution to this would be to broadcast all SDs one after the other, starting over at the beginning when the end is reached (called a flat broadcast). This, however, might not be the most efficient way. If a certain SD is required by terminals more frequently than another, then it makes sense to broadcast it more frequently. The priority should be to minimize the average access time of the entire system. That time, averaged across all required SDs by all terminals, is what should be minimized.

There are quite a few data scheduling algorithms, but most of these [c.f. 17–20] are pull-based, meaning that they rely on the clients making requests on a back channel for certain data items to be broadcast. These are not suitable for a WBC, as they are simplex broadcast channels. Hence, only push-based algorithms were included in our consideration. The following two push-based algorithms seem to be most promising for SDs scheduling on a WBC:

Broadcast Disks

This algorithm tries to minimize the average access time. It uses access probabilities of the data items to schedule their broadcasting. A data item that is accessed twice as frequently as another will have an access probability twice as high as the others. The algorithm works by assigning data items to virtual disks, which "spin" at different speeds. For each "rotation" of a disk, all the data items on that disk are broadcast once. If every data item is assigned to a disk with a speed relative to its access probability, then this algorithm is optimal among the push-based algorithms.

The main problem faced in using this algorithm is knowing the access probabilities of the SDs. The system will be ever-changing, so there will be no true numbers for probabilities. The challenge will be to find an accurate estimate of these.

Priority Index Policies

This algorithm also uses access probabilities to schedule the data items. Unlike broadcast disks, which is an off-line

TABLE 3

Attributes for the Voice-Call AS Service Type

	Attribute	Meaning	MT/MU use
For advertisement	Cost	Contains information about the cost of using the voice-call service (e.g., cost per minute to various recipients: VoIP, GSM, PSTN, different countries).	The MT can tell the MU how much each offered AS will cost for the call the MU wishes to make.
	Performance	Contains third-party measurements of the performance of this AS (e.g., calls dropped, sound quality, etc).	The MT can tell the MU which offered AS can be expected to perform best.
	Protocols	Which protocols are used by this AS.	The MT may not support the use of certain protocols.
	Features	A list of features apart from the basic that might differentiate one voice-call service from another (e.g., ability to send text during a call, music playing in the background, etc).	The MT can list these features to help the MU decide whether to use this AS.
For association	Software	An identifier that tells the MT what software is needed to access this service.	The MT can see if this software is already installed. If not, it can be downloaded from a software-download service provider.
	Supported versions	The version numbers of the software that are supported by this service.	The MT can see if the version of the currently installed software is up to date.
	Software parameters	Information about the service needed by the software (e.g., IP address and port numbers). This may have different fields for each value of the software attribute.	The software uses this information to associate with the service.

scheduling algorithm (i.e., it schedules the whole broadcast beforehand), this is an on-line one (i.e., it decides which data items to send in real time). At each slot n, it chooses which data item i to broadcast based on its priority index ÏiÁ.wi(n), where Ïi is the rate of client request generation for page i, wi(n) is the elapsed time since the last transmission of page i, and Á is an exponent that determines the relative importance of Ïi versus wi(n). It chooses the data item with the highest-priority index. By varying the value of Á, various results can be achieved, but study has shown that the value of Á = 0.5 has the best performance in most cases.

The advantages of this algorithm over the broadcast disks algorithm are: no need to perform any pre-computation before the broadcast; no need to store the whole broadcast; much less storage (as this is an on-line scheduling); and no need to wait until the end of the broadcast to incorporate a change in the data. However, there is one disadvantage that makes it less suited for use in a WBC: Since data is scheduled at every slot, the data is not predictable, so it would be hard to index.

Data Indexing

MTs have limited battery power. Therefore, any steps must be taken to reduce their power consumption, thereby extending their battery life. One of the biggest users of battery power is the MT's receiver. Therefore, it is advisable to minimize the time spent by terminals for listening to a WBC channel to receive the required SD (referred to as tuning time). Tuning time could be reduced by the use of indexing because without it, a terminal would have to tune in to a WBC and listen to the broadcast continuously until the required SD was transmitted. By adding indexing data to the broadcast, terminals can tune in, find out when the required SD will be transmitted, then tune out and wait until that time to tune back in again. By adding redundant data to the broadcast, however, the average access time will increase.

There are several indexing schemes that aim to provide a good tradeoff between a low tuning time and a low access time. All these schemes have several things in common: the attribute that the data items (records) are indexed by is called the key (in the case of a WBC, the key will be the service type); the order of the records in the broadcast is arranged by the value of their keys; the broadcast is divided into fixed size parts called buckets; and the indexing tells the client in which bucket to look to find the first record with the required key (as buckets are of a fixed size, the client will know the exact time a particular bucket will be broadcast).

Short description and comparison of the main indexing schemes are given in Appendix C. Note that flat-broadcasting schemes could be adapted to accommodate non-flat broadcasting as well.

To determine the best combination of data scheduling algorithm and data indexing scheme for use in a WBC, details need to be known about the intended coverage area of a particular WBC (i.e., national, regional), the number of SDs for broadcast, the number of service types, and an estimate of the number of terminals in the area and their access patterns. Once these details are known (or estimated), the best combination of schemes and algorithms to use in a WBC can be determined using simulations.

WBC Carrier Technologies

As a WBC is a simplex broadcast channel, it should be delivered by a broadcast technology. There are several properties that would determine how desirable a particular broadcast technology is for this purpose. It should be possible to achieve a very high level of coverage of a given area. Ideally, coverage should at least match that offered by the ANs (GSM, UMTS) in the area. Coverage should not be affected by terminal mobility. Reception should be available indoors and outdoors. The transmission power (and other running costs) to achieve this level of coverage should be as low as possible. Receivers should have minimal sizes, power requirements, and hardware manufacturing and integration costs so as not to adversely affect terminal size, stored energy requirements, and manufacturing costs. This means that WBC carrier technology will need to be or become a successful broadcasting (radio, TV) technology, because the more successful it is, the cheaper the equipment will become. If at all possible, the receiver technology will need to be able to handle all WBCs—local, regional/national, international, satellite, terrestrial, etc. Perhaps more than one technology could be used, e.g., one for international, national, and regional WBCs and another for local WBCs. There are several seemingly suitable broadcast technologies to consider, some already fully operational and some in the early stages of development. These fall into two main categories—terrestrial and satellite. Examples include the following:

Terrestrial Carrier Technologies

There are several terrestrial broadcast technologies that are designed to accommodate mobile reception. With most of these, coded orthogonal frequency division multiplexing (OFDM) with a guard interval between symbols ensures that the Doppler effect does not come into play when receiving the signal while the terminal is in motion.

Digital Audio Broadcast

DAB [28, 29] is a standard for digital radio developed by the Eureka 147 research project in the late 1980s that typically operates in Band III or L-Band. It delivers audio encoded using MPEG Layer 2 (MP2), but it can also carry data streams. Several radio channels are broadcast together in a 1.7MHz DAB multiplex. Each multiplex has a capacity of 1.184 Mbps when employing the most popular level of error protection. In the United Kingdom, Germany, and other countries, there are national multiplexes broadcasting national radio stations as well as regional and local multiplexes broadcasting local radio stations. DAB is well suited to provide large coverage, allowing six or more radio programs to be bundled together in a national multiplex broadcast on the same frequency all over the country, called a single-frequency network (SFN).

Services are up and running in about 20 countries, and others are having trial broadcasts or researching the idea. Receivers have been on the market for 10 years and recently receivers have been developed that are targeted for integration into mobile phones and personal digital assistants (PDAs). Developed in the 1980s, the original DAB technology, using MP2 audio codec, is now old. MP2 is very inefficient compared to the new high-efficiency advanced audio coding (HEAAC) used in the newer digital audio broadcasting standard. Countries that have yet to decide on digital audio broadcasting may choose one of the newer, more spectrally efficient technologies such as those mentioned below. While it would be preferable for WBCs to be broadcast over one technology, the realities in different countries may mean different technologies will be used.

Terrestrial Digital Multimedia Broadcast (TDMB)

TDMB, an improvement on the original DAB, may be used to broadcast data, audio, or video. In Europe, it has recently been approved as an ETSI standard [31, 32]. It can be broadcast using existing DAB transmitters with minor upgrades. It operates in the same bands as DAB and uses the same 1.7 MHz multiplexes. However, TDMB uses more efficient error protection to increase the size of a multiplex from 1.184 Mbps to 1.416 Mbps.

TDMB is fully operational in South Korea, and mobile phones are on the market that can receive TDMB. A technological upgrade on DAB, it uses the new HEAAC. It can also deliver TV. However, broadcasters in countries that have already implemented the original DAB are unlikely to change in the short term, given the heavy legacy receiver investment. However, suitable reductions in the cost of receivers could change this scenario.

Digital Radio Mondiale (DRM)

DRM [33] is a worldwide consortium promoting, developing, and implementing digital broadcasting using the analog AM radio spectrum allocations for digital sound and other services. Bands covered include short-wave, medium-wave, and long-wave bands. The design, development, and testing phases of the short-wave band are expected to be completed between 2007 and 2009. It uses channels of 9 kHz or 10 kHz, but channel extensions of up to 100 kHz are planned. A 100 kHz channel will be able to provide a data rate of 285 kbps. Smaller channels, rather than a large multiplex of several channels, would mean that a WBC would not have to rely on having other broadcasters to share a multiplex with before it could start operating. This could be very important for the regional versions of a WBC, as some local regions may not have many broadcasters. Capability of operation at lower-frequency bands means greater geographic coverage for the same transmit power. Transmissions in the short-wave band in particular are capable of covering huge geographical areas. The disadvantage of using these bands is that a larger antenna is required to receive the signal, and atmospheric conditions can cause interference problems.

Digital Video Broadcast–Handheld (DVB–H)

This is a new standard for broadcasting digital television to handheld devices [34]. It extends the DVB–Terrestrial (DVB–T) standard with time slicing and other additions to greatly reduce receiver power consumption. It is ideal for WBC use because it is designed for handheld devices and the low power consumption allows for high terminal mobility. It uses multiplexes of 5, 6, 7, or 8 MHz. The multiplexes are very large, and for the same reason that the small channels were an advantage for DRM, that is a disadvantage for using DVB–H as WBC technology.

It provides a higher data rate for the same amount of spectrum as DAB. Also, because it can transmit audio using the far more efficient HEAAC codec, it may be considered as a better option than DAB for digital radio.

Multimedia Broadcast/Multicast Service
This is a technology proposed by the 3GPP [35] to broadcast digital television on handheld devices over UMTS networks. As UMTS technology is being used, terminal mobility and indoor reception should not be problems. MBMS is still in the development stage.

Satellite Carrier Technologies
Broadcasting from a satellite has advantages and disadvantages when compared with terrestrial systems. A naturally large antenna pattern footprint is a big advantage when considering a WBC. Direct line of sight is normally required because of the high frequency allocations for satellite broadcasts. Thus for indoors, in tunnels, on city streets, etc., where reception would be weak or nonexistent, either repeater solutions or supplemental terrestrial systems would be required. The satellite system would not be attractive WBC advertisements of local significance, i.e., of generally small-footprint WBC communications. Some DAB–type satellite technologies are briefly described below.

Satellite DMB (SDMB)
SDMB [e.g., 40, 41, and 42] has the same purpose as TDMB, i.e., to broadcast data, audio, or video. It is broadcast from a satellite and operates in the S-band at much higher frequencies than TDMB. SDMB is operational in Japan and Korea, and mobile phones capable of receiving the broadcasts are on the market. "Gap filler" ground repeaters are used to enhance the signal in areas where line of sight with the satellite might be obstructed. A R&D group called MoDiS [36] looked into a European SDMB solution, under European Union (EU) funding (2002 to 2004). The present EU–funded research in this field is called the Mobile Applications & sErvices based on Satellite & Terrestrial inteRwOrking (MAESTRO) project, which aims at studying technical implementations of innovative mobile satellite systems concepts targeting close integration and interworking with 3G and beyond–3G mobile terrestrial networks.

Digital Audio Radio Satellite
There are several systems and standards for delivering digital radio by satellite. The most prominent are the WorldSpace system [37]—with geostationary satellites covering parts of Asia (AsiaStar), Europe, Africa (AfriStar), and one yet to be launched in South America (AmeriStar)—and XM Radio [38] and Sirius [39] (both operating in the United States). All these operate in the S-band. Formats vary. For instance, one generation of the WorldSpace system transmits up to 96 broadcast channels (BCs) in a time division multiplex (TDM) downlink beam with a frequency range of 1,453.384 MHz to 1,490.644 MHz. Each BC can contain up to eight service components (SCs), which are allowed to contain MPEG–coded audio or general data (e.g., JPEG).

Development of miniaturized receivers suitable for integration into a mobile terminal is well advanced. Reception is impaired in indirect line-of-sight situations. This can be counterbalanced by using ground repeaters in urban areas to enhance the signal.

Just as these systems have channel space dedicated to carrying services such as latest stock information, they also could act as a WBC carrying structured advertisements of SP ADA information.

Conclusion

Advertisement and discovery are part of an important R&D challenge to find solutions for the automated enabling of the entire process of ADA of 4G mobile services and for the evolution of UCWW. It will contribute significantly to continuity of connection to mobile services provided to MTs, that those MTs are able to discover the ANs and their attributes, if not in advance of entering the accessibility footprints of the ANs, at least when they have already entered (or are adjacent to) such footprints. To facilitate this, a wireless equivalent of the shop window and billboard advertising, called a WBC, was proposed in this paper. This would provide ANPs and ASPs with a very active dynamic means to advertise (i.e., proactively "push") their presence and services through a WBC operated by non–ANP service providers. The proactive push advertisement nature of a WBC would correspond well to ANP competitive desires to use all means to reach all MUs from their existing "loyal" consumers through to the impulse-buying ones. Advertisement and discovery of ANs, their wireless communication services and application services deployed and accessible in a particular area/location, and procedures for terminal association with them are defining characteristics of NGNs and especially a UCWW.

A WBC would also benefit the MUs because of automated discovering functionality of MTs scanning a WBC and updating service offerings and availability information, matching these against MU/MT profiles and proposing, and enabling casual or persistent consumer-type MT–AN association links for different user-desired services. The result of this process will be more up-to-date information for user-driven always best connected and served (ABC&S) decisions.

Implicit in this vision is the global standardization of these proposed WBCs as a vehicle and a wireless infrastructural support for ADA procedure. Practically it makes sense to have several WBCs directed at different geographic extensions—national, regional, etc.

This paper details the steps taken to come up with the best design for a WBC. An SD model, an SD format, and an ASN.1 SD encoding were elaborated for use in a WBC. Categorization of mobile services was provided and service templates were proposed. Steps were taken toward best data format on a WBC and the best technology to allow MTs to access a WBC. Critical review and comparison of possible carrier technologies for a WBC were also presented.

References

J. Pereira. "Optimising Spectrum Efficiency in a Heterogeneous Network Environment." In Wireless, Mobile and Always Best Connected, 1st International ANWIRE Workshop. Glasgow, Scotland. DVD Proceedings. Ed. M. S. O'Droma. ISBN 0-9545660-0-9. April 2003.

M. S. O'Droma and I. Ganchev. 2004. "Enabling an Always Best-Connected Defined 4G Wireless World," Annual Review of Communications, Volume 57 (Chicago, Ill.: International Engineering Consortium, 2004), ISBN: 1-931695-28-8, pp. 1,157–1,170.

IST-2001-38835 ANWIRE, Deliverable D1.2.1, "Always Best Connected" – Concepts & Definitions. www.anwire.org. February 2003.
IST-2001-38835 ANWIRE, Deliverable 1.1.3, "Report on the Coordination of Integrated Thematic Efforts Related to System Concepts." August, 2004.
T. H. Le, A. H. Aghvami. Performance of an accessing and allocation scheme for the download channel in software radio. Wireless Communications and Networking Conference, WCNC 2000. IEEE, Vol. 2, 517–521. 2000.
www.sun.com/software/jini.
IETF RFC 2608 – Service Location Protocol, version 2. www.ietf.org.
Salutation Consortium. www.salutation.org.
ITU–T Rec. X.680 Information technology – Abstract Syntax Notation One (ASN.1): Specification of basic notation. www.itu.int/ITU-T/studygroups/com17/languages.
ITU–T Rec. X.691 www.itu.int/ITU-T/studygroups/com17/languages.
3GPP User Equipment procedures in idle mode TS 25.304.
3GPP Non-Access-Stratum Functions Related to Mobile Station in Idle Mode TS 23.122. www.3gpp.org.
3GPP Radio Resource Control protocol specification TS 25.331 www.3gpp.org.
ANSI/IEEE Std. 802.11. www.ieee802.org/11.
Bluetooth Specification Core, Bluetooth Special Interest Group, Version 2.0. https://www.bluetooth.org.
M. Carugi, B. Hirschman, and A. Narita, "Introduction to the ITU–T NGN Focus Group Release 1: Target Environment, Services, and Capabilities." IEEE Communications Magazine, pp. 42–48, October 2005.
D. Aksoy and M. Franklin, "Scheduling for Large-Scale On-Demand Data Broadcasting," Proc of Infocom Conference, 1998.
S. Acharya, M. Franklin, and S. Zdonik, "Balancing Push and Pull for Data Broadcast," Proc of ACM SIGMOD Conference, May 1997.
J.-H. Hu, K L. Yeung, G. Feng and K F. Leung, "A Novel Push-and-Pull Hybrid Data Broadcast Scheme for Wireless Information Networks," 2000.
C.-L. Hu and M-S Chen, "Adaptive Information Dissemination: An Extended Wireless Data Broadcasting Scheme with Loan-Based Feedback Control," IEEE Transactions on Mobile Computing Vol.2, No.4, October–December 2003.
T. Imielinski, S. Viswanathan and B. R. Badrinath, "Energy Efficient Indexing on Air," Proc of ACM-SIGMOD, Intl Conference on Management of Data, Minnesota, May 1994.
D. Shin, D. Shin, E.-J. Jeong, J.-H. Kim and S. Kim, "Efficient Data Broadcast Scheme on Wireless Link Errors," IEEE Transactions on Consumer Electronics, vol. 46, November 2000.
T. Imielinski, S. Viswanathan, and B. R. Badrinath, "Power Efficient Filtering of Data or Air," Proc of 4th Intl. Conference on Extending Database Technology, Cambridge, March 1994.
T. Imielinski, S. Viswanathan, and B. R. Badrinath, "Data on Air: Organization and Access," IEEE Trans. Knowledge and Data Eng., vol. 9, May/June 1997.
K. L. Tan and J. X. Yu, "Energy Efficient Filtering of Nonuniform Broadcast," Proc of 16th ICDCS, 1996.
N. Shivakumar and S. Venkatasubramanian, "Efficient Indexing for Broadcast Based Wireless Systems," Mobile Network and Applications, Vol. 1, 1996.
K. F. Jea and M. H. Chen, "A Data Broadcast Scheme Based on Prediction for the Wireless Environment," Proc of 9th Intl. Conference on Parallel and Distributed Systems, Taiwan, December 2002.
World DAB. www.worlddab.org.
ETSI Radio broadcasting systems; Digital Audio Broadcasting (DAB) to mobile, portable and fixed receivers ETS 400 301.
ITU–T NGN Focus Group. www.itu.int/ITU-T/ngn/fgngn.
ETSI Standard TS 102 427.
ETSI Standard TS 102 428.
Digital Radio Mondiale. www.drm.org.
DVBH on-line. www.dvb-h-online.org.
3GPP Multimedia Broadcast/Multicast Service (MBMS); Architecture and functional description TS 23.246.
Mobile Digital Broadcast Satellite. www.ist-modis.org.
WorldSpace Satellite Radio Network. www.worldspace.com.
XM Satellite Radio. www.xmradio.com.
Sirius Satellite Radio. www.sirius.com.
ESA, Mobile Satellite Systems and Services, Satellite Digital Multimedia Broadcasting (SDMB) System. 3 Nov. 2005. telecom.esa.int/telecom/www/object/index.cfm?fobjectid=11985.
Mobile Applications & Services based on Satellite & Terrestrial Interworking project (MAESTRO). EU Project (IST-2003-507023). ist-maestro.dyndns.org/MAESTRO/index.htm.
Tanner, J.D., Editor, Telecom Asia, "T–DMB takes on S–DMB in Korea," 15 April 2005. www.telecomasia.net/telecomasia/article/articleDetail.jsp?id=158158.
M. O'Droma and I. Ganchev, "Ubiquitous Consumer Wireless World," at the IFIP/IEEE Mobile and Wireless Communication Networks Conference, MWCN'04, Paris, 25–27 Oct. 2004.

Notes

1. The term "mobile service" is used here to mean both access networks' communications services (ANCSs) and application services (ASs).
2. The list of possible services advertised on the WBC is by no means complete. Some of the listed service types may not be suitable for advertising on the WBC, and other (suitable) service types may have been omitted. Also, some of the service types belong to more than one service category.
3. Access time—The time from the moment a mobile terminal starts accessing a data item (SD) on the WBC until the moment it receives the full data item.
4. Access probability of a data item (SD)—The probability that a mobile terminal will try to access that data item (SD) at any time.

Appendix A. Service Discovery Protocols

JAVA Intelligent Network Infrastructure (Jini)

Jini [6] is a service advertisement and discovery system developed by Sun that is based on the Java programming language. Devices plug together to form a community, and each device provides services that other devices may use. It works by extending the Java application environment from a single virtual machine to a number of machines. It uses Java's remote method invocation (RMI) to move code around the network.

A Jini community uses lookup services. A lookup service is a directory for other services in the community. When a client or a service enters a network, its first step is to discover what lookup services are available to it. When it has discovered one or more lookup services, it can choose to join the corresponding communities. If the joining entity is a service, it registers with the lookup service, which then stores information about the service for prospective clients to see. If the joining entity is a client, it can then query the lookup services about available services in the community.

Service Location Protocol

SLP [7] is an IETF protocol for service discovery and advertisement. It is designed solely for IP–based networks. It uses service uniform resource locators (URLs), which define the service type and address for a particular service.

There are three entities in SLP: service agents (SAs), user agents (UAs), and directory agents (DAs). SAs advertise the locations and attributes of available services, while UAs discover those of services required by the client. DAs store information about available services. They are like lookup services in Jini. SLP can operate with or without DAs.

Salutation

Salutation [8] is architecture for service discovery developed by the Salutation Consortium, which includes industry and academia members. It provides a standard method for devices to describe and advertise their capabilities and the capabilities of the services they offer to other devices.

The Salutation manager (SLM) is similar to the lookup service in Jini. Services register their capabilities with an SLM as functional units. Clients can search for services using the SLM. During a service search, an SLM will work with other SLMs to find available services. When a client discovers a service, the SLM also manages the connection between client and service.

Appendix B. Service Description Formats

Jini SD Format

An SD is stored as an instance of the Java class ServiceItem. The class has the following fields:

- *Service ID*: A unique 128-bit identifier for the service
- *Service object*: A Java object used to provide service interface to clients
- *Attribute sets*: An array of instances of the entry class, each specifying certain attributes of the service

SLP SD Format

A SD is stored as a cached SrvReg message. All fields (below) in a SrvReg message are represented as text strings:

- *URL entry*: With two subfields, a service URL and the lifetime that URL can be assumed to be valid for. A service URL is a string with the following format:

"service:"<srvtype>"://"<addrspec>

where <srvtype> is the service type; <addrspec> is the hostname (which should be used if possible) or dotted decimal notation for a hostname, followed by an optional `:` and the port number.

- *Service type*: The same as <srvtype> in the service URL. Each type of service has a unique string that specifies the service type.

- *Scope list*: Scopes are used to administratively group services together. All services with the same scope belong to the same grouping. A service has one or more scopes, which may include the default scope.

- *Attribute list*: A list of attribute name and value pairs. Each service type has a service template that specifies those attributes that must be included in this list and those that are optional. A SD must follow the service template of its own service type. There are four possible attribute types: Boolean, integer, string, and opaque (represented as text in the attribute list).

Salutation SD format

A SD is stored as a functional unit description record. A functional unit description record is some encoding of an ASN.1 [9] sequence, with the following fields:

- *Functional unit ID*: Like service type in SLP, this is an identifier unique to the type of service. Unlike in SLP, though, this is an integer and not a string.

- *Set of attribute records*: A list of zero or more attribute records. Like in SLP, each type of service has a template that specifies those attribute records that must be included and those that may be included. An attribute record has the following fields:

- *Attribute ID*: This identifies the attribute. It is similar to attribute name in SLP but is an integer, not a string.

- *Compare function*: A function (e.g., isLessThan) used to compare the value of an attribute to the value of the corresponding attribute in a query.
- *Value*: The value of the attribute. Attributes can be any standard ASN.1 type. The attribute type is specified in the template.

Appendix C. Data Indexing Schemes

(1,m) Indexing

(1,m) indexing [21] is based on a multilevel tree-based index. It is for a flat broadcast. All the records are arranged by their key values. The index is broadcast periodically, m times during each broadcast. All buckets contain an offset to the beginning of the next index segment. Clients tune in and find out when the next index segment will be broadcast. They tune in again at that time and find out when the first record with the required key will be broadcast. They tune in once more at that time and read all the records with the required key until a record with a different key is broadcast. The optimal value of m can be calculated for both error-free and error-prone links [22].

Distributed Indexing

Distributed indexing is similar to (1,m) indexing except that only part of the index is replicated. The index tree is divided into two parts: a replicated part (the top r levels) and a non-replicated part (the bottom k–r levels, where k is the total number of levels in the tree). The nodes in the tree in the non-replicated part are only broadcast once in each broadcast, directly before the data segment that they index. The optimal value of r can be calculated. For large index trees, distributed indexing provides a better access time than (1,m) indexing.

Hashing

This scheme [23] uses hashing, rather than an index, to index the data. Like (1,m) indexing, the algorithm is for a flat broadcast and all the records are arranged by their key values. Each bucket contains a hash function that maps key values onto buckets.

This is a flexible scheme, meaning that by varying certain parameters of the scheme (e.g., the hash function), better results can be obtained for access time or tuning time, depending on the needs.

Flexible Indexing

This scheme [23] maintains the flexibility of the hashing scheme but tries to improve on the access time and tuning time. It is also for a flat broadcast, and all the records are arranged by their key values. The broadcast is divided into p data segments, and every bucket has an offset to the beginning of the next segment. The first bucket in each segment contains a data part and a control part. The control part has a control index, which can be divided into a binary control index and a local index. The binary control index tells clients in which segment the first record with a particular key can be found, and the local index specifies in which bucket it may be found. This scheme is flexible by varying the value of p. Generally, a larger p gives a better tuning time, whereas a smaller p gives a better access time.

Non-Clustered Distributed Indexing

This scheme [24] is the same as the distributed indexing scheme except that it has a modification so that it is not necessary for all records to be arranged by their key values.

Flexible Indexing for Non-Uniform Broadcast

This algorithm [25] is specifically designed for a data broadcast that has been scheduled using the broadcast disks scheduling algorithm. It is similar to the flexible indexing algorithm, but it has some modifications to allow for a non-flat broadcast.

Huffman-Based Indexing

The index in this scheme [26] is based on alphabetic Huffman trees. It works when the popularity patterns of individual data items (an indication of the expected number of access to the items) are known before the broadcast. It works for non-flat broadcasts.

Predictive Indexing

This is an index scheme that aims to reduce tuning time by allowing clients to access data by predicting locations [27]. Instead of an index tree, the scheme only fills in the broadcast channel a few parameters for location prediction. It is a flexible scheme and can be optimized to outperform flexible indexing, hashing, and (1,m) indexing on tuning time and access time.

Leveraging MIMO in Wide-Area Networks

Steven Glapa
Director of Marketing
ArrayComm LLC

A growing number of current and prospective wireless wide-area network (WAN) operators are adopting strategies that include mobile broadband access and rich multimedia services. These strategies present significant challenges to their wireless networks, including requiring large improvements in network capacity, subscriber data rates, range, and coverage quality, in order to build and sustain viable business models. The potential performance gains offered by smart antenna technologies such as multiple inputs, multiple outputs (MIMO) are of increasing interest to operators as they grapple with these challenges to network economics. The promise MIMO has shown in the wireless local-are network (LAN) field, combined with recent advances in client device technologies, have prompted much exploration of wide-area MIMO application.

Many inherent characteristics of LANs that have driven MIMO success in that domain differ substantially in wide-area environments, so the technology transfer must be handled with care. In the following brief overview of wide-area MIMO application, we highlight interference and limited scattering as the most important of these differences and recommend key considerations in implementation. The good news for wireless operators is that a large portion of the theoretical gains from MIMO can indeed be achieved in the wide area by network-aware solutions designed to minimize interference in multi-cell environments and maintain robust operation in limited-scattering situations. We also note that as these performance gains can be achieved without changes in underlying wireless protocols, MIMO for WANs may be closer to realization than generally understood.

Defining MIMO Technology

Smart antenna deployments in commercial WANs to date have used multiple antennas only on the base station side of the link, with a single antenna on the client device, largely because of significant cost sensitivity on the subscriber end. As pressure to improve WAN economics has continued to rise, and as increased chipset integration for client devices has reduced the marginal cost of adding smart antenna processing there, interest in solutions with smart antennas on both ends of the link has increased as well. Multiple antennas at both ends allow for many new transmission techniques that are not possible in systems with multiple antennas at only one end, with additional performance gains in many situations. The industry's current dialog on smart antennas includes widely varying definitions of the terms used for a range of implementations, so it's worth stepping through a brief overview of how the taxonomy fits together.

We start with the simplest case. Consider a system with a single antenna at each end of the link. Although the signal is transmitted in all directions (typically within a 120° sector), a particular wireless channel may only have two dominant paths, as illustrated in *Figure 1*. We show here an example of an elevated base station communicating with a mobile handset (or more generically, a "client device," since it could be a mobile computing platform) down at street level, where the bulk of the received signal comes from reflections off neighboring buildings. This corresponds to a single input, single output (SISO) channel. [Note that in the industry's chosen nomenclature, "input" and "output" are defined in these terms from the frame of reference of the channel itself, rather than the perspective of the devices on either end.]

Here we can introduce the simplest and currently most common form of smart antennas. If the receiver has more than one antenna, it can intelligently combine the signals from the antennas and recognize that the signal indeed is arriving from two main directions. It can do this because the two paths have different spatial characteristics or different spatial signatures. Since the receiver recognizes that there are two different spatial signatures, it can combine the signals from the two antennas so they add coherently resulting in a stronger combined signal. This corresponds to a single-input (to the channel), multiple-output (from the channel) (SIMO) scenario, and this is the well-known case of receiver diversity. Receive diversity is used widely in 2G and now 3G cellular networks on the base station side of the link.

If instead the transmitter has multiple antennas while the receiver has only one antenna, the signal still travels along the same paths since the physics are the same (the buildings are still there). This corresponds to a multiple-input, single-output (MISO) scenario. The main difference compared to

FIGURE 1

A Wireless Channel with Two Dominant Propagation Paths between a Base Station and Client Device, Represented by the Arrows, Overlaid on the Base Station's Nominal 120° Sector Transmission Pattern

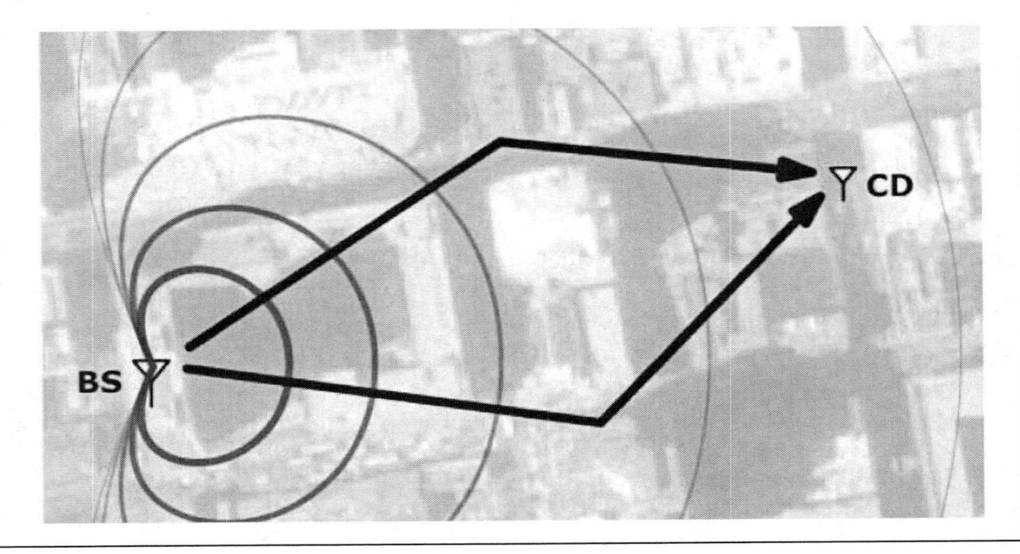

SIMO is that the combining has to be done at the transmitter instead of the receiver. By weighting the transmit antennas appropriately, the two paths can be made to add coherently in the same way as for the SIMO case. This approach is used widely in personal handset systems (PHSes) and high capacity–spatial division multiple access (HC–SDMA) systems with multiple antennas on the base station side, for both receive (working in SIMO mode) and transmit (working in MISO mode).

Providing multiple antennas at both ends of the link corresponds to a MIMO scenario. In this case, we can exploit the two paths much more efficiently, as *Figure 2* illustrates. The transmitter can weight its antennas so that one stream of information, shown in blue, is sent along the first path (i.e., spatial signature) and another stream of information, shown in orange, on the other path. Since the receiver also has multiple antennas, it can separate the two streams by detecting that they have different spatial signatures. In this case, two data streams can be sent, potentially doubling the data rate seen by the subscriber. This offers a material advantage in the best case over what can be achieved with MISO or SIMO processing alone. This MIMO advantage can be achieved without requiring extra bandwidth or power. The multipath propagation that generally impairs the performance of single-antenna links is instead exploited in the MIMO case to increase the channel efficiency and quality.

It is important to understand that MIMO systems exploit multipath propagation provided that these spatial dimensions exist in the propagation environment. In the figures above, there are four antennas and only two dominant paths. Hence, in this case, only two data streams can be formed, even though there are four antennas. Therefore, the performance of MIMO is closely related to the multipath richness of the environment where the system is employed. Fortunately, in many environments there is enough scattering and multipath to support several parallel data streams.

Results from information theory suggest that the system capacity, which represents an upper bound on the data rate, grows linearly with the number of antennas (under certain channel assumptions and holding total power constant) if multiple antennas are used at both ends. In *Figure 3*, the theoretical capacity for different MIMO systems with equal numbers of transmit and receive antennas is shown. The capacity for the 8x8 MIMO system (i.e., with eight antennas on each end of the link) can be up to eight times the capacity of a single antenna system. When all network operational expenses (OPEX) and capital expenses (CAPEX) are taken into account, MIMO techniques offer substantial net performance and economic benefits over single-antenna systems. Especially for high-data-rate services such as true broadband access, Internet protocol TV (IPTV), and large file transfers, where limited bandwidth poses a severe problem, MIMO techniques are a promising solution. Increasing levels of component integration in client devices of the past few years have significantly reduced the marginal cost of adding multi-antenna processing on the subscriber end of the link, making consideration of MIMO applications more practical in many situations.

The predictions in *Figure 3* only characterize the performance limits of an ideal system. Information theory does not provide a great deal of practical guidance for how to achieve these limits. Practical systems face the challenge of how to use the spatial dimensions provided by the channel to best advantage. In principle, three main ways of exploiting the channel have been proposed, the first two focused on individual link performance and the third on overall network performance. Those ways are as follows:

- *Increased data rate:* The previously described technique (illustrated in Figure 2) is usually referred to as spatial multiplexing. For channels with a rich scattering environment, it is possible to increase the data rate by transmitting separate information streams on each

FIGURE 2

A Communication Channel with Two Dominant Propagation Paths Being Used to Double the Subscriber's Data Rate through MIMO (Note the Beam Shaping Enabled by Multi-Antenna Processing that Directs Signal along the Channel of Interest and the Absence of Signal along the Other Dominant Channel)

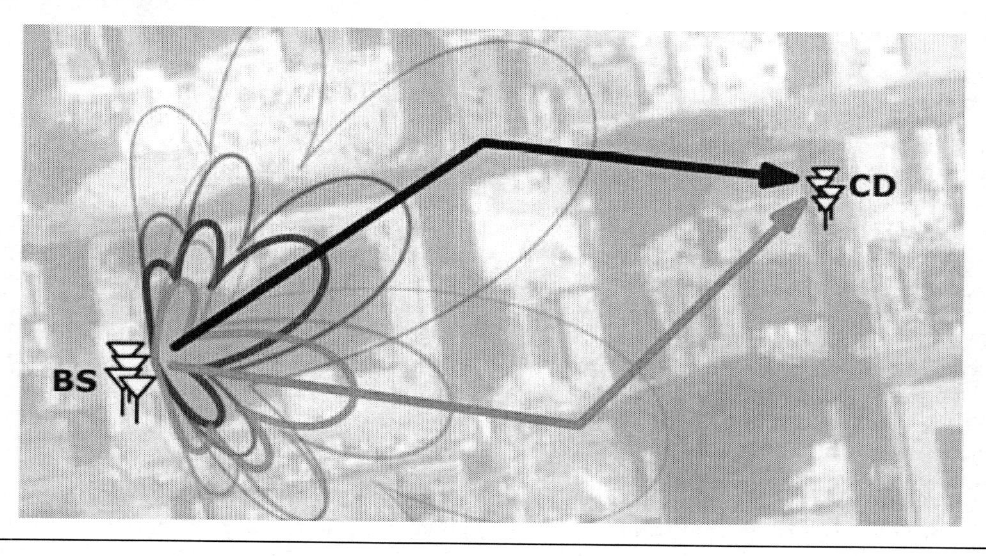

FIGURE 3

Theoretical Average Capacity versus Signal-to-Noise Ratio for MIMO Systems with N Transmit and Receive Antennas, Holding Total Transmit Power Constant

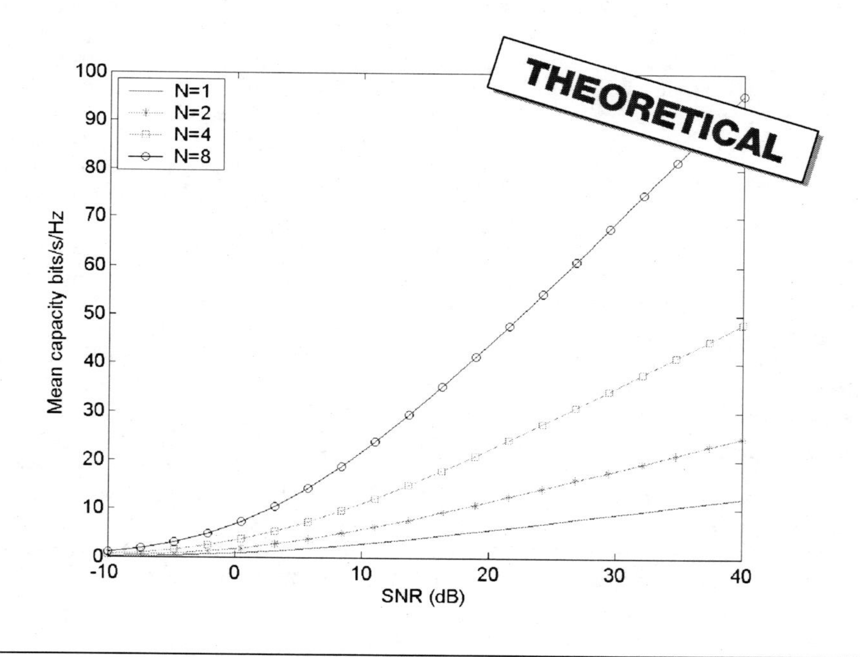

antenna. Using sophisticated receiver technology, the streams can be separated and decoded. For example, using four transmit and four receive antennas, four times the capacity of a single antenna system can be achieved.

- *Improved quality of service (QoS) through diversity*: Conversely, by transmitting the same signal over multiple antennas over multiple symbols, the reliability of the transmission can be improved instead of the data rate. Essentially, this technique provides space-time diversity by sending multiple copies of a signal over different antennas and time instants. This technique of spreading or coding the information symbols over space and time is called space-time coding.

- *Achieving a mix of higher data rate and improved QoS through interference mitigation*: Another alternative exploitation of the spatial dimensions in a MIMO system, suitable to more interference-rich environments, is optimizing the distribution of radio energy in the system overall, to minimize creation of and sensitivity to co-channel interference in the network. Described in more detail in our final section, this solution meets the objectives of higher data rates and more robust links through higher signal to interference-plus-noise ratio (SINR) (which enables higher modulation classes and, therefore, higher data rates in the link) and classic diversity (which increases link stability). As with the MISO system, the multiple spatial channels are used by the base station to create a coherent combination of energy at the client device. They are used by the client device to improve its effective sensitivity in those spatial "directions" (as in a SIMO system), reducing the power required on transmit by the base station. The reverse is done on uplink. The base station and client device work automatically in concert to reduce the level of interference in the system. As we will discuss in the next sections, overall network performance is a critical dimension for wide-area system optimization, and interference is a dominant driver of broadband network performance.

Research labs throughout the world have demonstrated the practical feasibility of MIMO techniques in the initial and most promising application, wireless LANs, by achieving capacities close to those predicted theoretically in their labs using both spatial multiplexing and space-time coding. Because of the large performance gains shown in that initial application, MIMO techniques have quickly left the research labs to be implemented into WLAN products.

Early MIMO Success with Wi-Fi

The most widely publicized implementations of MIMO have been in fixed, wireless local-area environments where the primary benefit of MIMO is to increase individual client device throughput. In particular, WLANs (a.k.a. Wi-Fi networks) in home and enterprise applications exhibit several characteristics, including the following, that make them ideal candidates for initial adoption of MIMO:

- *Rich scattering*: Most Wi-Fi systems are deployed in environments with rich scattering such as indoor or dense urban settings. In these environments, there are usually several propagation paths or spatial dimensions that can be exploited to form multiple streams. In fact, the indoor environment closely matches the conditions necessary to achieve the linear increase in capacity with the increased number of antennas shown in *Figure 3*.

- *Independent deployment*: Another key element for rapid deployment is that Wi-Fi equipment usually is purchased by the end users and deployed independently in their own network. Interoperability of different MIMO Wi-Fi solutions is not critical, as indicated by the success of IEEE 802.11n products, before a common MIMO standard has been agreed upon. This has allowed a very rapid deployment of MIMO without waiting for a standard to be formed.

- *Limited interference*: Also key is that the Wi-Fi environment is closely matched to the theoretical assumptions under which MIMO techniques have been studied. Because of the short range of Wi-Fi networks, MIMO receivers generally operate without significant levels of co-channel interference. The performance of these solutions degrades quickly if operated with significant levels of uncompensated co-channel interference in a network of multiple access points.

The successful deployment of MIMO in Wi-Fi demonstrates that the potential performance improvements made possible by MIMO are real. The fact that it has moved from promising lab results to Wi-Fi products in just a few short years suggests a tempting opportunity for operators of wireless WANs to replicate this success.

The Challenge for WANs

The same performance benefits that enabled the successful adoption of MIMO in Wi-Fi products also make MIMO a good potential candidate for the mobile environment in wide-area wireless. However, the mobile, multi-cell environment is fundamentally different from the Wi-Fi radio environment in some ways, including the following, that make straightforward deployment challenging:

- *Interference*: The wide-area environment experiences significant levels of interference because of dense deployments and large cells. In such environments, rejection of interference and high throughput are both necessary. Hence, to extend MIMO's WLAN success to WANs and mobile broadband data services, new MIMO solutions that take into account both interference and data rate are necessary.

- *Limited scattering*: The wide-area scattering environment can in some cases be limited, with only one or two dominant paths. For instance, if there is line of sight (LOS), there is only one dominant propagation path, which limits the use of spatial multiplexing techniques.

- *Interoperability*: In a WAN, all the users need to communicate with the base stations seamlessly throughout a large network (spanning both geographies and operators), which makes interoperability a necessity. Solutions such as those outlined above that use spatial multiplexing or space-time codes require protocol changes that can increase the time to market for MIMO solutions significantly in the wide-area domain. For example, the receiver needs to know the space-time code the transmitter is using for it to decode the data. Work is already under way in several standardization bodies to include MIMO in mobile systems such as the IEEE 802.16e standard, but there is still some time before robust commercial products will begin to appear.

These factors make adoption of MIMO more challenging in a WAN than in Wi-Fi. New solutions that account for the specific properties of a large, multi-cell network are necessary. Successful implementations of MIMO in WANs will share two key attributes:

- *Interference suppression*: In the wide-area case, the additional degrees of freedom available through the antenna arrays at both ends of the link should be used at least in part to reduce interference. Interference mitigation at both the transmitter and receiver can dramatically reduce network interference compared to systems with only interference rejection at one end.

- *Robust solutions*: Solutions can be developed that account for a limited number of dominant propagation paths. Significant performance gains can still be achieved through coherent combining at both the transmitter and receiver, even in channels with limited scattering. Recent research shows that even for channels with only one dominant propagation path—the so-called keyhole channels—substantial gains can be achieved using smart antenna techniques at both ends of the link.

MIMO for WANs

A significant portion of the ultimate MIMO gains can be achieved now in existing WANs without changes to existing protocols and without waiting for completion of new protocols. Substantial performance improvements can be achieved with adaptive array-processing techniques at the base station combined with similar processing at the mobile terminal—the third basic MIMO approach outlined above. In fact, theory indicates that this is the optimal approach under many of the channel conditions common in WANs. These gains, both in terms of signal strength and interference suppression, are critical for the evolution of WANs, in support of operators' increasingly broadband and multimedia service objectives. As Robert Syputa, senior analyst at Maravedis Telecommunications Research, recently said, "To live up to the promise of providing metro/wide-area network coverage at true broadband data rates, both WiMAX and future versions of 3GPP must adopt MIMO and AAS technologies."

A solution that balances interference rejection and throughput can be achieved in the following way. The base station calculates combining weights for its antenna array that minimize interference from the base station point of view. Similarly, the mobile terminal uses its antenna array to minimize interference from the handset point of view. Since no special coding in the link is required at either the base station or the client device, the MIMO processing can be implemented and operated completely independently on each device. The result is a self-organizing and self-optimizing system that continually adapts to the changing interference environment and to the changing service needs of the user. The independence of each end of the link in this MIMO approach provides robust performance even in the case of a heterogeneous network or one in an upgrade transition—where not all base stations and client devices are equipped with multiple antennas. Single-antenna endpoints simply participate using SIMO (in transmit) or MISO (in receive) channels with their multi-antenna counterparts. The overall network benefit of this interference-minimizing MIMO approach would be an increasing function of the penetration of multi-antenna devices in the system.

In addition to maintaining more robust performance in loaded networks than would be possible with "plain vanilla" MIMO techniques adopted directly from the Wi-Fi domain, the interference-optimized MIMO approach we describe has an additional benefit in wide-area application. In these networks, with relatively large cells and stringent indoor-coverage service objectives, there are many cases at cell edges and deep indoor locations where client devices must operate with very poor channels—where signal-to-noise ratio (SNR) is too low to employ the more complex MIMO techniques. The adaptive antenna processing inherent to an interference-optimized MIMO approach will allow the network to serve users more effectively in these low–SNR cases by "devolving" from the complex MIMO modes to the more classic adaptive antenna approach (often referred to by the oversimplified term "beamforming").

Conclusion

The performance gains provided by MIMO techniques show great promise of fueling the next chapter in the ongoing revolution in wireless communications. MIMO equipment is already providing performance gains to the Wi-Fi market, and WANs will soon be next. However, the radio environment in a wide-area, mobile wireless system is very different from the Wi-Fi domain, with interference posing the most significant challenge. Fortunately, there are wide-area MIMO solutions available now based on adaptive antenna processing that provide tremendous gains over single-antenna systems. These solutions balance interference and throughput requirements by exploiting through multiple antennas the spatial dimensions inherent in the channel. A significant portion of these gains can be achieved without changes to protocols and can be deployed in the near future. Hence, wide-area MIMO applications may be closer to realization than is generally understood.

Revolution by Satellite

Pradman Kaul

Chairman and Chief Executive Officer

Hughes Network Systems

Introduction

In the past decade, the telecommunications industry has undergone revolutionary changes, driven by the world's insatiable demand for information and connectivity via the Internet and pushed by major developments in system and component technologies. Indeed, the universal availability of broadband access is now understood to be an economic imperative by most nations.[i]

Throughout this period, satellite has played a significant role in expanding the addressable market for broadband services, yet it remains hardly known or poorly understood compared to wireless and terrestrial alternatives. Ten years ago, satellite represented a relatively small, fragmented part of the telecom world, dominated by government or quasi-governmental operators and a handful of proprietary equipment suppliers. Today it is poised to enter the telecom mainstream.

This paper outlines the evolution of the satellite industry and its significant role in building the broadband economy. Major market and technology trends are explored that indicate an even greater role for satellite in the decades to come.

History of Satellite Communications

The Beginning

Sir Arthur Clarke is rightfully credited as the father of communications satellites since, in October 1945, as a little-known pilot for the British Air Force, he published his now famous article in *Wireless World* entitled, "Extra Terrestrial Relays: Can Rocket Stations Give World Wide Coverage?"[ii] Clarke combined the nascent technologies of wireless, rocketry, and radar to envision a system of three geo-stationary space stations that would ensure complete coverage of radio communications around the globe.

His conclusions about the advantages of such architecture were the following:

- Ubiquity for all types of services
- Unrestricted use of at least 100GHz of bandwidth, which, combined with use of beams, would yield an almost unlimited number of channels
- Low power requirements, because of high illumination (transmission) efficiency
- Costs of operating would be incomparably less than terrestrial world networks, however great the initial expense

Only the last conclusion could be excessive, based on what we know today about the economics of different communications technologies, but we can certainly confirm the first three. Indeed, continent-wide coverage and cost-effective delivery of high-bandwidth video services remain the primary advantages of satellite communications technology over terrestrial alternatives. Even though Clarke's thesis raised only a few eyebrows at the time, it was only a dozen years later that the launch of Sputnik electrified the world and space very much became the new frontier.

A New Industry Is Born

Government and military interests dominated the early days of satellite development. Although the commercial promise of satellite technology was great, in the early days satellites were both expensive and largely unproven. Satellite transmission also carried with it legal considerations in using spectrum across sovereign geographies. For this and a host of other reasons, development of the satellite infrastructure was largely driven through government-backed international organizations.

Perhaps the most important of these was the International Telecommunications Satellite Organization (INTELSAT)[iii], which was created in 1964 by the United States and 84 other nations to implement a worldwide commercial satellite-based telecommunications system. In April 1965, the Early Bird was placed into geo-stationary orbit, which heralded the launch of the commercial satellite telecom industry. Interestingly, while most countries designated state-owned telecommunications companies as partners or signatories to INTELSAT, the United States designated a private company, Comsat Corporation, to be its INTELSAT representative. The United States' penchant toward private-sector approaches to satellite—even in its early days—would continue to be an important element as the industry moved from quasi-government entities and control to its current free-market form.

Later in 1965, the American Broadcast Company (ABC) first proposed to use satellites to distribute television signals, beyond the initial long-distance voice and data services. By the mid-1970s, television broadcasting, aided by significant technological breakthroughs, became the single largest user of satellite bandwidth, which led to highly publicized satellite capabilities such as the live broadcast of the Tokyo Olympics. Indeed, it was the emergence of satellite as the most cost-effective backbone to deliver movie channels and super-station content virtually anywhere that, in turn, spawned the modern terrestrial-based cable television industry.

A significant turning point for the satellite industry came in the 1980s, when the U.S. government began to actively encourage the development of commercial satellite communications systems that could compete with INTELSAT. This drive resulted in the creation of PanAmSat, the first private company that launched satellites to compete as a global satellite system with INTELSAT. The push toward private, commercial competition in the 1980s and 1990s led to an acceleration of new entrants into the satellite market not only in the launching and manufacture of satellites, but also in the area of ground equipment manufacturing and satellite services. In the past decade, the satellite communications market overall has grown more than threefold to an approximately $100 billion industry.

The VSAT and Private Networks

As satellite backbone operators were entering commercial maturity in the early to mid-1980s spurred by growth of television services, it was the invention of the very-small–aperture terminal (VSAT) that ushered in the concept of satellite private networks for enterprise. VSAT technology enabled a dramatic decline in the size, complexity, and cost of satellite earth stations and satellite transmission, making it cost-effective for business to implement private satellite networks. For example, in the early days of INTELSAT, earth stations were 100-foot dishes with cryogenically cooled maser amplifiers costing as much as the equivalent to $10 million in the 1960s to build. Cost to carriers per circuit was approximately $100,000. Today, antennas for normal satellite services are typically 15-foot dish reflectors costing the equivalent to $30,000 in the 1990s. Direct-broadcast antennas are as small as a foot in diameter, and receivers cost a few hundred dollars.[iv] Compared to the mid-1960s, today's cost has dropped by a factor of 100.[v]

It is noteworthy that the first VSAT customer has now become the largest market cap business in the world, namely Wal-Mart. In 1984, Sam Walton was looking for a technology that could solve two problems. The first was data and inventory management. To maintain a competitive advantage, he believed that all Wal-Mart stores needed to have a constant fix on inventory and sales figures. However, telephone lines could not handle the immense data needs for his nearly 1,000 stores across the United States. The second challenge could be called people management. Sam Walton insisted on personally visiting every store and giving his employees personal pep talks. However, this required videoconferencing, which at the time was not available in all locations by terrestrial means. It was his gamble to implement a satellite-based solution to solve both problems that suddenly launched the VSAT enterprise networks industry. Fortune magazine has called it one of the 20 decisions that have "shaped the modern world of business."[vi]

FIGURE 1

VSAT vs. Competing Technologies for the Enterprise Segment

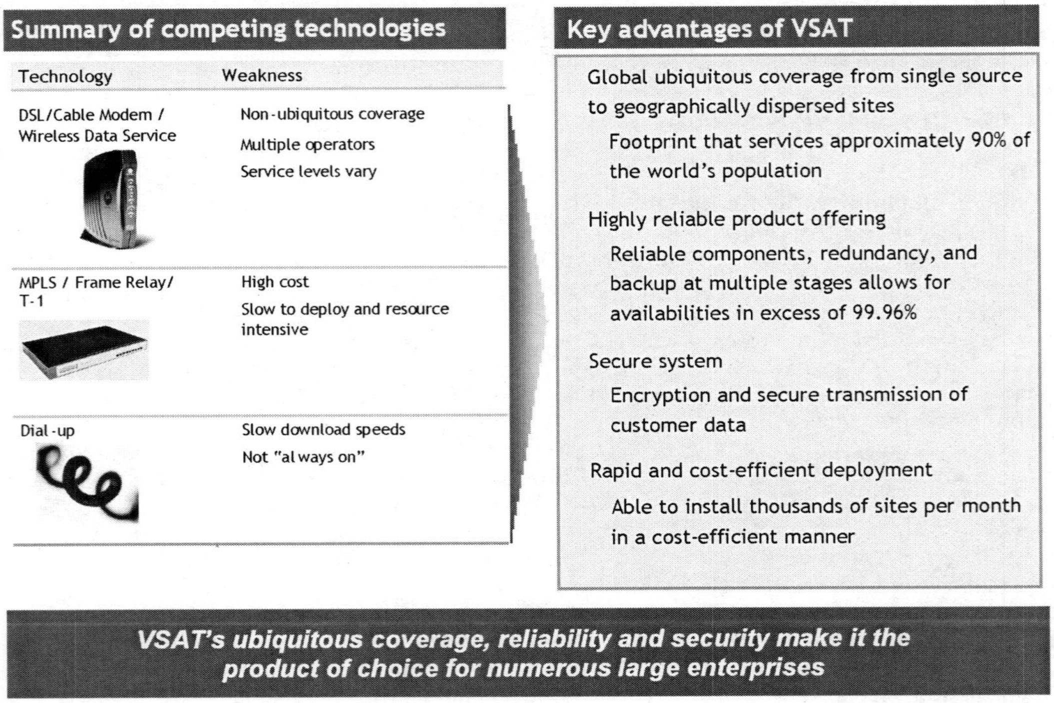

Summary of competing technologies

Technology	Weakness
DSL/Cable Modem / Wireless Data Service	Non-ubiquitous coverage Multiple operators Service levels vary
MPLS / Frame Relay/ T-1	High cost Slow to deploy and resource intensive
Dial-up	Slow download speeds Not "always on"

Key advantages of VSAT

- Global ubiquitous coverage from single source to geographically dispersed sites
  - Footprint that services approximately 90% of the world's population
- Highly reliable product offering
  - Reliable components, redundancy, and backup at multiple stages allows for availabilities in excess of 99.96%
- Secure system
  - Encryption and secure transmission of customer data
- Rapid and cost-efficient deployment
  - Able to install thousands of sites per month in a cost-efficient manner

VSAT's ubiquitous coverage, reliability and security make it the product of choice for numerous large enterprises

Since its inception 20 years ago, the market for VSAT private networks itself has grown to be approximately $4 billion per year with approximately 1 million sites around the world, serving not only major enterprises, but also a growing list of small-to-medium enterprises (SMEs), small offices/home offices (SOHO), and consumer customers.[vii]

DBS and DARS

By the mid-1990s, the cost of two-way satellite receivers and service had dropped to about $1,000, which was affordable to SMEs,. Not surprisingly, when the cost of receive-only units correspondingly dropped below a few hundred dollars, the so-called direct broadcast satellite (DBS), or personal entertainment business, was born. Tracking the VSAT revolution, the DBS started slowly and then grew exponentially. In just 10 years, this technology has rocketed U.S. providers DirecTV and Echostar to success, today collectively claiming more than 30 million subscribers.[viii] Globally, there are more than 350 million DBS service subscribers.

In fact, digital audio radio services (DARS) appeared quickly on the heels of DBS. The growth in DARS services has been arguably even more explosive than that of satellite-based television. Only four years old, satellite radio already commands 5 percent of the U.S. radio market with 9 million subscribers.[ix]

Learning from Mistakes

As with most emerging technologies, satellite communications had its bumps in the road. In the heady days of the Internet boom, there were several efforts to challenge terrestrial-based mobile cellular networks with satellite-based networks. For the two most publicized of these projects—Iridium and Teledesic—this involved building and launching hundreds of low-earth-orbit (LEO) satellites. Although these efforts were backed by significant money and noted personalities, both failed, though Iridium survives as a much smaller and more focused enterprise.[x] Clearly, satellite cannot compete with terrestrial cellular networks in delivering primarily mobile voice services.

While these highly publicized efforts encountered their difficulties, satellite infrastructure overall has continued to grow in size and sophistication for primarily fixed services. Today, estimates indicate there are more than 1,000 satellites orbiting the earth, all of which play a vital role in enabling a wide range of networks and services for governments, businesses, and individual consumers.[xi] Today, satellite technology is opening new vistas and new business models based on its ability to efficiently broadcast or multicast video, data, and voice services. Even certain mobile satellite services are proving successful when targeted to particular market sectors, such as in the cases of Thuraya and Mobile Satellite Ventures (MSV). Thuraya's success is based on lower rates for Global System for Mobile Communications (GSM) customers who roam outside their home areas, whereas MSV's success is based on specific mobile data services for targeted customer sectors such as public safety.

Future growth of the satellite industry is being accelerated by three major trends in the industry and three shifts in market demand. The three trends in the industry are investment, standards, and technology. The three shifts in market

FIGURE 2

Global Site Forecasts

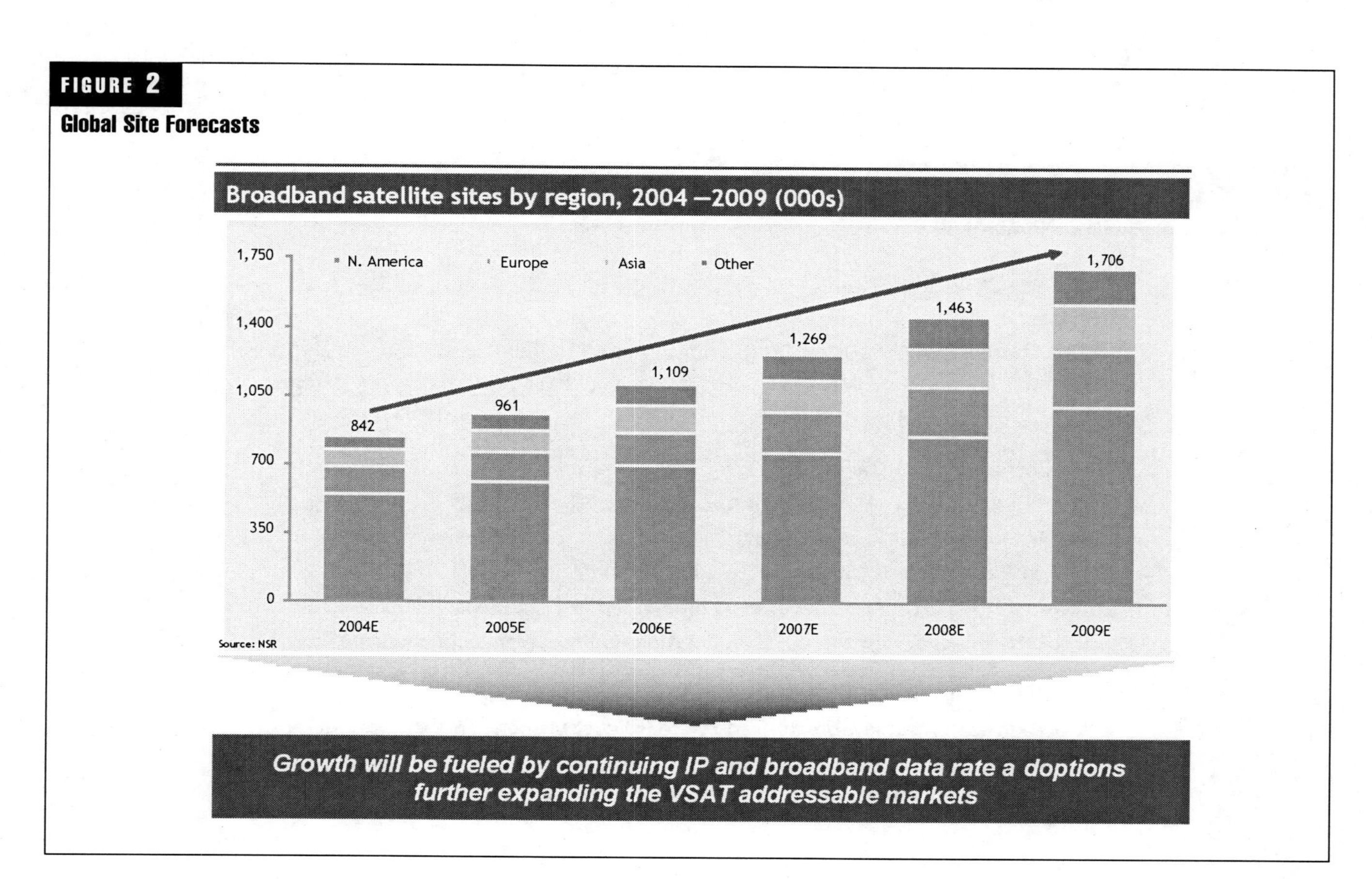

demand are multipoint access, the demand for video, and emergency communications.

Privatization and Consolidation
The influx of private equity in the satellite industry has been impressive by any standard. Few within the satellite world have not caught their eye. In fact, private equity investors have recently acquired the major satellite operators, including Inmarsat (December 2003, $1.5 billion) NSS (January 2004, $956 million), PanAmSat (August 2004, $4.2 billion) and INTELSAT (October 2004, $5 billion). Eutelsat SA, the European consortium, has also recently begun the process of privatization.[xii]

After many of these newly restructured private equity holdings came several sizable initial public offerings (IPOs). Examples include PanAmSat (March, $900 million IPO), NSS (May, $180 million), Inmarsat (June, $699 million) and WorldSpace (August, $249 million). This cash infusion into the industry has increased the satellite industry's access to capital while imposing discipline and rationalization into the industry's growth.

One of the outcomes of that discipline and rationalization is industry consolidation. As new private equity owners push the industry toward a leaner and more efficient private-enterprise model, major players are naturally seeking to find economies of scale.[xiii] A few examples include the acquisition of GE Americom by SES Group, INTELSAT's acquisition of Loral, and, most recently, INTELSAT's acquisition of PanAmSat.[xiv] The new influx of private equity capital, combined with consolidation, is making for a much healthier and synergistic satellite industry.

Satellite Standards
Another important driver of future satellite growth is the coalescence of the industry around a select set of standards and protocols. In the past, each satellite service and equipment provider developed proprietary products and services. This happened despite history indicating how agreement on standards has propelled growth. For example, in video, it was the emergence of the VHS standard and later the DVD standard. For computers, it was MS-DOS and then Windows. For cell phones, it was GSM.

So a very positive market force yet to be unleashed is that the satellite industry has now settled on primarily two technology standards, namely Internet protocol over satellite (IPoS)[xv] and digital video broadcasting–return channel satellite (DVB–RCS). Each has its proponents, but IPoS has emerged as the only true international standard, having been approved by the European Telecommunications Standards Institute (ETSI), the International Telecommunication Union (ITU), and the Telecommunications Industry Association (TIA) standards.

Standardization within the industry is already attracting new entrants for equipment manufacturing and value-added applications. Adoption of a standard will undoubtedly yield manufacturing and service economies of scale to accelerate cost reduction, technology advancement, and fuel market growth.

New-Generation Satellites and Terminals
As noted previously, technological advances such as VSAT dramatically decreased the cost and increased the flexibility of ground-based terminals.

Today's advances are occurring at three levels. First, there is a new generation of satellites that will soon enter the market. These satellites are not just mirrors that bounce signals from one place on the planet to the other as originally envisaged by Clarke. The new generation of smart satellites incorporates routing, switching, and beam-form-

FIGURE 3
Telecommunication Standards Activities

IPoS and DVB-RCS have achieved global or regional ratification as standards f or FSS services

- IPoS approved as TIA (US), ETSI (Europe), and ITU (global) stand ard. In addition, recently endorsed by INTELSAT. It is the only global standard for VSATs today
- DVB-RCS approved by ETSI (Europe)

HNS contributions to Standards

- IPoS: DVB-S compatible standard for two-way broadband services - over 500K terminals
- GMR-I: Geo-mobile ETSI standard - 250K terminals
- RSM-A, ETSI TS102: Next generation broadband multimedia services
- DVB-S2: HNS provided the LDPC modulation and receiver technology

HNS strongly supports standardization

- Standardization cannot be allowed to create cost burdens on the equipment or service
- Standards must achieve manufacturing scale

ing technologies on board, which will enable them to function more flexibly and efficiently, thereby significantly increasing the satellite's addressable market. Just as the VSAT fundamentally changed the economics of satellite-based networks, the new generation of smart satellites will fundamentally change the economics of satellite capacity. For example, the new generation of SPACEWAY satellites developed by Hughes Network Systems operates in the globally assigned Ka-band spectrum, and will yield 10 times greater throughput and capacity than today's Ku-band satellites. Each satellite is capable of up to 10Gbps of throughput, equivalent to 5,000 E1 terrestrial links. Three such satellites will approximate the total capacity of the 27 Ku-band satellites orbiting the United States. By conservative estimates, the SPACEWAY technology will position the industry to expand its addressable market by four times or more.

A second technological change is taking place for land-based terminals. New compression and component technologies are solving the problem of slow uplink speeds. Satellites have historically been an ideal platform for broadcast communications but have been plagued by slow uplink speeds. Indeed, the latest technology developments will enable two-way satellite broadband throughputs of multi-megabits per second or more, speeds that are comparable or greater than DSL and cable modems.

Finally, and perhaps most important, the satellite industry is beginning to take a more systems-based approach to providing managed networks. This new managed services approach positions satellite technology at the hub of a much more complex and integrated system that incorporates the best of satellite and terrestrial network offerings. Boeing's Patrick Ryan has put it this way:

> We are not just building satellite platforms anymore—we're supplying satellite systems that unite elements in space, in the air, on the ground, and at sea to make network-enabled operations possible. These are effectively an integrated architecture forming a network-enabled system that provides users the critical information they need, in real time, to make decisions and take action.[xvi]

Satellite is uniquely suited for this role. This is very likely the glue that brings together the best of network technologies into one integrated architecture.

Demand for Broadband and the Limitations of Terrestrial Connectivity

In addition to the internal dynamics of the satellite industry, there are external connectivity demands that satellite is uniquely suited to fill.

Multipoint Access

In his recent book, *A Flat World*, a New York Times columnist outlines the new economic reality where broadband connectivity enables work from multiple geographies in real time. At the heart of the system, he places satellite because its ubiquity allows multipoint access to the Internet whether a person is in a remote village in Mongolia or amidst the tens of millions of people in Mexico City.

Integration of Video

Throughout the 1980s and 1990s, voice and data were king. However, the future of subsequent generations of communication systems rests in the cost-effective delivery of video, data, and voice. Land-based systems are not efficient delivery systems for video.

FIGURE 4

Fast Packet Switch in the Sky

The SPACEWAY satellite provides high quality, flexible packet-switched network in the sky

- 10 Gbps capacity

Combines spot beams and onboard processing (demod/remod) for interbeam routing with multicast/broadcast capability

- Enables both high speed interactive applications and multicast/broadcast applications

Provides bandwidth-on-demand

- Interactive digital communications
- Services and high-speed Internet access
- Supports data, voice, images, and video
- Low-cost, easy-to-install terminal

FIGURE 5

Key SPACEWAY Advantages

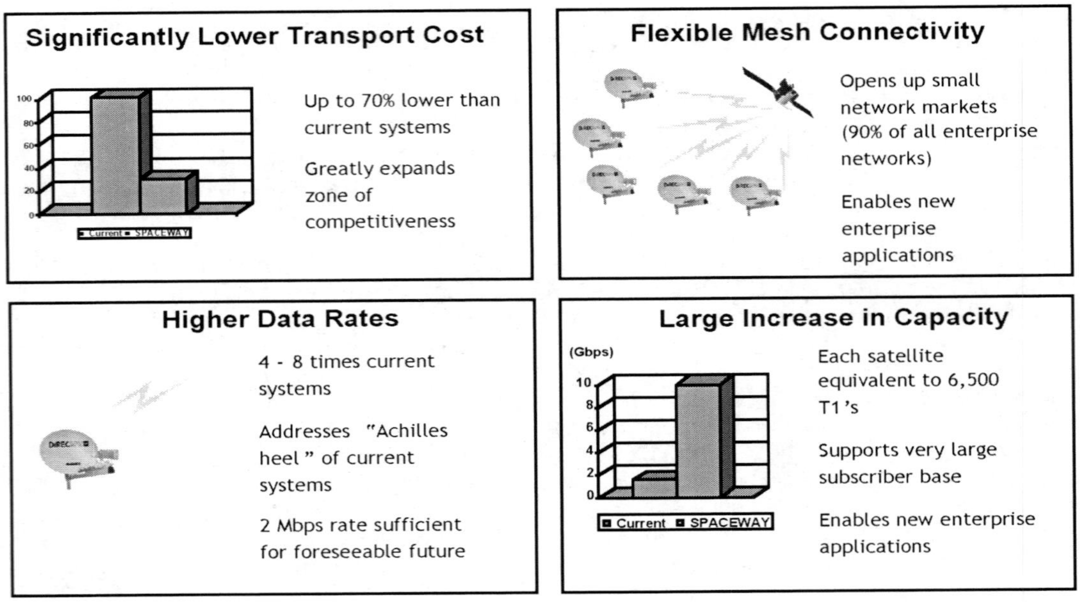

Significantly Lower Transport Cost

100 80 60 40 20 0

Current SPACEWAY

Up to 70% lower than current systems

Greatly expands zone of competitiveness

Flexible Mesh Connectivity

Opens up small network markets (90% of all enterprise networks)

Enables new enterprise applications

Higher Data Rates

4 - 8 times current systems

Addresses "Achilles heel" of current systems

2 Mbps rate sufficient for foreseeable future

Large Increase in Capacity

(Gbps)

10 8 6 4 2 0

Current SPACEWAY

Each satellite equivalent to 6,500 T1's

Supports very large subscriber base

Enables new enterprise applications

According to Vinton Cerf, "the Internet is not ready to be a true entertainment medium. As a result of its architecture, the Internet cannot cater to a vast number of people simultaneously asking for large files such as movies."[xvii] Cerf, in fact, refers to the terrestrial-based Internet. These land-based Internet service providers (ISPs) notably assume that consumers will only use a fraction of the bandwidth available to them at any given time, because doing otherwise would be too taxing on the system. Cerf proceeded to say that a big part of the Internet's future is multicasting and spot-casting.

With the advent of new smart satellites such as SPACEWAY, satellite will become the preferred platform for the integration of video into whatever communications application is demanded—from videoconferencing to distance learning to delivering the content a customer wants, whenever and wherever it may be.

Emergency Communications

In 2004–2005, the world experienced a series of natural disasters, including tsunamis in the South Pacific, hurricanes and floods in the United States, and earthquakes in Southern Asia. One of the many lessons the world has learned is the importance of communications for emergency preparedness and disaster recovery. In every case, satellite proved to be a vital link that enabled the world to respond and assist in a time of emergency. Beyond being the go-to technology in coordinating assistance after natural disasters, satellite is also poised to be a key component for emergency access for business to fast reliable broadband connectivity for both voice and data.

Indeed, virtually all the early communications efforts after the disasters that befell the world in 2004–2005 were satellite- and wireless-based. This is a lesson for government, business, and society's general communication needs. Fiber, cable, satellite, and wireless all have their strengths and weaknesses. No single approach can efficiently and effectively serve every communication requirement. Moreover, reliance on a single communications platform is unwise in today's uncertain world.

Conclusion

Based on this assessment, satellite networks will enter the telecom mainstream and be seamlessly integrated with terrestrial fixed and mobile networks to deliver an expanding universe of personal broadband services that will available globally on demand by all customers, including enterprises, governments, and consumers.

In the words of the first satellite visionary, Sir Arthur C. Clarke:

> There is no real technology threat to satellites on the horizon because even if or when a totally integrated and seamless terrestrial network is created, there will still be a requirement for independent redundancy and backup. In addition, the unquenchable thirst for bandwidth makes all sources of supply relevant and attractive.

Notes

1.
2. Slekys, Arunas, "Broadband Teledensity: The New Paradigm for Economic Growth," IEC Annual Review of Communications, Vol. 58.
3. Clarke, Arthur C. "Extra-Terrestrial Relays," Wireless World, October 1945.
4. Hecker, JayEtta Z., Director, Physical Infrastructure, Government Accountability Office (GAO), testimony before the House Energy and Commerce Subcommittee on Telecommunications and the Internet, April 14, 2005.
5. Whalen, David J., "Communications Satellites: Making the Global Village Possible," SatCom History, National Aeronautics and Space Administration.
6. Whalen, ibid.
7. Hajin, Corey, "Sam Walton Explores the Final Frontier," Fortune, July 27, 2005.
8. Broadband Satellite Markets, Fourth edition, Northern Sky Research, January, 2005.
9. "Satellite Industry History," Satellite Broadcasting & Communications Association.
10. Dalton, Richard, J., "Satellite: Up, up and away," Newsday, December 11, 2005.
11. Jung, Helen, "Neither McCaw nor Gates could keep Internet-in-the-sky alive," Associated Press, October 14, 2002.
12. Toto, Christian, "Eyes, ears in orbit," Washington Times, December 1, 2005.
13. Lardier, Christian, "Comsat market pickup ahead," Interavia, October 1, 2005.
14. Pultz, Jay E, Rali, Patti A., "INTELSAT's PanAmSat Buy to Continue Satellite Consolidation," Gartner Research, September 1, 2005.
15. Villian, Rachel, "Key Trends for the Satellite Industry." Space News, September 5, 2005.
16. TR 101 985, TC-SES; Broadband Satellite Multimedia; IP over Satellite; Standardization Issues, ETSI.
17. Chase, Scott, "Satellite Manufacturing: Building a Strong First Link," Via Satellite, December 1, 2004.
18. Cerf, Vinton, "Broadband Dreams, Multicast Beams," Perspectives, Cnet, September 29, 2004

Overview of a Personal Network Prototype

Dimitris M. Kyriazanos
Research Engineer, Ph.D. Candidate, Communication, Electronic, and Information Engineering
National Technical University of Athens, Greece

Michael Argyropoulos
Research Engineer, Ph.D. Candidate, Communication, Electronic, and Information Engineering
National Technical University of Athens, Greece

Luis Sánchez
Associate Professor, Communications Engineering Department
Universidad de Cantabria, Spain

Jorge Lanza
Associate Professor, Communications Engineering Department
Universidad de Cantabria, Spain

Mikko Alutoin
Research Scientist, Adaptive Networks
VTT Technical Research Centre of Finland

Jeroen Hoebeke
Researcher, Department of Information Technology (INTEC)
Ghent University, Belgium

Charalampos Z. Patrikakis
Senior Research Associate, Communication, Electronic, and Information Engineering
National Technical University of Athens, Greece

Abstract

Ubiquitous connectivity and access to services is a challenging task as today's users move through heterogeneous networks and technologies while using a wide selection of devices. Personal networks (PNs) aim to provide a unified overlay network in a transparent and seamless way. In this paper, an overall view of a PN prototype is provided.

The Personal Network Concept

As emerging technologies offer an even wider variety of capabilities to users, the need for convergence and homogenization also rises. Moreover, the real needs of the user should also be taken into consideration to guarantee that provided solutions are more tailored to users' needs. The current sociotechnological status quo raises the demand for shifting environments to become smarter, more responsive, and more accommodating to the needs of the individual without jeopardizing privacy and security. These challenging requirements are encompassed by the PN concept [1], offering a self-configuring and self-organizing network of personal devices, irrespective of geographical location, communication capabilities, and mobility.

A PN is the set of all networking-capable devices that someone uses for personal purposes, including communication, financial transactions, information, and entertainment. A PN may be geographically distributed (e.g., home cluster, car cluster, office cluster) with the clusters interconnected by means of virtual private networks (VPNs) building a "personal overlay network." Despite its very dynamic nature, a PN must be strictly guarded, since the resources it interconnects contain a significant amount of personal information such as contact lists, bank accounts, passwords, or various preferences. The physical layout of a PN is presented in *Figure 1*.

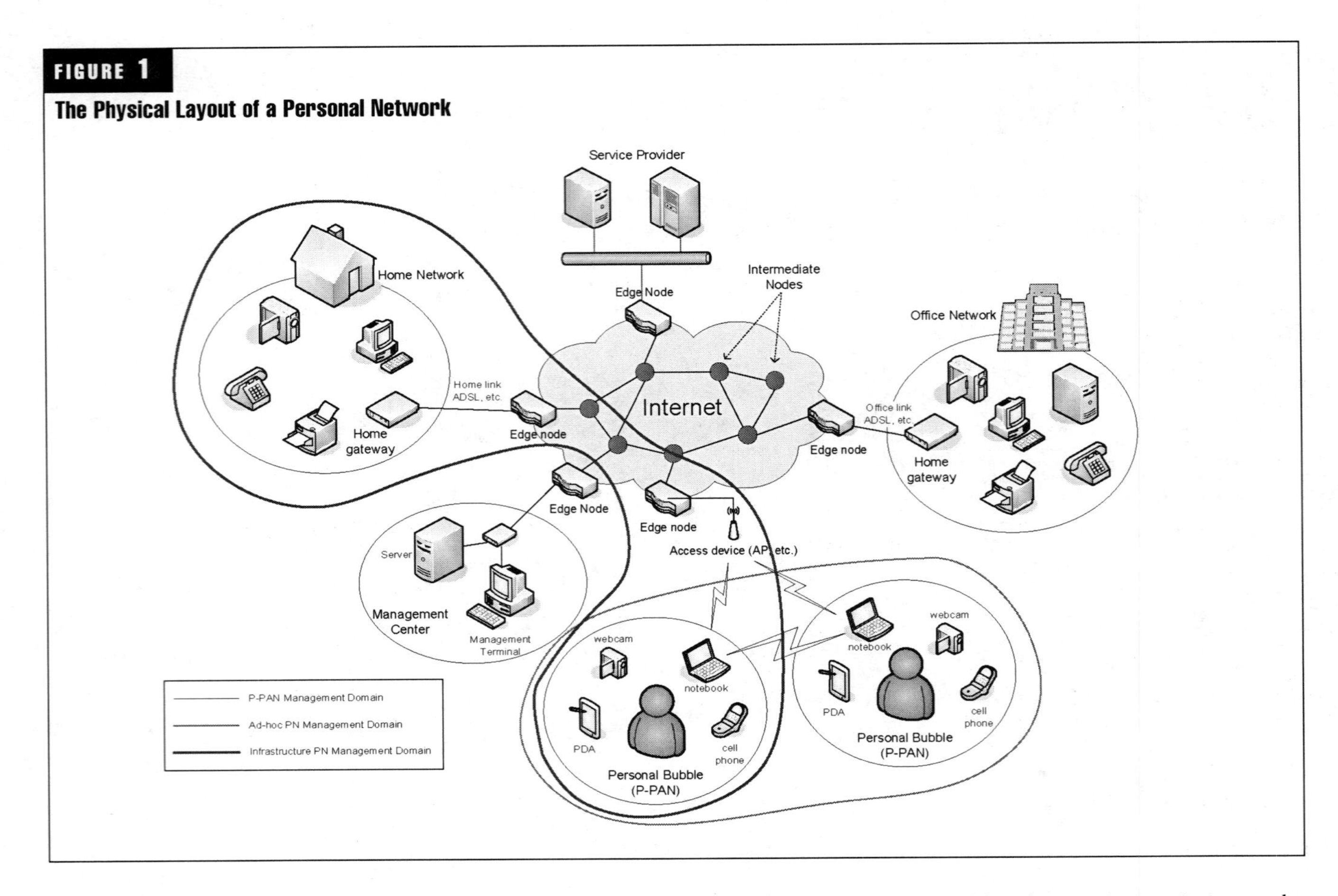

FIGURE 1

The Physical Layout of a Personal Network

In this paper, we present a PN prototype—implemented during the Information Society Technologies' (IST) My Personal Adaptive Global Net (MAGNET) project [2]—that illustrates the self-configuring and self-organizing capabilities [3] required for PN networking and can also serve as an enabler for personal services. The resulting PN architecture paves the way toward PN federations, where multiple PN users with common interests and tasks can communicate and cooperate.

PN Framework and Requirements

Platforms

Before introducing the different technical solutions integrated in the PN prototype, this section reports on the platforms that make up the integrated solution, covering both hardware technologies and software framework. It is important to note that all the communications are both Internet protocol version 4 (IPv4)– and IPv6–based, depending on the final addressing schema defined and the applications in use. The integrated components support both network layer protocols for this purpose.

To assist in better understanding the platform and its components, the presentation will be based on a demonstration setup, including different scenarios that accentuate the full functionality of the PN prototype.

Hardware Platforms

- *Computing devices*: To set up the different clusters and the private personal area network (P–PAN), which is the "bubble" created by devices in proximity to the user, laptops are used to mimic high-capability devices that can be part of the users' PN (*Figure 1*). Personal digital assistants (PDAs) are also used as personal nodes to come closer to real-life situations. Summarizing, the equipment is made up of a set of laptops and PDAs forming the P–PAN and the other PN's clusters, and some PCs acting as edge devices used at the borders of the interconnecting structures. Furthermore, the functionality of the foreseen applications is provided by additional equipment (e.g., cameras, additional PCs acting as servers, a printer), used wherever necessary.

- *Communication equipment*: To emulate the wireless communications in the P–PAN, widely adopted wireless technologies (e.g., wireless PANs [WPANs], wireless local-area networks [WLANs]) have been selected. Institute of Electrical and Electronics Engineers (IEEE) 802.11b/a/g and Bluetooth technologies are the choices for the demonstration. Compliant PC cards, universal serial bus (USB) dongles, or built-in systems are used as the wireless access method. As the prototype is not only focusing on the P–PAN, but also on the PN, a communication with foreign nodes has to be established as well. Gateway nodes have been used for that purpose. For this, they include interfaces both for internal P–PAN communication (Bluetooth, IEEE 802.11 a/b/g) and interconnecting infrastructures featuring wired connections (Ethernet) as a minimum requirement.

Software Environment
The software modules used have been implemented on (and for) the aforementioned hardware platforms. As has been mentioned, the prototype is based on the use of PCs, laptops, and PDAs. The PDAs run embedded Linux OS, while the PCs are based on Linux OS, running kernel 2.6.x family. Linux OS has been chosen for the development framework, as it offers an open-source license with access not only to operating system source code, but also to free drivers for all the communication equipment.

System Requirements
In this section, the authors present a set of base requirements, following related research concerning all layers of a PN. A user-centric approach for PN design, which was adopted by the authors, can be clearly seen in some of the following requirements:

- P–PAN architecture must accommodate devices that implement different types of PAN radios. A PAN radio refers to a radio interface specification that has been defined under the auspices of IEEE 802.15 or 802.11 or in a research project such as MAGNET.

- Service and context discovery must be scalable in order to meet the definition of the potential sizes of a PN. Since PNs can potentially become rather large, the number of services and the volume of context information can also grow rapidly. The service discovery (SD) system must be able to cope with such huge amounts of data.

- P–PAN specific security mechanisms must be hidden from the end user or from applications. The only exception is the authentication phase, where the owner of a P–PAN decides which PAN devices can join his P–PAN.

- Keys and/or passwords used in authentication, integrity, and data origin protection and encryption must be installed in a PAN device so that they are not accessible from unauthorized people.

- A P–PAN must provide a mechanism that hides private services from external nodes. The owner of a P–PAN may want to not only limit access to some particular services within a P–PAN, but also limit the visibility of these services.

- Different wireless access technologies must be able to coexist within a single device. The address allocation for the nodes in the P–PAN should be automatic without the need for user intervention.

- A P–PAN is a family of IP and non–IP devices. Proxy functionality for the non–IP devices is required.

- SD must be able to function without infrastructure, i.e., in a true point-to-point way, and in a heterogeneous link technology environment.

- The SD mechanism must guarantee the integrity of SD information.

- The interconnecting structure must provide secure means for communication between nodes in a PN over insecure network architectures.

- The interconnecting structure must support P–PAN (cluster) movement, causing minimal effects to its nodes, and must not require any user intervention.

- There should be naming and addressing support for the PN devices, while address allocation for the nodes in the PN should be automatic, without the need for user intervention.

- The PN should have a routing mechanism that enables seamless PN connectivity over heterogeneous wired and wireless networks.

- The PN should be established without user intervention.

- Naming schemes must accept updates and should support local name spaces, which differ from one user to another.

Services and Applications
The end-user services used in the PN prototype are a subset of the services that a fully implemented PN system can offer. Many of the end-user services are based on access to files in different nodes and devices such as Web-based file access. Therefore, together with the end-user application programs, dynamically controlled file access has been used in the prototype to implement a user environment that covers many of the services needed in a personal network. Among others, video and audio streaming is supported and has been tested, together with other important services such as real-time surveillance monitoring and remote control of home functions and appliances (lights, washing machine, etc.).

PNs: Challenges and Prototype-Provided Solutions

In this section, the main challenges and requirements that have been tackled by the prototype are summarized. The structure follows the three abstraction levels in which the PN has been divided in the MAGNET project, namely connectivity, network, and service levels. Additionally, the mobility management is presented, since it gathers many of the requirements imposed by PN users.

Convergence of Heterogeneous Interfaces on the Connectivity Level
Connectivity level requirements appear at the cluster level. The main challenge here is the provision of mechanisms that cope with heterogeneity in the cluster. Multiple radio domains coexist in a PN cluster. To have full connectivity among them, multimodal personal nodes supply the mechanisms for handling frames coming from heterogeneous air interfaces, as required by the peculiarities of each radio domain (i.e., point-to-point connection technologies), and for selecting the most appropriate output device in case the nodes are part of several radio domains. The solution implemented for the prototype has been the universal convergence layer (UCL) [4], designed to support cluster formation and maintenance. Neighbor discovery is another issue that

has been also resolved at the connectivity level. A personal node is aware of personal nodes and devices within the same radio domain and inside its coverage area.

Finally, security is present at all abstraction levels defined in the MAGNET PN architecture [5]. Link-level security mechanisms (i.e., encryption of link-layer frames) have been exploited as far as possible during the cluster formation by providing a safe communication channel between trusted nodes on a hop-by-hop fashion. This has been accomplished by the establishment of link-level secure sessions on detection of new neighbors. Keys for these sessions derive from the PN long-term pair-wise keys.

Self-Configurability, Addressing, and Routing Challenges on Network Level

Once the nodes have established the communication paths at the connectivity level, there is a need to select the way secure and efficient information exchange is achieved. Every time nodes have to communicate with each other, they need to know both their peer names and locations. Therefore, the first task to be accomplished within P–PAN formation is the assignment of a name and an address that univocally identify the node. After the node has self-configured its basic network parameters, the next step is to find the way to interact with other personal nodes in the neighborhood as well as with foreign nodes through appropriate gateways.

The diversity of devices requires the definition of a set of solutions allowing every personal device, independently from its characteristics, to access any other peer resource. At connectivity level the cooperation among different radio domains has been solved: all P–PAN air interfaces are grouped under a unique identifier and the link-layer mechanism chooses the appropriate interface for establishing communication with a peer. However, personal nodes are unreachable in a single hop if they are out of the coverage range. To achieve end-to-end communication, this connectivity needs to be solved at the network level. Some already known solutions are ad hoc routing protocols that enable multi-hop environments through cooperative use of resources. *Figure 2* shows a situation where link-layer solutions are not enough and a network approach is necessary.

A P–PAN can be considered a network entity by itself, but in the PN scope, it is a requirement that communication with external nodes, either in its neighborhood or by means of infrastructure, can be established. With connectivity acquired, services can also be discovered and accessed or provided. A gateway enables connectivity to nodes outside the P–PAN while a delegated node known as the service management node (SMN) acts as a repository of the services offered by P–PAN nodes and devices. The service discovery must be as lightweight as possible so as to be feasible for any P–PAN node to make use of it (smartphones, PDAs). Delegation of the SMN is a result of a selection procedure, which occurs at the moment when connectivity among all P–PAN nodes is supplied by the naming and addressing procedure (i.e., the moment when the P–PAN is formed). Once the SMN is selected, every node is able to discover any registered service, including the gateway functionality service.

The selection procedure involves a sequence of tests in which the strongest device—from a computational and networking point of view—becomes the SMN and is based on the request parameters that fit the source peer capabilities. In the P–PAN, different paths to an outside network could be configured. Security is also a main concern in the PN scope and plays an important role in the whole architecture. From the connectivity level, there is an inherited security for each link. Securing the communication on the connectivity level may not always be feasible; therefore, nodes must be enabled with network-level solutions to guarantee the level of security needed within the P–PAN. There are many security mechanisms that can be used to protect communication. In this case, unlike in the connectivity level, the channel will be secured end to end. This would be sufficient, provided that both peers have previously identified themselves. This identification procedure is conducted at the initial steps of the P–PAN formation through the exchange of keys and identities.

Extending all above, when a foreign node or device and their provided services are going to interact with a MAGNET P–PAN or PN cluster, long-term trust relationships must be examined and secure communications between nodes should be enabled. The connectivity with a local but

FIGURE 2

Network-Level Communication

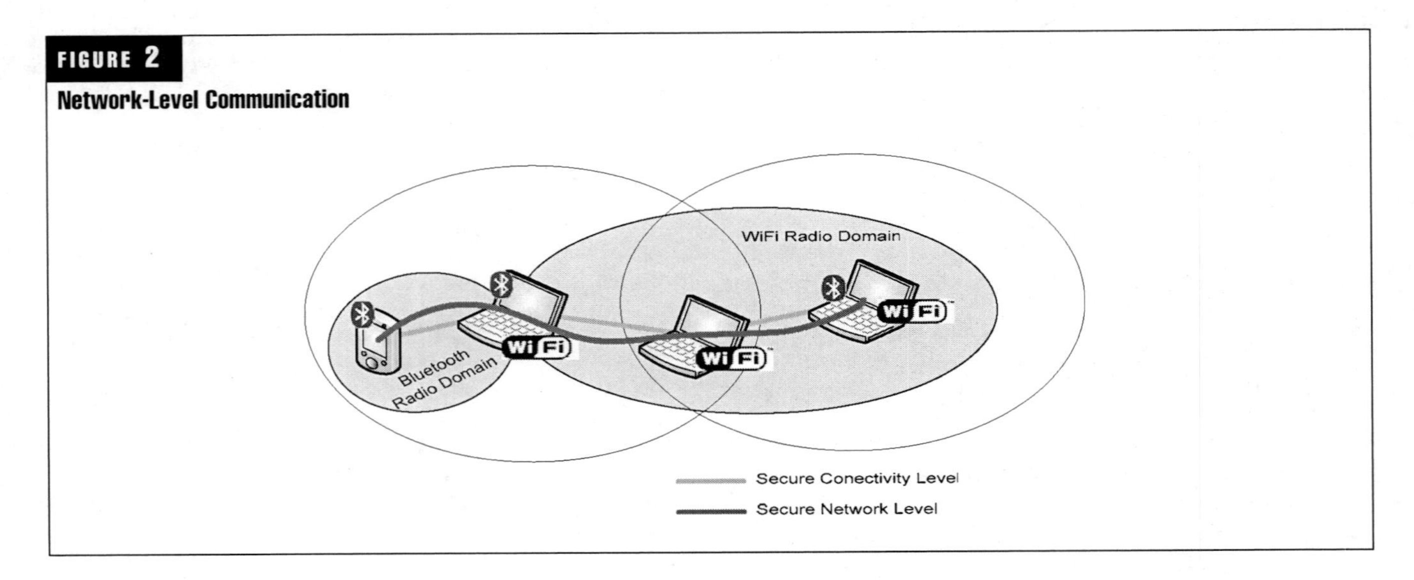

foreign node or device to the P–PAN node will depend on the security profiles available at the service and connectivity level. For the PN-to-foreign-user requests, profile mechanisms and service access control by light AA components can be used.

Unique PN prefixes are envisaged advocating the use of private or local IPv6 addresses that are different from those used by public foreign nodes as well as other private nodes. To communicate with the local-foreign node, a network address translation (NAT) function would typically be needed for Layer 3 (L3) communications. Thanks to the use of IPv6 addressing, the probability of colliding addresses is very low and there is no need for a NAT function. In the other direction, for foreign requests, the P–PAN is in need of a special authentication. This can be based on public key infrastructures (PKIs) or prior symmetric key exchange similar to the PN imprinting. An ephemeral exchange of secrets can be used to temporarily allow access to the P–PAN and PN.

In this point, we have to go one step beyond. In this scenario, we suppose that a node of the P–PAN uses a public service outside our PN. This public service may be located practically anywhere on the Internet, but not in the vicinity of the P–PAN. The prerequisite for this scenario is that P–PAN has been formed utilizing the pre-shared PN keys. Furthermore, one or more gateway nodes are set up to provide connectivity to the outside world. The P–PAN may or may not have edge routers (ERs) available. If there is no ER available, the gateway node handles the required functionalities itself. So when establishing the connection to the remote public service, gateway nodes must conclude the most suitable media for the connection. In a general case, this means that all the connection types provided by gateway nodes must be evaluated with respect to the quality of service (QoS) requirements of the service. The gateway node must provide the necessary network-layer connection functionality for the service. If the PN address space is private, then NAT needs to be performed by the gateway node or ER. Other services may additionally require proxy mechanisms from the gateway node such as port forwarding for active FTP. The gateway node provides these functionalities for the user application or delegates them to a nearby ER. However, the gateway node must not engage in the NAT function using a temporary Internet address, because then the established connection is broken once the address becomes invalid (e.g., due to mobility). In this case, the ER provides the NAT function for the cluster. This also applies if the access network is using private IP addresses.

Finally, the PN technical scenario extends all technical cases described before, not only for personal nodes and devices in the local vicinity of the user, but also for those that are farther away (at home, at work, in the car, etc.) and it describes how these personal nodes in different clusters securely communicate with each other over a fixed interconnecting structure such as the Internet. To this end, the remote clusters have established a secure communication channel between each other, transparent for the end user and applications. To the end user, the PN is seen as one virtual network offering a plethora of personal services, but at the same time hiding the details of the interconnection structure, tunneling mechanisms, PN formation, and maintenance. *Figure 3* shows the deployment for visualizing this PN technical scenario.

ER and Cluster Gateway Interaction

Any node in a PN cluster can become a gateway node. As a result of the successful completion of the ER discovery process in a PN node, an IP address of an ER is retrieved. This event is triggering an IPSec key negotiation mechanism in order to set up a tunnel between the public interface of this PN node and the discovered ER. The direct consequence of this IPSec tunnel establishment is that the PN node becomes a PN gateway node. Once the node becomes a PN gateway node, it informs the rest of the nodes in the cluster about this new capability.

PN Agent Creation/Construction and Cluster Registration

As the establishment of overlay networks to provide intra–PN connectivity relies primarily on the concept of the PN agent—whose role is to maintain key PN information for networking purposes—a description of the PN agent construction is first presented to ease understanding of intra–PN self-configuration and networking. In the PN agent concept, the following two steps occur:

- The PN agent server stores the PN agent information. If the PN agent is centralized, there is a unique server. Otherwise, the distributed servers collaborate to disseminate PN agent information and resolve queries.
- The PN agent client asks the PN agent to perform the cluster registration and deregistration actions.

When a cluster connects to an edge node, the gateway passes its cluster name to the PN agent client. The incoming cluster name is concatenated with the location information of the ER (e.g., the ER's Internet address) into a cluster name record that will be advertised to the name resolution network (i.e., to the PN agent). These records actually form the PN agent within the naming system. In this manner, the PN agent will keep the information about the cluster gateway and its attachment point IP address. The name resolvers (NRs) point-to-point overlay network exchanges the name records so that the PN agent can be accessed from any intentional NR (INR). At the time of cluster registration, the cluster nodes announce their names to the naming system. As it is shown in *Figure 4*, in order to tackle the aforementioned procedures, the PN agent has been implemented using the intentional naming system (INS)/Twine framework, where an INS application in the ERs takes the PN agent client role, while the point-to-point network formed by the INR offers the PN agent server functionality in a distributed manner.

Name-Based Tunnel Establishment

When establishment of tunnels between PN nodes is based on policies relying on names, we talk about name-based tunnel establishment. A flexible and scalable naming system such as INS/Twine has been selected to conduct the tunnel establishment and, as previously mentioned, embed the PN agent framework.

Cluster Mobility

The PN agent, using the INS/Twine framework, is capable of assisting mobility management in the PN architecture. When a cluster moves, it de-registers itself from the PN agent by removing its corresponding name record from the INRs. It then passes its name to the new visited ER, which will announce this name to the naming system. Therefore, the mapping between the cluster name and its associated

FIGURE 3

PN Technical Scenario

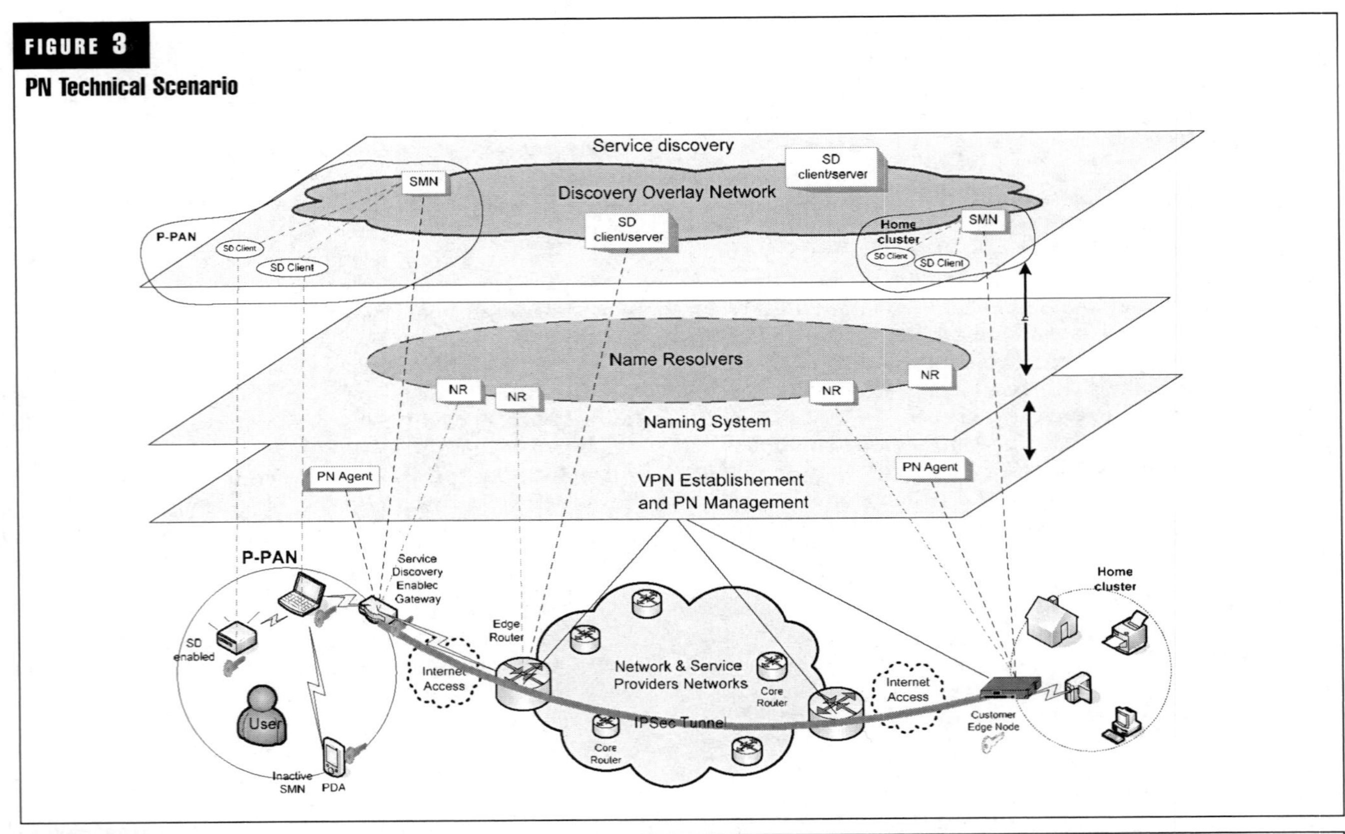

FIGURE 4

PN Agent Concept within Naming Systems

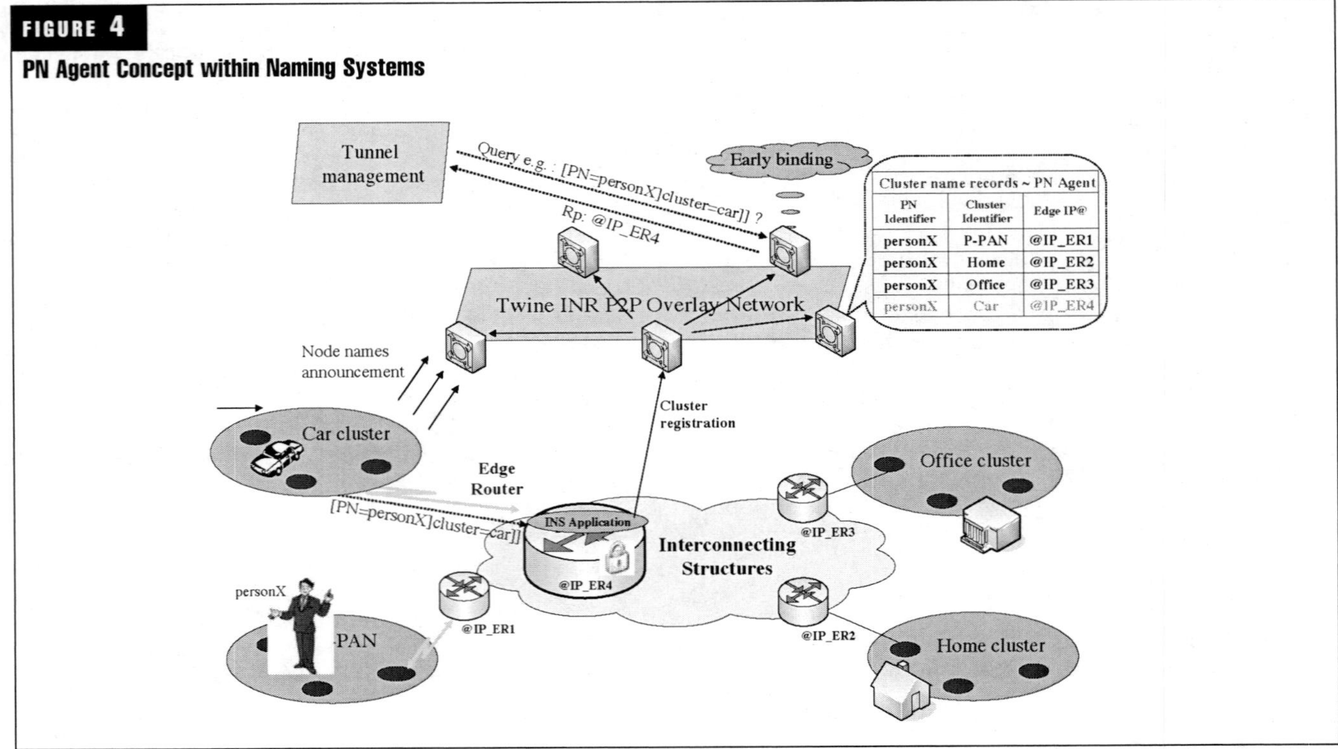

serving ER IP address will be updated. This procedure is done dynamically without user intervention and assures an always-connected behavior since the PN gateway nodes will be ready to support PN connectivity even under mobility conditions.

Virtual Router and PN Agent Interaction

As depicted in *Figure 5*, the PN agent interacts with provider edge (PE) nodes that support PN services to achieve PN networking, according to the dynamic changes in the clusters and the P–PAN. The PN agent maintains a table of registered clusters and the IP addresses of the edge nodes that

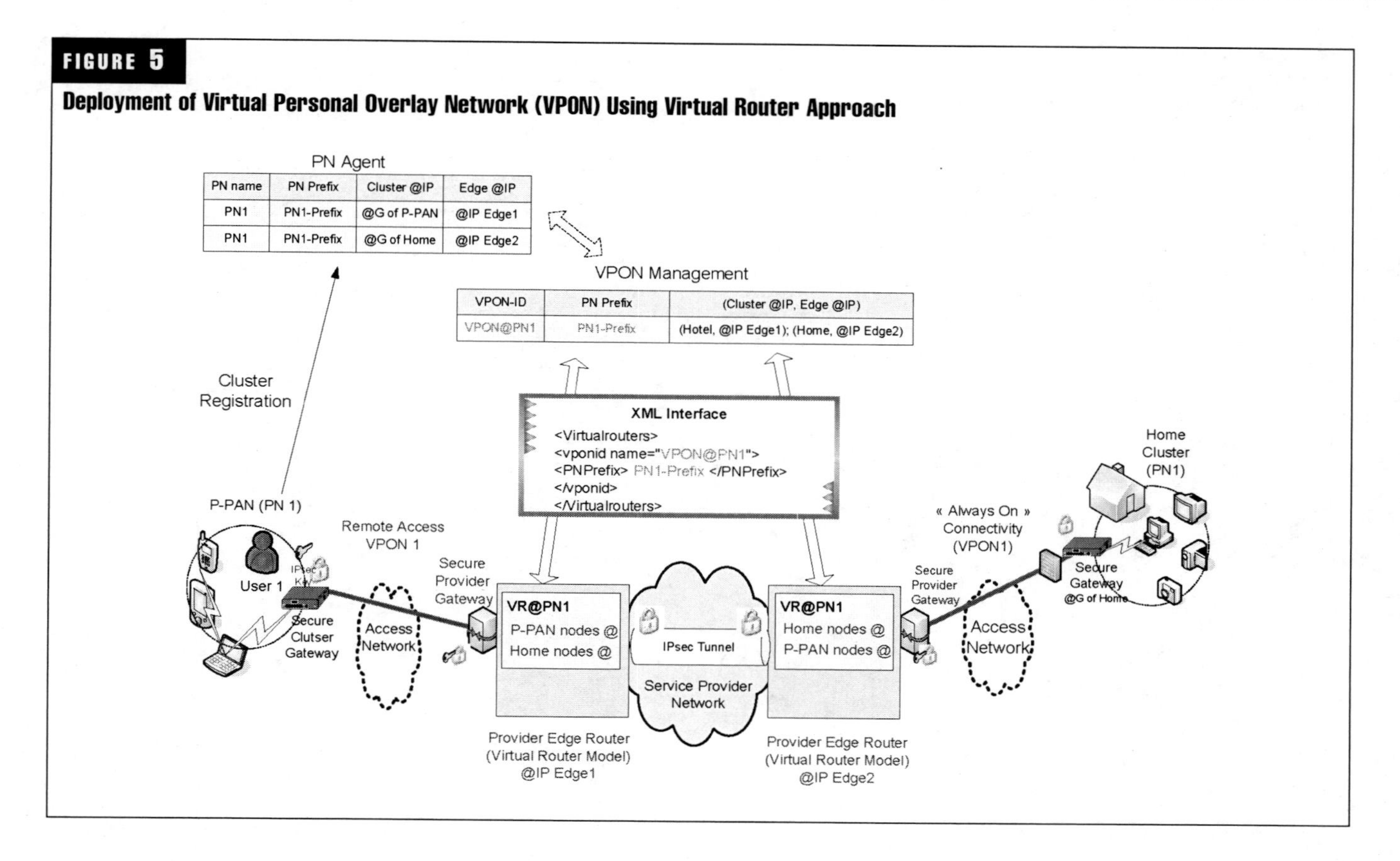

FIGURE 5

Deployment of Virtual Personal Overlay Network (VPON) Using Virtual Router Approach

are serving as their ingress and egress tunnel endpoints. It partakes in PN establishment, maintenance, and management by interacting with the naming system, addressing and routing, PN management, mobility management, and the security framework.

Cluster Mobility Management Using a Dynamic Tunneling Approach

The previous sections described cluster registration and the creation of tunnels to establish secure intra–PN communication. This section explores the mobility aspect of a roaming cluster using dynamic tunneling mechanisms. This extended section describes also the interactions between proactive routing in the P–PAN and the interfaces and information exchange with the virtual router (VR) instances [6, 7].

Figure 6 depicts a scenario where a moving P–PAN is changing its attachment point to the network. In this scenario the mobility results in a change in the PE. To maintain connectivity, management of cluster mobility must be combined with the dynamic tunneling framework. This change in the attachment point is reflected by an update in the PN agent that consequently adapts PN networking according to the dynamic changes in the clusters.

When roaming, clusters change their point of attachment (as a result of Layer-1 [L1] monitoring and the proactive routing framework in the P–PAN), the following subsequent actions take place in the VPON nodes:

- Old remote access through PE1 is canceled and a new remote access is established by the P–PAN through PE3.
- The same actions described before are repeated to establish a new inter-cluster connectivity. The PN agent and the VPON membership table are updated at run time with the new IP address of the gateway node as well as the IP address of the PE router 3.
- A new VR instance is created in the PE3, which interacts with PE2 through a static tunnel across the service provider backbone. All PEs collaborate to achieve fast-forwarding of context data and pending packets in the old PE1.

Wide-Area Service Discovery, Overlay Access Control, and Policy Management on Service Level

The service discovery architecture inside the P–PAN is based on a centralized approach, that is, SMN will federate all service-related needs. It acts as a repository of services registered within the P–PAN and also enables searches outside the P–PAN in case services are outside the P–PAN.

The PN–wide service discovery is achieved through a point-to-point overlay network of SMNs, as depicted in *Figure 3*. The SMN acts as a super-peer for its cluster and is part of the PN SMN point-to-point overlay network that includes all cluster SMNs as members. The P–PAN/cluster SMN super-peer is added to the PN point-to-point overlay as soon as the external connectivity is available through any of its cluster gateways.

When a P–PAN/cluster SMN receives, from its cluster service discovery client, an SD request that does not match any local services, it propagates this service discovery request within the PN super-peer overlay network to all the other

FIGURE 6

Cluster Mobility Management Using Dynamic Tunneling Approach

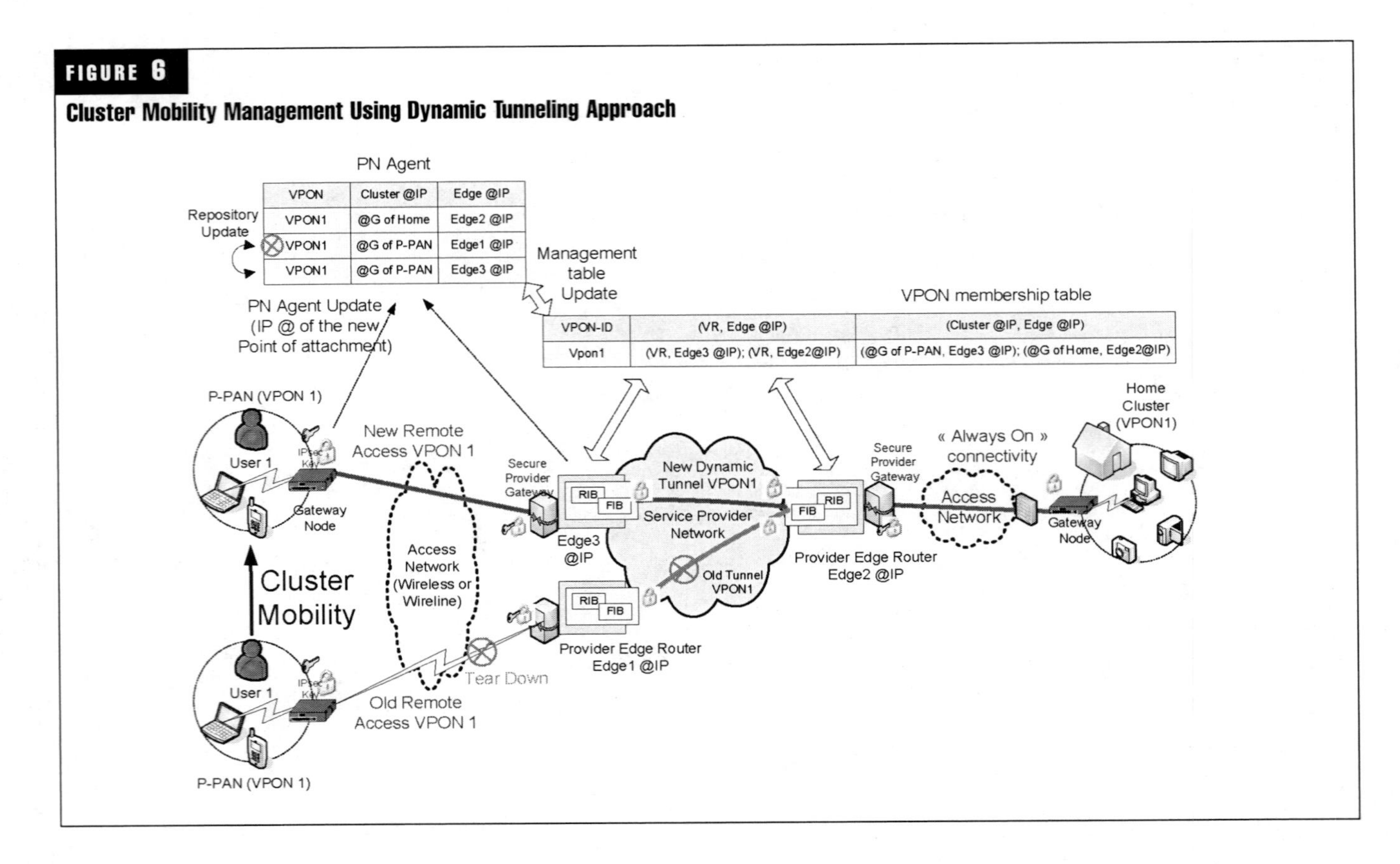

PN cluster SMNs. Each of these distant SMN super-peers then performs the following operations:

- Extracts search attributes contained in the received SD request
- Searches in its cluster service repository for local services that match the search attributes
- Sends an SD response for all retrieved matching local services to the SMN that initiated the SD request

The P–PAN/cluster SMN (the search initiator) gathers all the SD responses that are sent by the other PN SMNs within the super-peer overlay network in order to form the global SD response. Finally, this SMN sends back this global SD response, in the appropriate SD framework format, to its cluster service discovery client that initiated the search [8].

Access control throughout the PN is a challenging task. Besides the lack of a centralized authority due to connectivity restrictions, connectivity is again not guaranteed to every cluster inside the PN. Therefore, certain clusters might be out of reach of the owner-administrator. With respect to these problems, we present here a solution for a distributed and decentralized access control system [9].

Based on the security profiles, access control modules and profile repositories were designed and developed to function in a distributed way throughout the overlay access control system formed by the interconnected set of SMNs. In this way, each SMN acts independently as an access policy officer for all the devices and access requests under its jurisdiction. Moreover, devices that are capable of holding profile repositories and access control modules subsequently perform access control in a decentralized way. For devices incapable of holding such modules, including sensors, proxies are needed.

On the other side, the user performs policy management over the entire PN using a proper administration tool. Even in case of a cluster losing connectivity to the rest of the PN, the corresponding master device is expected to block any unauthorized access requests. However, in situations of an isolated cluster, access rights and certificate revocations issued by the PN administrator will fail to reach the SMN. In order to minimize the impact of such unfortunate situations, the granting of privileges and access rights below a default high level of security should at all times be ephemeral and stamped with an expiration time. Such time stamps reduce the risk and the extent of damage caused by exploiting a lack of connectivity to the PN administration, since any expired granted rights can be revoked, even by isolated nodes, equipped with the proper modules. As a future work, time-stamping techniques and modules will be developed for enhancing the PN access control system.

Finally, any communication between modules and the policy management administration tool remains secure since it is based on the underlying security infrastructure created from the imprinting procedure, trust establishment, and subsequent key and certificate generation.

Policy management throughout the PN is achieved by properly propagating profile information updates by the PN administrator. For this purpose, a tool is provided for the policy administrator through which policies can be issued regarding entities placed anywhere inside the PN, providing policy management for the overlay network through a tree-like structure graphical user interface (GUI), shown in *Figure 7*.

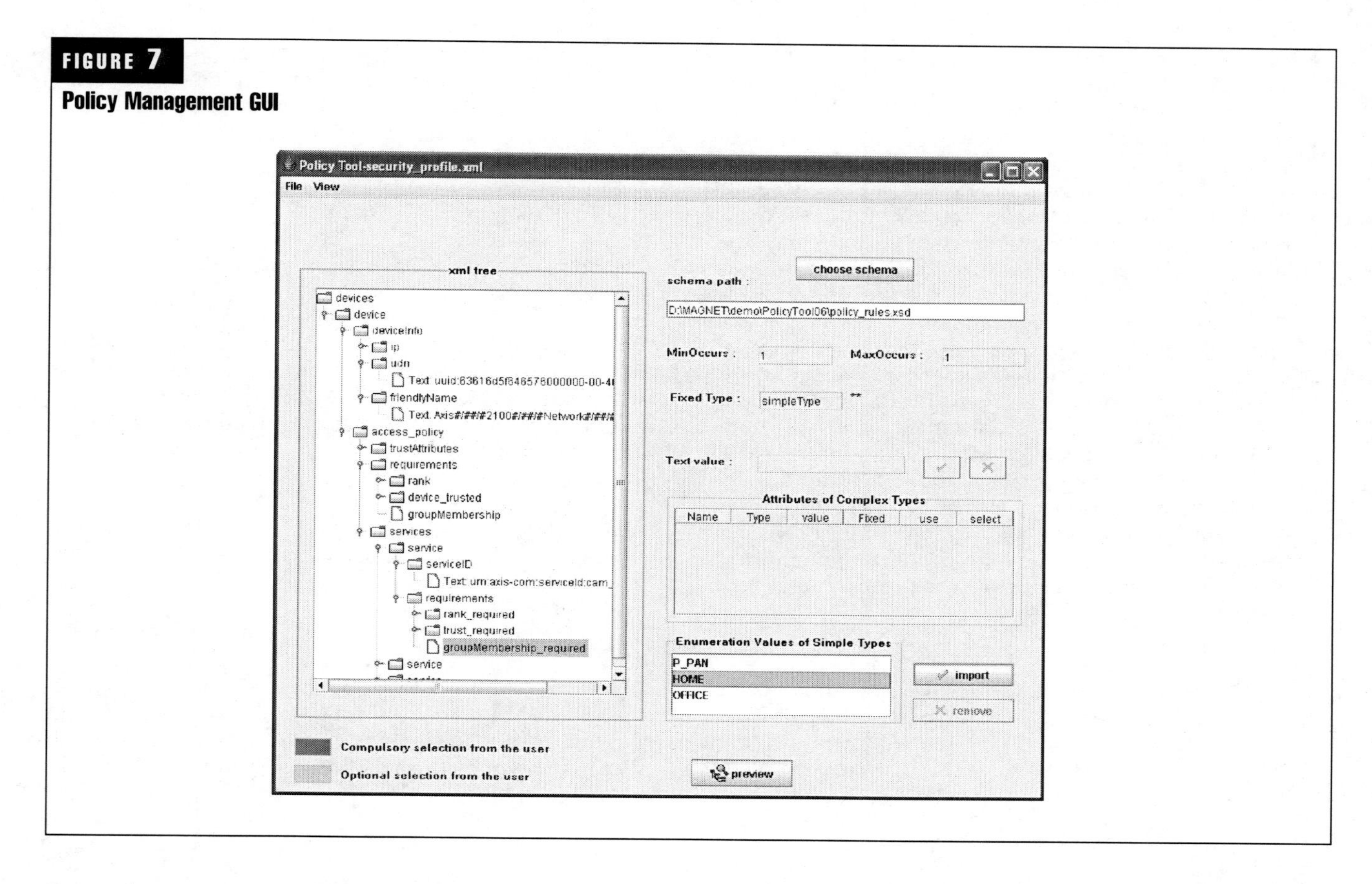

FIGURE 7

Policy Management GUI

Once policies are personalized according to user interaction, propagation of policy updates throughout the PN is achieved by utilizing the SMN functionality inside the PN, properly notifying responsible SMNs for any changes in their local policy settings.

Mobility Management

The mobility scenarios described in this section explain how the PN organizes and maintains itself in the light of dynamics such as changes in gateway nodes and ERs. The way the PN protocol solutions are able to deal with these dynamics in a timely manner, while maintaining session continuity, will be demonstrated. The behavior and performance of the PN architecture will be demonstrated for three types of dynamics, which will be described in more detail.

Mobility Scenario 1

In the first scenario (see *Figure 8*), we assume a cluster that has two gateway nodes, connected to the same ER (e.g., one node connected via 802.11 and one using an Ethernet cable). One of the gateways is used to establish a connection to a remote personal node at home. Suddenly, the connection of that gateway node to the ER breaks. At that point, traffic should be re-routed by the intra-cluster routing protocol. This does not have any implications on the ongoing communication session.

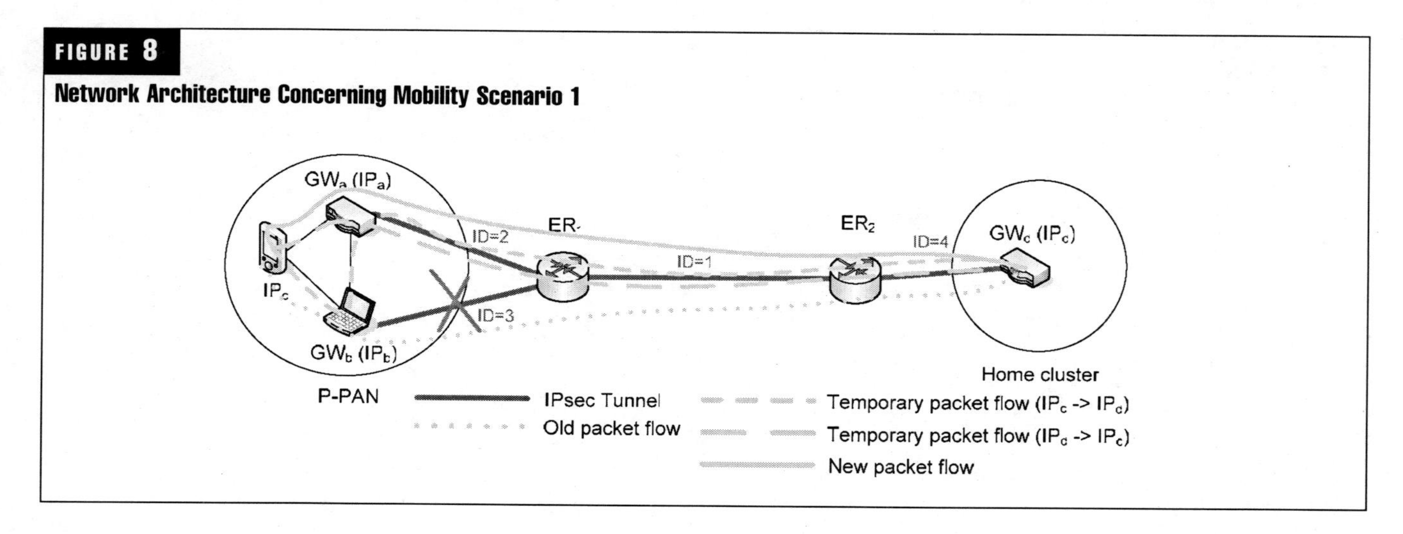

FIGURE 8

Network Architecture Concerning Mobility Scenario 1

Mobility Scenario 2

In the second scenario, the roaming P–PAN (see *Figure 9*), we assume a P–PAN that is connected to an ER in the interconnecting structure through a gateway node. A communication session with a remote personal node is ongoing, thereby using a dynamic tunnel established between the local and remote ER. Due to the mobility of the P–PAN, the P–PAN looses its connectivity to the interconnecting structure and thus the current ER. Next, the P–PAN rediscovers connectivity to the interconnecting structure and an associated ER (e.g., via another access point). When connectivity is regained, the P–PAN, which has a new point of attachment to the interconnecting structure, discovers a new ER and registers itself to the new ER and the PN agent. This registration will trigger the establishment of a new tunnel between the new and remote ER. In addition, the old ER should detect that connectivity to the PN cluster has been lost and tear down the tunnel that had been established, cleaning up the existing information in the PN agent. During the whole procedure, the ongoing communication session does not break but, in the worst case, simply experiences a short delay.

Mobility Scenario 3

Finally, the last mobility scenario (see *Figure 10*) is a combination of the two previous scenarios. Now the P–PAN has two gateways (e.g., WLAN and Universal Mobile Telecommunications System [UMTS]) that are connected to different ERs. One of the gateway nodes is used for an ongoing bidirectional communication session. At a certain moment, one of the ERs becomes unavailable. The PN protocol solutions reroute the traffic so that ongoing communication session is not interrupted.

PN Prototype: A Storyline

The previous sections have described the different technical solutions implemented and integrated in the PN prototype. In this one, a unified vision will be given by defining a storyline in which different scenarios will occur following a timed sequence of events, thus resulting in the demonstrator of the prototype described in this paper. In this sense, the four main scenarios into which the prototype can be divided are as follows:

- PN formation
- P–PAN formation
- Introducing a new node in the PN
- Discovering services
- Within the P–PAN
    - Outside the P–PAN

The demonstration illustrates the situation of a journalist who needs to attend the next European Union leaders' summit in Brussels. The same day of the meeting, the journalist flies from his hometown to Brussels. He travels with his lap-

FIGURE 9

Network Architecture Concerning Mobility Scenario 2

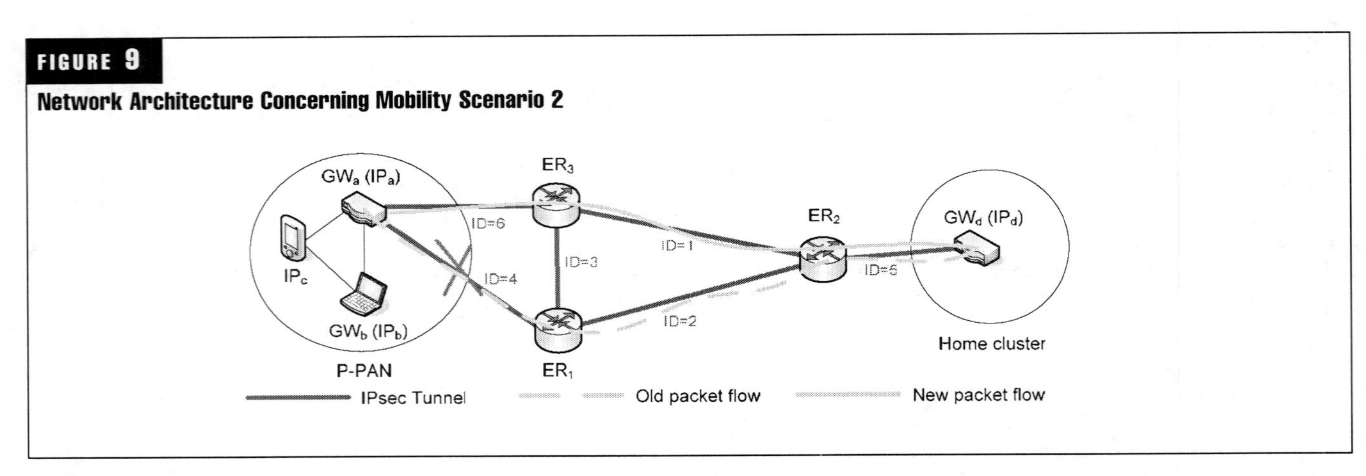

FIGURE 10

Network Architecture Concerning Mobility Scenario 3

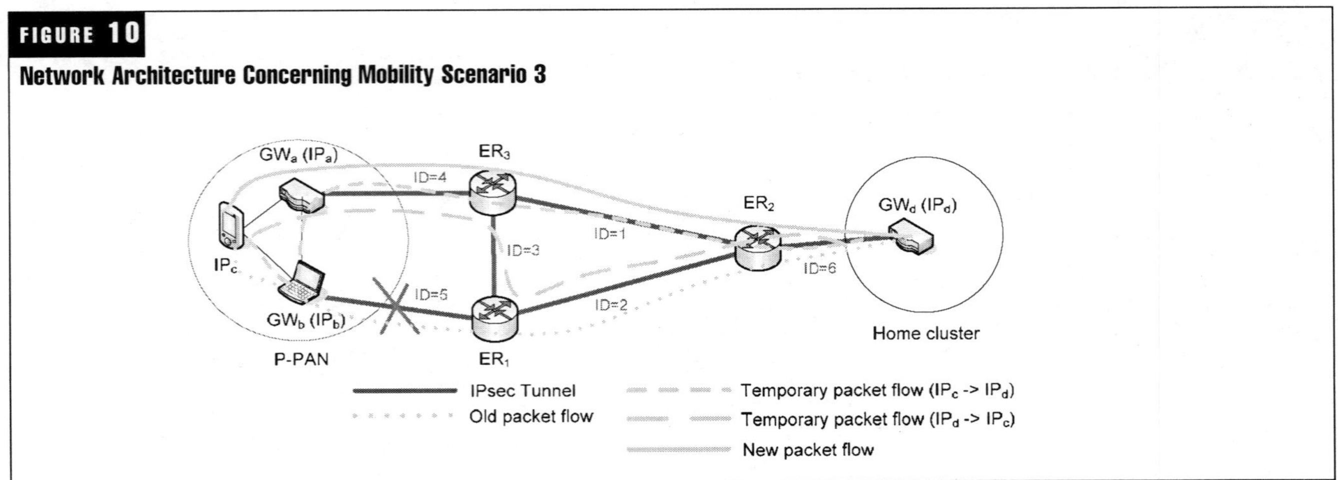

top, mobile phone, camera, and new PDA (in the demonstration, these devices are three laptops and one PDA, respectively). As he boards the plane, he has to switch off all his electronic devices, but the rest of his personal devices are still on. At work, he has a couple more devices (two laptops—GW–O, SMN–O) and at home he has his home network with remote control and leisure facilities (another two laptops in the demonstration—GW–H, SMN–H). This situation is shown in *Figure 11*.

As can be seen, both clusters are interconnected through secure tunnels. As soon as the journalist arrives as his destination, he starts switching on the personal devices he is carrying. As shown in *Figure 12*, he first switches on his mobile phone (GW–P). As this device is able to connect to the Internet, it automatically registers himself in the PN agent to allow the establishment of the secure tunnels with the remote clusters, both at his home and at the office. The tunnels are established and the PN enlarged with a new cluster, this time the P–PAN.

Any node in a PN cluster can become a gateway node for the other cluster members, enabling remote intra–PN communication. As a result of the successful completion of the

FIGURE 11

Journalist's PN Composed of Home and Office Clusters

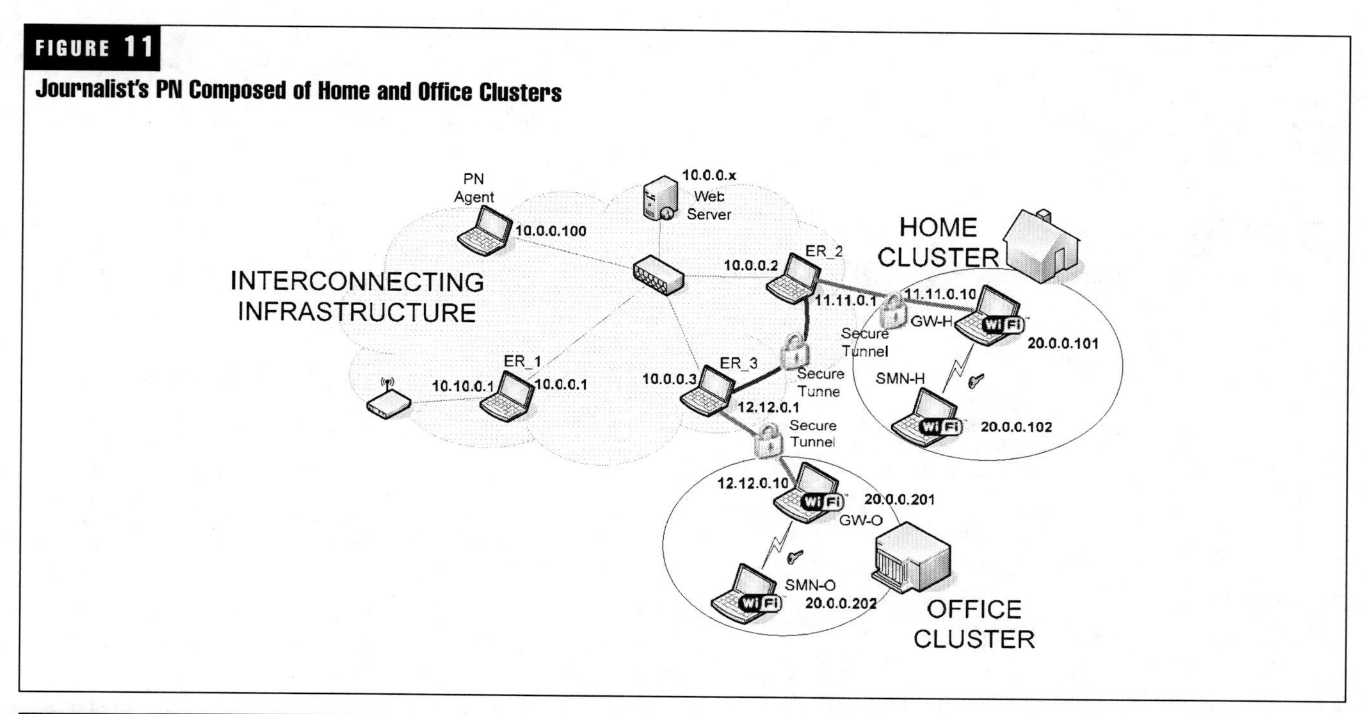

FIGURE 12

P–PAN Registration in the Journalist's PN

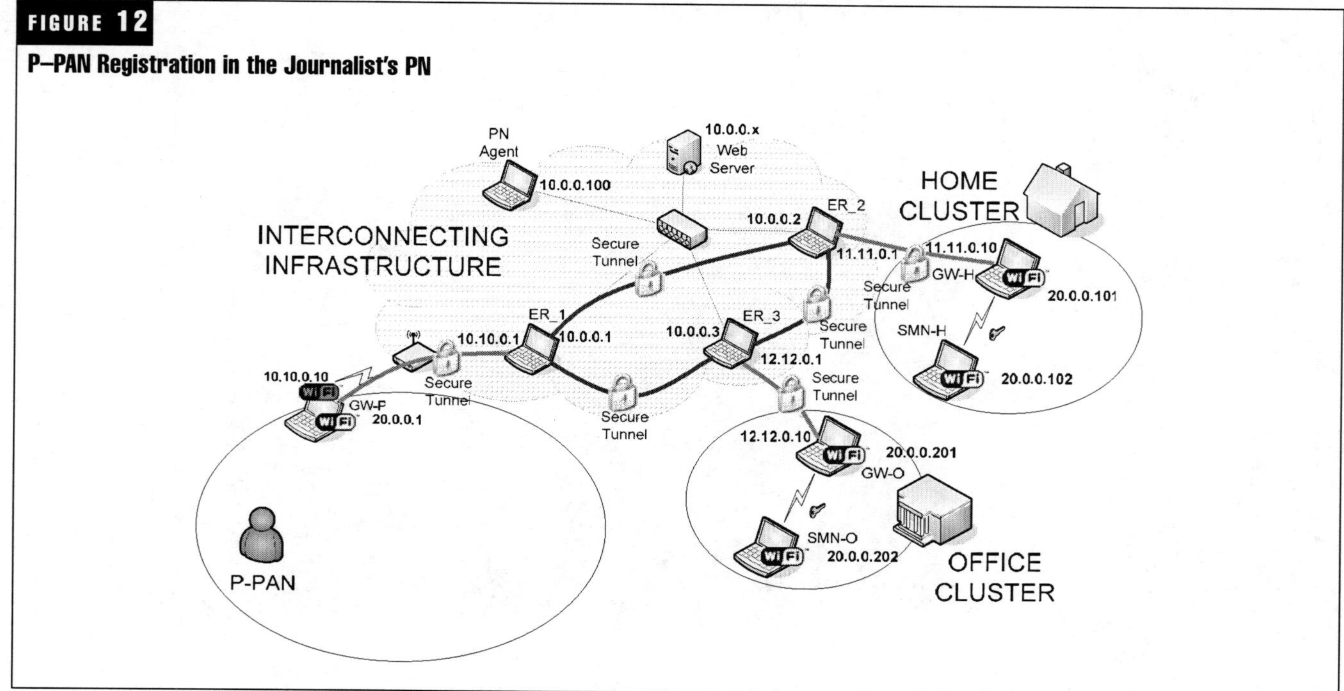

ER discovery process in a PN node, an IP address of an ER is retrieved. This event will trigger Internet protocol security (IPSec) key negotiation mechanism in order to set up a tunnel between the public interface of this PN node and the discovered ER. The direct consequence of this IPSec tunnel establishment is that the PN node becomes a PN gateway node. Once the node becomes a PN gateway node, it informs the rest of the nodes in the cluster about this new capability.

All the journalist's devices discover each other and form the P–PAN by exchanging the session keys that will allow them to communicate in a private and secure way. Finally, the rest of nodes discover the mobile phone (GW–P) as their gateway to the Internet and they are registered in the PN agent so that the rest of clusters have knowledge of the exact composition of the P–PAN. As shown in *Figure 13*, a GUI was implemented to allow the user managing the P–PAN composition. In this GUI, personal nodes appear in blue while foreign ones appear in red. Thanks to this GUI, the user realizes that he has not yet imprinted (i.e., established the long-term secrets that bootstraps the trust relationship between personal nodes) his PDA with his camera. *Figure 14* shows how the journalist can press a button on this GUI to start this imprinting procedure.

The result of this process is the exchange of long-term cryptographic secrets that will be used to derive session keys and authentication mechanisms in order to protect the communications between this pair of nodes. As shown in *Figure 15*, autonomously and transparently to the user, all his devices have formed his PN: a protected secure person-centric network that connects all his active personal devices, including personal devices at remote locations such as the home network, office network, and car network.

Once the PN is formed, the journalist can start accessing the services provided by all his personal devices on a secure and

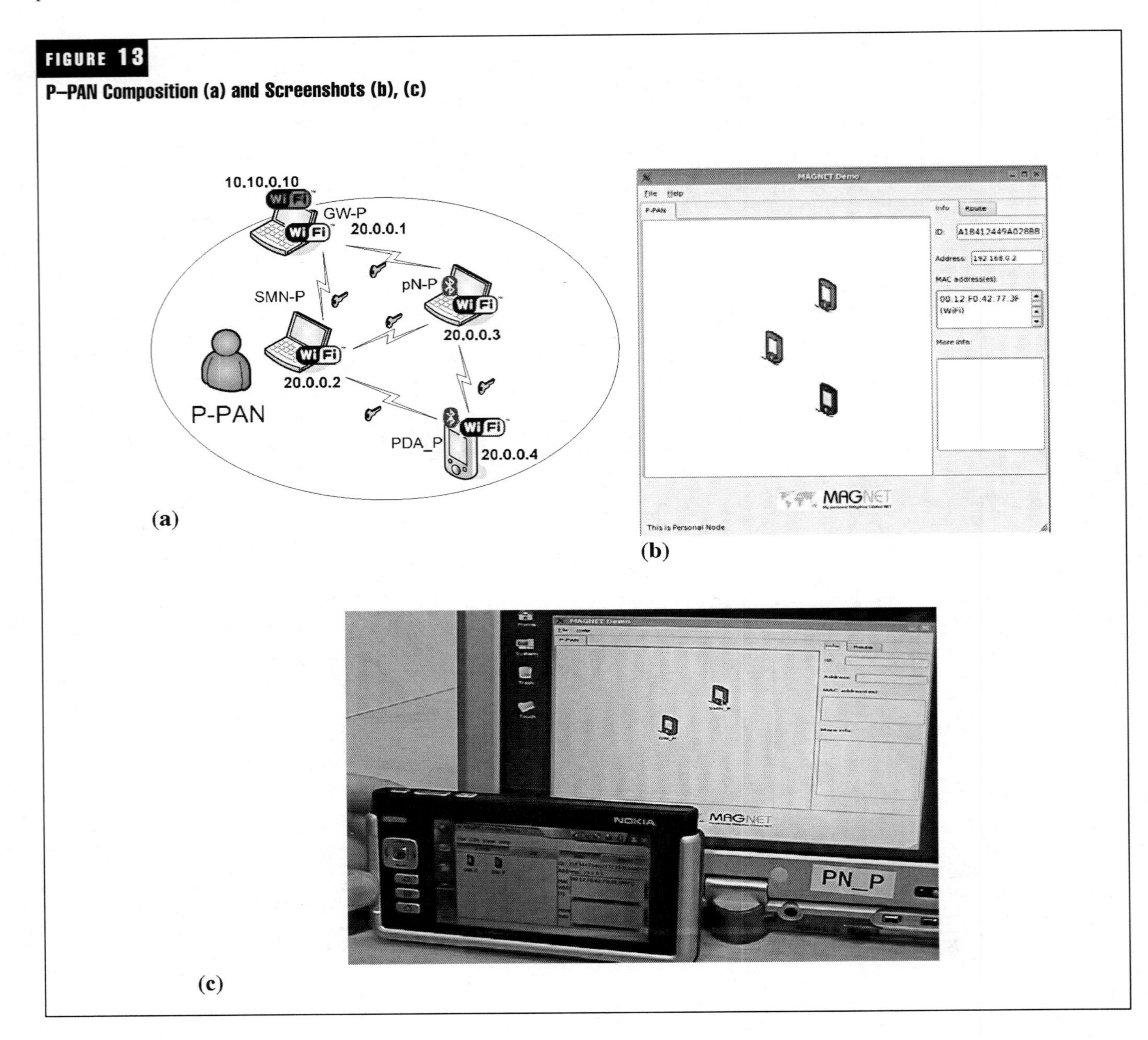

FIGURE 13

P–PAN Composition (a) and Screenshots (b), (c)

FIGURE 14
Imprinting Process

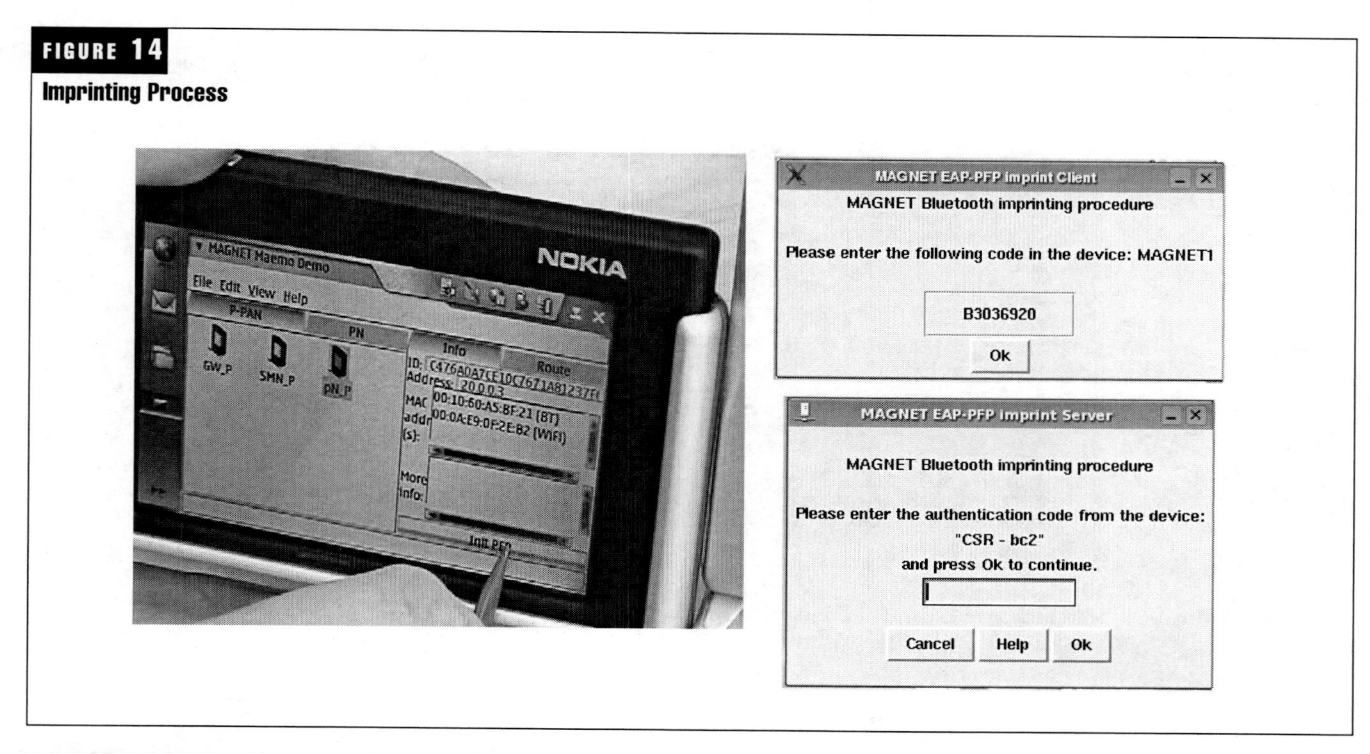

FIGURE 15
PN Composition

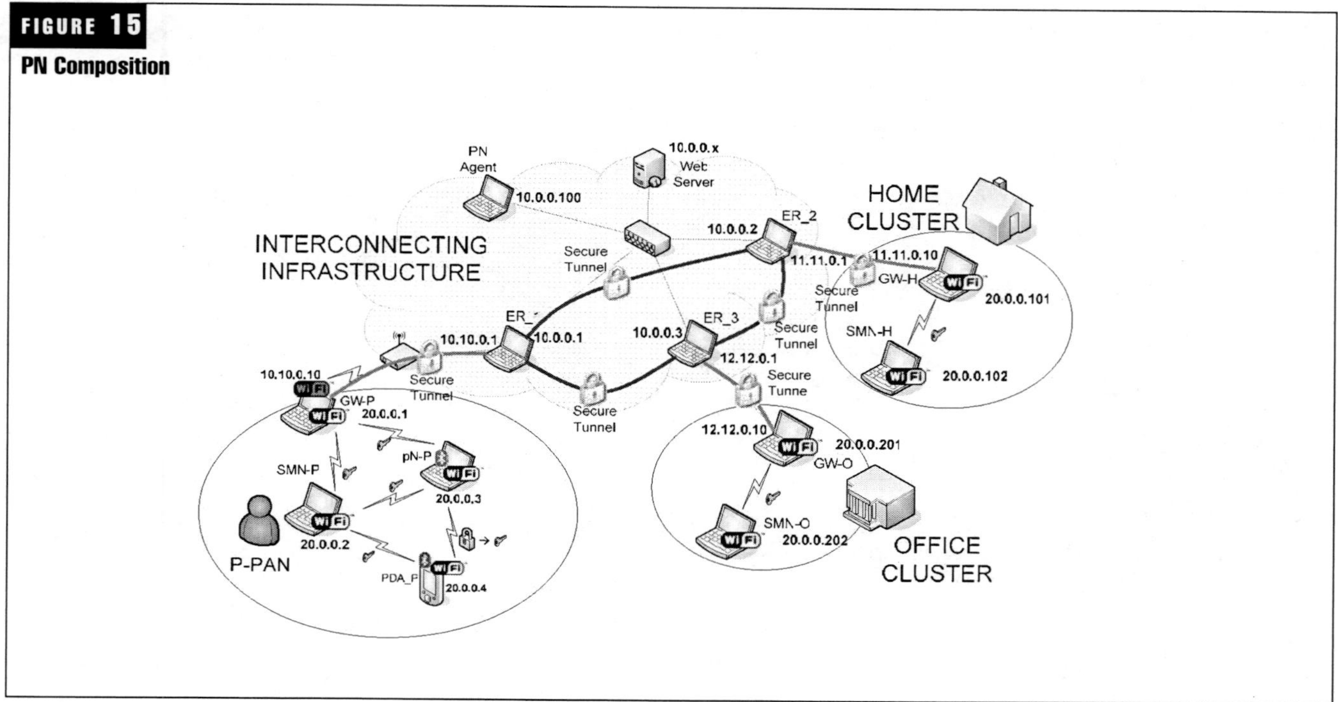

private way. The journalist is able to discover all the available services within the P–PAN as shown in *Figure 16*.

The journalist is now able to access, from his laptop, the video and audio stream his camera is generating. With his video edition application in his laptop, he starts to record the stream adding comments and excerpts. He realizes that some documentation he has in his computer in the office would be really helpful to finalize his report of the meeting. Using the same GUI shown in *Figure 16*, he discovers his content access service at his office. He is able to access all the content of his computer just by clicking a couple of buttons. Once he has finalized the report, he uploads it on a shared folder so that his colleagues are able to review it. He also starts a videoconference with his colleagues at the office to discuss the report. All the traffic is securely transported over the Internet so nobody can have access to this material.

Once he has finished his work and while waiting on the boarding gate, he takes advantage of his PN to remotely control his home appliances (he starts his washing machine and checks if he left the lights on, etc.). Finally, he has some

Service Discovery

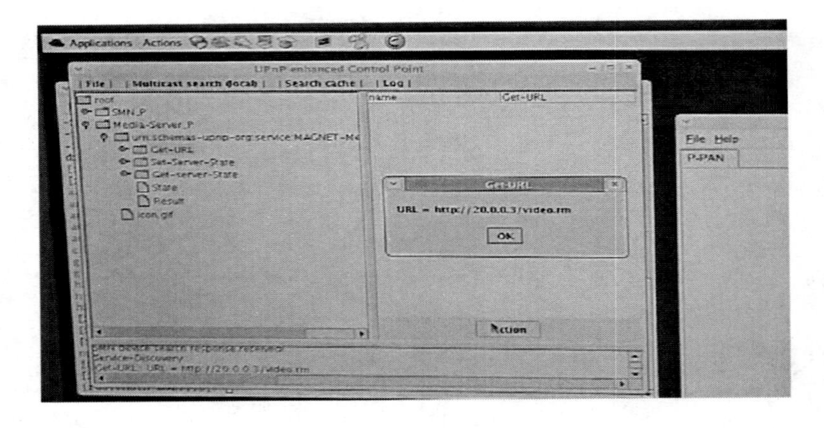

time left and relaxes watching a movie he had recorded the previous night that is stored on his home digital video recorder.

Conclusions

The developed PN architecture and implemented and integrated software components have led to the development of a working PN prototype, enabling secure communication between personal nodes independent of their location and personal services. Currently, this work serves as a basis to develop real pilot PN services and as an enabler for PN federations.

Acknowledgement

This work has been performed in the framework of the IST projects MAGNET and MAGNET Beyond [2], partly funded by the European Union. The authors would like to acknowledge the contribution of their colleagues from the consortium.

References

1. I.G. Niemegeers and S. Heemstra de Groot, "From Personal Area Networks to Personal Networks: A user oriented approach," Journal on Wireless and Personal Communications 22 (2002), pp. 175–186.
2. IST-MAGNET consortium, www.ist-magnet.org.
3. L. Munoz, et al., "A Proposal for Self-Organizing Networks," Wireless World Research Forum Meeting 15 (SIG 3), Dec. 8–9, Paris, France.
4. L. Sanchez, J. Lanza, L. Muñoz, J. Perez, "Enabling Secure Communications over Heterogeneous Air Interfaces: Building Private Personal Area Networks," Wireless Personal Multimedia Communications – Aalborg, September 2005, pp. 1,963–67.
5. M. Petrova et al., MAGNET Public Deliverable D2.1.2 "Overall secure PN architecture."
6. W. Louati and D. Zeghlache, "Network based Virtual Personal Overlay Networks using Programmable Virtual Routers," IEEE Communications Magazine, Vol. 43, No. 8, Aug. 2005, pp. 86–94.
7. W. Louati and D. Zeghlache, "Virtual Router Concept for Communications between Personal Networks," eighth International Symposium on Wireless Personal Multimedia Communications, 2005, Aalborg, Denmark, Sept. 18–22, 2005.
8. W. Louati, M. Girod Genet, and D. Zeghlache, "Implementation of UPnP and INS/Twine interworking for scalable wide-area service discovery," eighth International Symposium on Wireless Personal Multimedia Communications 2005, Aalborg, Denmark, Sept. 18–22, 2005.
9. D. Kyriazanos, et al., "MAGNET Personal Network Security Model: Trust Establishment, Policy Management and AAA Infrastructure," Wireless World Research Forum Meeting 15 (SIG 3), Dec. 8–9, Paris, France.

Smooth Integration of Mobile Video Telephony to Windows Mobile

Tsahi Levent-Levi

Products Manager, Technology Business Unit
RADVISION

Abstract

Third generation (3G) video telephony usually uses the 3G–324M protocol. Today, handset manufacturers that wish to roll out a 3G Universal Mobile Telecommunications System (UMTS) or time division–synchronous code division multiple access (TD–SCDMA) handset understand that 3G–324M support is mandatory. As Windows Mobile gains popularity and becomes more widely deployed in smartphones and personal digital assistants (PDAs), the smooth integration of 3G–324M in Windows Mobile is a prerequisite to success. How do you go about doing that?

3G–324M Overview

3G–324M is based on the International Telecommunication Union (ITU) H.324 standard and is designed to send several channels over low-bit-rate networks. In 3G, this rate is 64 kbps over a circuit-switched connection (UMTS or TD–SCDMA).

This technique for mobile video telephony is the best solution available today for handling latency and round-trip delay issues, implementing authentication and billing mechanisms, and minimizing equipment upgrade costs for operators.

From the standard design above, we can see that there are the following issues 3G–324M application developers must deal with once a 3G–324M stack implementation is available:

1. *Integrating with the network, through the baseband:* This includes dialing and tearing down calls as well as sending and receiving 3G–324M–related bit streams (using the H.223 component).

2. *Integrating with the media codecs*: This includes audio and video codecs, including optimization when required. In order to achieve the best quality, auxiliary algorithms such as acoustic echo canceler (AEC) and noise reduction (NR) are also required.

3. *Building the application itself.*

This article discusses these three issues, focusing on the integration with Windows Mobile 5.0.

Windows Mobile Integration Challenges

Properly integrating video telephony with the Windows Mobile platform involves more than simply writing an application. Windows Mobile 5.0 integrates all the mobile elements incorporated in the handset in order to ensure that users receive a unified look and feel that is intuitive and easy for them to learn and master. Customers expect video telephony to be an integral part of that look and feel and the functionality of the handset, not another third-party, stand-alone application that needs to be installed. Windows Mobile comes with a set of interfaces that should facilitate integration of video telephony applications. Some are related to the network and the media codecs, but there are also application-specific interfaces that are required.

The rationale behind doing integration is the poor user experience that would result if integration was not carried out. In addition, technically, if the video telephony application is part of the operating system image, it is loaded automatically each time the handset is powered on. This means that the application is loaded with no user intervention.

Chipset Architecture

Usually, a handset using Windows Mobile will include two or more central processing units (CPUs), including the following:

1. *Application chip*: This chip is the one that actually runs Windows Mobile. It is used by the operating system as well as the various applications and serves as the actual user interface for the end customer.

2. *The baseband*: This chip is responsible for communication with the baseband by using radio frequencies (RF). It usually uses a proprietary operating system or a scaled-down real-time operating system.

Sometimes, additional chips handle video and audio coding to reduce complexity and increase battery life.

FIGURE 1

3G–324M Standard Design

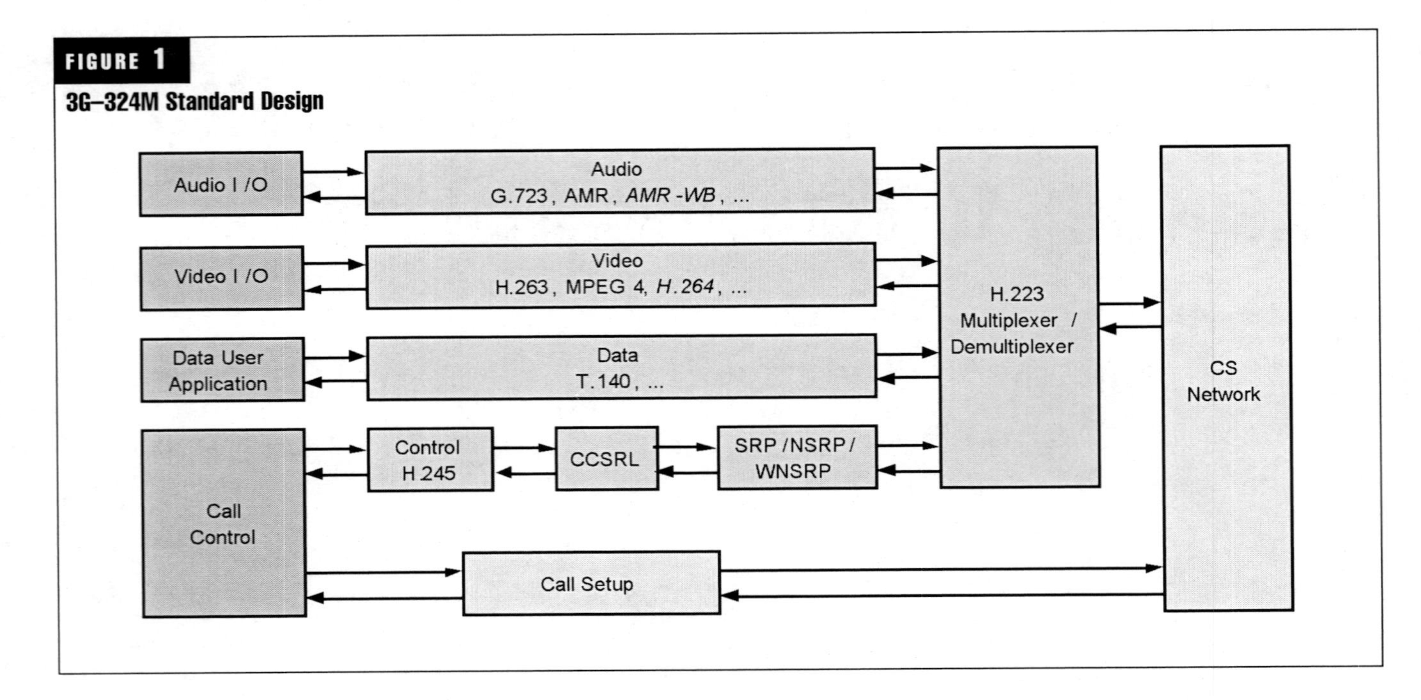

In these standard architectures, 3G–324M is usually optimally located on the application chip for several reasons, including the following:

1. The application chip runs Windows Mobile. Because the goal is to achieve the best user experience possible—maximum integration with the user's operating system, in this case—Windows Mobile is preferred.

2. Windows Mobile interacts with the baseband using a layer called telephony application programming interface (TAPI), and it is easier to handle this layer when 3G–324M is on the application chip itself.

3. Development and debugging on the baseband usually requires a lot of time and expertise. Development tools for these platforms are scarce and less advanced that those found on Windows Mobile. Since time to market is of prime importance, developing more on the application chip and less on the baseband is preferable.

4. Developers have more control over the application chip's resources and peripherals than they do the baseband's. Any "bells and whistles" required can be implemented on the application chip but might not be possible on the baseband.

Baseband and TAPI

Due to the fact that video telephony is a networking application, it should be connected to Windows Mobile through TAPI. Usually, a Windows Mobile handset will be based on hardware architecture with two or more CPUs, where one is responsible for networking (also known as the baseband) and the other is responsible for the rest—in our case, the application chip running the Windows Mobile operating system itself. The actual interface between the application chip and the baseband is done using a radio interface layer (RIL). This layer is not developed by Microsoft, but rather by the baseband provider or the integrator of the handset, or, in other words, the party responsible for mixing between the application chip, the baseband, and the operating system.

Since the RIL is implemented for each baseband or platform independently, different basebands behave differently and require additional investment.

Media Codecs and Direct Show

When deploying video telephony, we would like to see both the audio and video channels opened and properly utilized, despite bandwidth limitations. Doing that requires codecs that compress and decompress the audio and video signals so they can be transmitted over 64 kbps—the gross bandwidth available. Effectively, the application is left with 60 kbps or less dedicated for both media types. This is not very much and presents the following issues that must be dealt with:

- *Optimization:* A good video codec (Moving Pictures Experts Group [MPEG]–4 or H.264) is CPU–intensive. By definition, it will take over the application chip's resources, leaving you with one of the following choices: optimize it for your CPU; use a more powerful CPU; or use hardware acceleration. My choice would be hardware acceleration. In addition to leaving more room for the CPU to breathe and allow other applications to run simultaneously and smoothly, it is also likely to extend the battery life of the handset—no easy feat when deploying video telephony.

- *Direct show:* This issue is closely related to Windows Mobile. Everything needs to be properly integrated with direct show. If it is not done correctly, end users will curse the day they purchased the phone every time there is an incoming video call while they are listening to their favorite MP3. And that will be the least of your problems. Improper integration could lead to inferior media quality or a low frame rate, which may cause customers to run and buy a different handset from another vendor.

Remember that mobile video telephony is all about having video in the call. You should get the best platform with the best codecs and then integrate them in a manner that will make your customers satisfied with a job well done.

Phone Canvas

Windows Mobile 5.0 supports an advanced level of customization of the phone's look and feel. This allows differentiation among manufacturers using this operating system. To take full advantage of this component, developers should plan 3G–324M integration into the platform with these issues in mind. Correct integration requires adding configuration parameters to the phone settings, defaults, reusing configuration/data from voice-call settings, changing the menus, interface to hard-buttons, and sometimes getting events from other operating system [OS]/phone modules.

Fully integrating the application will allow for unified branding by operators and enhancement of the entire user experience.

Contacts and Call History

Using the same features for voice calls enhances the user experience. Users expect to have the same experience for video calls as they have for audio calls and, of course, enjoy their 3G handset more than their "old" 2G handset. This means using the call history feature most mobile handsets have had for years, with lists of missed calls, dialed calls, incoming calls, dialing from an address book, etc.

Call history is the Windows Mobile component that logs missed, dialed, and/or incoming voice and video calls. An implementation of video telephony should take care of this aspect as well, enabling the user to dial video calls from these lists directly.

In addition, Windows Mobile enables users to easily synchronize with their desktop Microsoft Outlook Contacts list, effectively creating a single address book for both the desktop and mobile handset. This powerful feature allows one-button dialing directly from the contact information. Users should expect no less for video calls. Easy dialing and a user-friendly interface for video calls will eliminate complexity and should lead to increased usage of video services.

The integration of the application with these features must be done on the operating system level in order to have a single, unified look and feel for audio and video calls.

Conclusion

Windows Mobile is an advanced operating system that is quite suitable for the 3G arena. As we move toward total convergence, mobile handsets are becoming more complex, with a wide range of applications integrated within. In order to take full advantage of Windows Mobile, an application as important as video telephony must be implemented with extreme care. The integration must take into consideration every aspect of the application's requirements, as well as Windows Mobile architecture and interface. Exploiting the full power of Windows Mobile will guarantee a high-quality product that delivers a superior user experience. Failing to do so will surely result in less appealing handsets for consumers.

The Law of Mobility

Russ McGuire
Director of Business Strategy
Sprint

An article on page B1 of the September 22, 2005, issue of *The Wall Street Journal* began with this observation: "Alan Foster learned about Hurricane Katrina's landfall while watching news channel MSNBC on the small color screen of his Sprint cell phone, while waiting for his wife in a shopping mall near Los Angeles." The article went on to report that "in the week that followed, he kept tuning into his cellular TV whenever he was away from a TV set. At work, colleagues gathered around his cell phone to watch live television updates on the hurricane's devastating impact."

Anyone who read this article had to ask the question, "Why would anyone want to watch television on a two-inch screen?"

The answer is that Mr. Foster and his co-workers want to watch cellular TV because television has become the latest product to experience "the law of mobility."

How valuable is a two-inch TV with a somewhat jerky frame rate? A 50-inch flat panel set with high definition television (HDTV) can easily cost $4,000 or more. Lower your appetite to a 32-inch tube set and you're down in the $400 range. If you can settle for a 13-inch portable, you're down to $70, and if you're really cheap, you can pick up a Sylvania five-inch black-and-white set with AM/FM radio and interchangeable faceplates from Best Buy for less than $30. That would imply that a two-inch set probably shouldn't cost more than $20.

So, why is Mr. Foster probably paying that much every month for the opportunity to watch his mini set? The key phrase in the WSJ article is "whenever he was away." Mobility exponentially increases the value of any product. That is the essence of the law of mobility.

Over the course of a year, Mr. Foster will pay approximately twelve times the value of a two-inch TV simply because he can watch it anytime and anywhere. I once bought a small portable television, thinking that I would then be able to watch shows that I otherwise would have missed. I never took it with me to the mall. I never took it to work. In fact, within a couple of months, I never took it anywhere. Mr. Foster takes his two-inch television everywhere. He has it with him all the time. Anytime he has a few minutes that otherwise would be wasted, he can pull out his television and get caught up on the latest news.

Why is this true? Because his television has been built into a device that Mr. Foster carries with him nearly all the time and everywhere he goes—his cell phone. This simple reality makes this otherwise tiny excuse for a television many times more valuable than anyone would have ever imagined.

Mobilized Products

I mentioned that the television is only the latest product to experience the law of mobility. A growing list of mobilized products enjoys the increased value promised by this nascent reality. Camera phones and mobile e-mail are two of the most prominent examples.

Think about the camera phone. The Samsung model A560 is very similar to the Samsung model PM-A740 camera phone, except that the A560 doesn't have a camera. The PM-A740 has a $60 higher list price. Being able to send photos from the camera phone also requires a data plan that's likely in the $10-per-month range, so the first-year cost of adding a camera to your phone can easily exceed $150. That must be a valuable camera!

At Best Buy, for $150, I can buy the Fuji FinePix A345 digital camera with 10.8x zoom and picture resolution of up to 2304x1728 pixels. The Samsung camera phone doesn't zoom and only takes pictures at 640x480. Best Buy doesn't carry digital cameras with those lowly specs, but Wal-Mart offers a Vivitar camera with similar specs plus a flash for less than $20.

But all of us with camera phones understand why they are worth many times the value that the quality of the product would imply. Camera phones are valuable because you have a camera with you everywhere you go. Every time I capture a photo (even with poor resolution) of an event that otherwise would have been lost, I experience the law of mobility.

Every Blackberry addict fully understands the law of mobility. Does mobile e-mail offer the richness of Outlook or Notes on the desktop? No. Yet we are willing to pay a premium so that we can access our e-mail in every otherwise wasted moment everywhere we go.

What are the implications of the law of mobility on the telecom industry? Quite simply that mobility will be built into every product and every process, and that mobility will rely on mobile and fixed telecom networks.

Three Laws

The phrase "the law of mobility" probably brings to mind other technology "laws" such as Moore's Law and Metcalfe's Law. It's worthwhile to consider these historical examples.

Moore's Law, as illustrated in *Figure 1*, predicted that the effective cost of computing power would be cut in half every couple of years. The most obvious effect of Moore's Law has been the introduction and mass adoption of personal computers. Moore's Law effectively predicted that there would come a time when it would make economic sense for computing power to move out of the data center and onto the desktop. Once that moment occurred, personal computers (PCs) exploded onto the scene in homes and offices around the world. Moore's Law hasn't stopped, resulting in computing power being built into virtually every kind of product with the power to feed a microprocessor, from $2 toys to $60,000 cars and every kind of item in between.

FIGURE 1

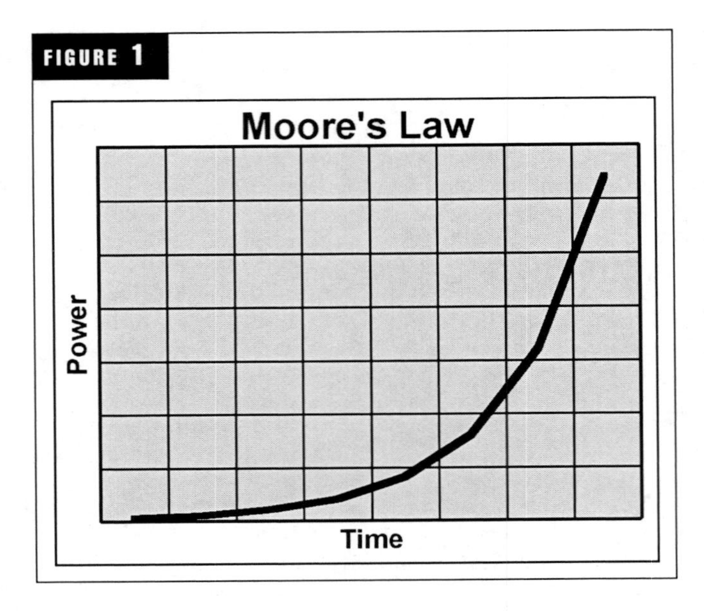

Metcalfe's Law, as illustrated in *Figure 2*, observed that the value of a network is exponentially related to the number of users of that network. The most obvious effect of Metcalfe's Law has been the emergence of the Internet. This global network plugged away in obscurity until a moment in the mid-1990s when suddenly the number of users of the network translated into value in the network that outweighed the cost and trouble of implementing it. From that moment forward, a chain reaction of exploding value rapidly resulted in it being unimaginable for individuals or businesses to not be connected to the Internet. The Internet is now being built into a vast array of products, anything that has information that could be shared or that would benefit from information beyond it, including games, telephones, televisions, encyclopedias, magazines, maps, utility meters, cameras, and watches.

FIGURE 2

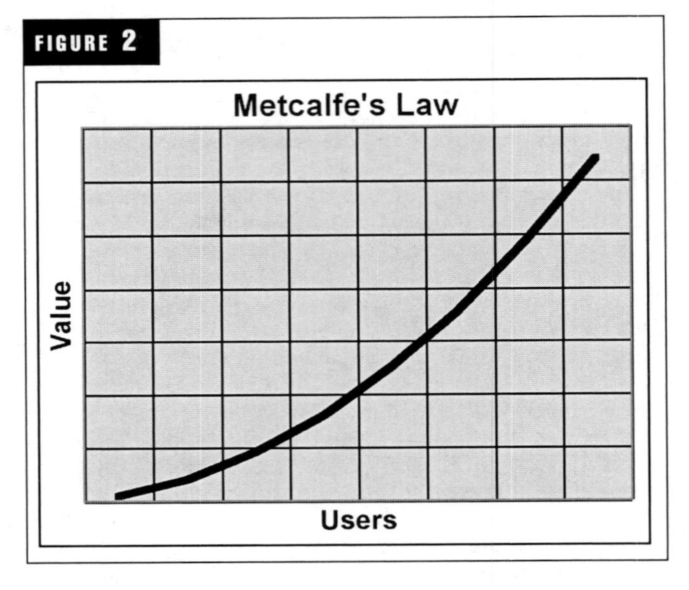

The law of mobility observes that the value of a product increases with mobility. However, until recently, the cost of adding mobility to any product has outweighed that increased value. A simple measure of mobility is the percent of time that the product is available for your use. Thanks to a combination of Moore's Law, scalability resulting from Metcalfe's Law, device convergence, and the increasing ubiquity of third-generation (3G) wireless networks, the cost of making any product (especially one involving information) available all the time is plummeting. Therefore, just as computing power and the Internet have been built into virtually every product, mobility is beginning to be built into every product.

FIGURE 3

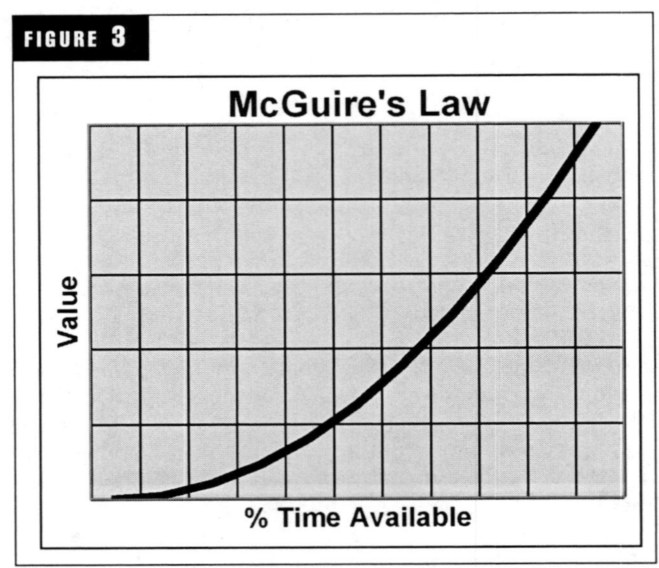

Power and Danger

The implications for business customers are also significant. Again, it's worth considering the effect that the PC era, predicted by Moore's Law, and the effect that the Internet era, predicted by Metcalfe's Law, had on businesses.

From the mid-1980s to the mid-1990s, business information technology (IT) activities were dominated by the effect of the PC. On one hand, the PC unleashed tremendous power for businesses. Spreadsheets enabled every employee to quickly perform complex financial analysis. Word processing and even early forms of e-mail greatly accelerated and improved internal communications. In short, the PC allowed businesses to make better decisions faster and communicate those decisions more clearly and broadly.

However, the PC also introduced tremendous danger to businesses, and especially to IT departments. Critical company information could easily be lost because of a hardware failure, malicious software, or simple human error. IT budgets exploded with increased capital spending for rapidly evolving hardware and software and the staffing of armies of desktop support. As information became decentralized and open to easy analysis from multiple angles, counterproductive political battles were armed with carefully selected and creatively presented "facts."

The Internet era similarly resulted in IT scrambles to capture the power and manage the danger of this new technology. From the mid-1990s to today, IT departments have been consumed with the implications of the global connectivity of information.

The Internet unleashed tremendous power through increased velocity of information flow through and between companies. Power was also released through the use of the Web browser as a universal user interface and the ability to access applications and data from any location with an IP address. The dangers of the Internet to business are similarly well documented and well understood.

We are already beginning to see business IT departments wrestle with the power and danger of mobility. It's reasonable to guess that the next ten years will be dominated by these opportunities and challenges.

Telecom Opportunity

The telecom industry needs to partner with our business customers to help them capture the power of mobility. We must provide ubiquitous mobile broadband services so that anyplace can be a workplace. We must deliver on the promises of network convergence so that our customers can focus on their business while counting on voice-data and fixed-mobile networks to simply work together. We must enable business applications to leverage identity, location, and presence information to operate with meaningful context and appropriate security.

We must help businesses become mobile businesses so that they can grow into new markets and drive efficiencies into their operations. The service providers that help business customers capture the power and manage the danger of mobility will capture inordinate value.

Arguably, the biggest winner in the PC era was Microsoft. DOS and then Windows provided a standard platform on which businesses could deploy multiple value-creating applications. By providing a standard platform that could be deployed across the enterprise, Microsoft helped IT departments minimize user training and, therefore, user error, reduce supply chain and support complexity, and leverage an accelerating pipeline of new applications and tools.

Similarly, the biggest business-to-business winner in the Internet era arguably was IBM. The eBusiness campaign clearly positioned IBM as the partner businesses could trust to capture the power of the Internet while managing the danger. At the beginning of the Internet era, IBM was a commoditized product company. Today, IBM is a highly valued solutions company.

Who can play a similar role as business customers wrestle with the power and the danger of the mobility era?

Measuring Value

In the mobile industry, the most basic measure of value creation is average revenue per user (ARPU). Customers' willingness to pay for mobile services is a reliable proxy for how much value those services are perceived to create. Across the industry, ARPU has been relatively flat throughout this decade, staying in the $50 to $52 range. However, two carriers have had significantly higher ARPU than their peers—Nextel and Sprint (now combined as Sprint). The outperformance of these two companies can be directly attributed to their ability to leverage the law of mobility in combination with Metcalfe's Law (for Nextel) and Moore's Law (for Sprint).

Arguably, all mobile companies are benefiting from the law of mobility. As mentioned above, the industry average ARPU for mobile service is about $50. The industry average monthly revenue per residential phone line is about $32. Given that a single residential phone line typically has multiple telephone handsets connected to it, used by multiple people, this probably translates to an ARPU in the $15 range. Therefore, as expected, mobile telephone service creates a multiple of value relative to fixed telephone service.

However, Sprint has been able to command an even higher multiple with ARPU in the neighborhood of $60—about 20 percent higher than the industry average. Sprint has clearly communicated that much of this ARPU lift is attributable to data services. Sprint has led the industry in introducing

FIGURE 4

Monthly Revenue per User

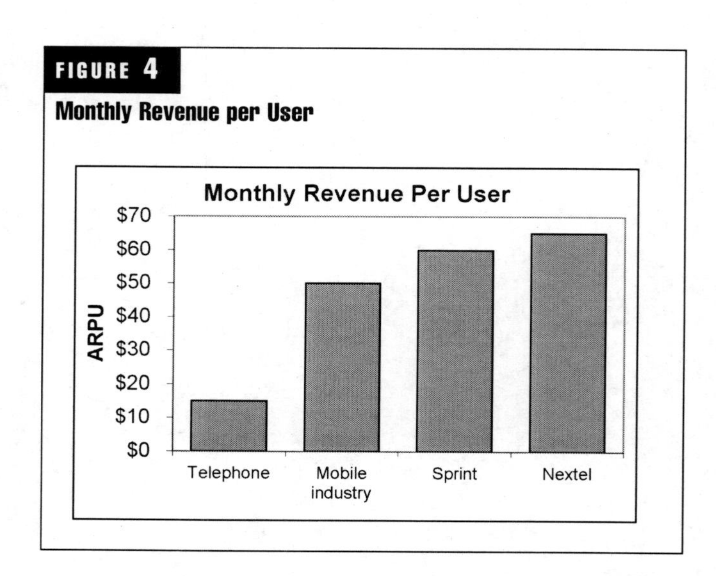

devices and services that combine the power of Moore's Law (computing power in the handset translating into valuable features for customers) with the power of the law of mobility.

Nextel has been able to command an even higher multiple, with ARPU typically in the $65 to $70 range—more than 30 percent higher than the industry average. Industry observers attribute this premium to Nextel's DirectConnect push-to-talk service. Highly productive ecosystems have been built within specific industries using the Nextel service. Because Nextel's push-to-talk service has been built on proprietary technology, these have been closed networks. Once these networks reached critical mass, anyone wanting to participate in the ecosystem had to become a Nextel customer to connect with others using the DirectConnect network. This is classic Metcalfe's Law economics being borne out in real-world performance.

Those of us at the combined Sprint have the enviable challenge and opportunity of determining how to leverage these strong starting points to win the industry race to the applications that can combine the power of Moore's Law, Metcalfe's Law, and the law of mobility to create unique value for our customers. How can we reach the Alan Fosters of the world in new ways that translate mobility into real value? Stay tuned. I'm sure we'll watch the mystery and drama play out together.

Detecting Packet Mishandling in Mobile Ad Hoc Networks

Sirisha Medidi

Assistant Professor, School of Electrical Engineering and Computer Science
Washington State University

Muralidhar Medidi

Associate Professor, School of Electrical Engineering and Computer Science
Washington State University

Sireesh Gavini

Associate Software Engineer, Marketing, Research, and Development
Schweitzer Engineering Laboratories, Inc.

Abstract

Mobile ad hoc networks are inherently prone to security attacks, with node mobility being most likely to allow security breaches. This makes the network susceptible to Byzantyne faults with packets getting misrouted or dropped. We propose solutions using an unobtrusive monitoring technique to locate malicious packet dropping. This technique uses information from different network layers to detect malicious activity. Any single node can use unobtrusive monitoring without relying on cooperation from other nodes, making this technique easy to implement and deploy. We implemented our technique using an ns-2 simulator and conducted extensive experiments using several mobility models. Results show that this technique was effective in detecting malicious behavior with a low false positive rate.

Introduction

Ad hoc networking opens up a host of security issues: use of wireless links makes eavesdropping easy (this may give an adversary access to secret information). Establishing any sort of trust among communicating parties is difficult, as there is no centralized infrastructure to manage and/or to certify trust relationships. This is compounded by the fact that these networks are often very dynamic—with nodes free to join and leave at will—and thus network topology and traffic change dynamically. Malicious activity is difficult to identify by behavior alone because many perfectly legitimate behaviors in wireless networking may seem like an attack. Selfish behavior or node misbehavior is also likely.

In mobile ad hoc networks, information on mobile nodes needs to be updated continuously, making them more susceptible to Byzantyne faults such as misrouting, corrupting, and dropping packets. The nature of these attacks is such that they consume resources associated with various network elements. This impedes efficient functioning and provision of services in accordance with their intended purpose [9, 11]. Unfortunately, current protocol architectures do not provide adequate support or protection against such malicious behavior. This paper takes a fresh look at security management from a real-time perspective and proposes solutions using an unobtrusive monitoring technique that does not require modification of all nodes in the network and relies on readily available information at different network levels to detect malicious nodes. The technique is designed with the following principles in mind:

- Single-node operation, which requires modification only to the node it runs on
- Compatibility with existing protocols
- Battery conservation through the use of readily available network information
- No security associations required
- Detection of malicious packet dropping and misrouting

The rest of the paper is organized as follows: we review the related work; introduce the unobtrusive monitoring technique; describe the experimental design and present our performance results; and present conclusions and future directions of research.

Related Work

Several techniques have been proposed for the detection, identification, and isolation of malicious nodes in wired networks. These techniques are either not adaptable to ad hoc networks or are not practical to implement. Perfect ingress filtering, one of the techniques proposed for wireless networks, places filters at every node [10]. In route-based distributed packet filtering [23], the algorithm performs routability checks on incoming packets. Filters need to be placed at key points in a network; hence it would be difficult to deploy since some nodes may have limited processing capability. "Watchdog" [21] is a technique in which each node "snoops" the retransmission of every packet it forwards. If the watchdog detects that the node has not correctly retransmitted the packet, it raises a warning. This requires omnidirectional antennas. Another variation of this technique, "nodes bearing grudges," proposed in [5] requires security associations between nodes to authenticate messages. The Internet packet dropping mechanism proposed in [34] is based on statistical analysis. The intrusion detection system proposed in [35] uses both local and shared data to detect malicious behavior. It also requires security associations for message authentication. "Resurrecting duckling" [32] technique establishes security associations where a new device can imprint on another device; the imprinting takes place either through close-range transmission or by direct contact between the two devices.

The audit trail approach facilitates tracing via traffic logs at routers and gateways [20, 28]. It is suitable for off-line traceback of denial of service (DoS) attacks but requires significant storage and processing overhead at routers. In behavioral monitoring [22], the likely behavior of a malicious node during a DoS attack is monitored to identify the source. For example, the malicious node may perform domain name server (DNS) requests to resolve the name of the target host that may not be resident in its local name server's cache. During a DoS attack, the malicious (attacking) node may try to gauge the impact of the attack using various service requests, including Web and Internet control message protocol (ICMP) echo requests, so logging those events and activities may reveal the malicious node. IP traceback [3, 29, 30, 31] is similar to the "trace route" command. In IP traceback, packets are sent out with ever-increasing time to live (TTL) values, and the node waits for returning ICMP time-exceeded packets. Traceroute [17, 25] is useful for finding the actual route to a node with a non–source-routed protocol. These techniques are used after the attack has happened, not to identify the attack while it is happening. They are not as useful for source-routed protocols that are likely to be used in ad hoc networks, since the packets already contain the route that the packet traveled. The distributed network monitoring approach to detecting disruptive routers proposed in [4] is based on conservation of flow but still needs to verify assumptions of conservation of flow, according to [16]. A theoretical framework is presented in [26] to detect malicious packet dropping in ad hoc networks with unlimited bandwidth, but no implementation was provided. In [27], a technique for detecting malicious packet dropping by bottleneck nodes is presented. This technique requires the collaboration of the destination node with the sender, which is expensive to implement. Several techniques were proposed for intrusion detection in fixed networks; a survey of those intrusion detection techniques and approaches is presented in [33].

Whenever a node identifies a malicious node, it broadcasts a special blacklist message [21] identifying the malicious node. When a node receives a blacklist message, it removes the blacklisted node from all of its routes, effectively isolating the bad node. "Voting" can also be used—each blacklist message counts as a vote, and a node is only blacklisted after receiving a certain number of votes. A major disadvantage of blacklisting is that it makes the network vulnerable to DoS attacks. With pathrating [21], each node maintains ratings of all paths to all other nodes it knows about. The rating for each path increases with each good transmission and decreases each time a broken link is detected. When a malicious node is identified in a path, its rating is lowered. This pathrating is an extension of blacklisting.

The solutions proposed so far either cannot be adapted to mobile ad hoc networks or are expensive to implement and require modification of all nodes in the network. Moreover, it is important to develop solutions that are scalable, implementable, and capable of detecting malicious behavior while communication is in progress. Techniques such as nodes bearing grudges rely on security associations between nodes in the network. This can be accomplished through the use of a certificate authority, which requires infrastructure support that mobile ad hoc networks lack. The resurrecting duckling technique can be used to establish a security association. However, secure imprinting involves physical contact between the devices, and this may not be possible. In addition, they consume the battery life of the mobile devices to perform the overhead of cryptographic operations. Another challenge in implementing these techniques is that they involve modification of software on most or all of the nodes in the network. For some techniques, all nodes in the network would have to run the modified software to participate. This means devices lacking the necessary capabilities, or devices where the necessary software modifications are not available, would be left out. Finally, all devices in the network may not be under the control of the same person or organization. An ad hoc wireless network at an informal gathering or conference may have tens or hundreds of devices, each owned by a different person [13].

Unobtrusive Monitoring

We propose an unobtrusive monitoring technique to overcome some of the problems associated with existing techniques. The technique can be used to detect Byzantyne faults such as dropping or misrouting packets. In this paper we focus on malicious packet dropping. Details for the case of misrouting can be found in [13, 14].

The unobtrusive monitoring technique relies on readily available information at different network levels to detect the presence of malicious nodes. This technique involves collecting and analyzing locally available data. It is implemented on the source nodes requesting service. The data collection and analyzer components form the core of the detection technique. Local data such as route request and route error messages, ICMP time exceeded and destination unreachable messages, and transport control protocol (TCP) time-outs are used to detect misbehavior. The data collection component gathers this information received within a certain interval of time called the detection interval (explained later in this paper). Any information older than the detection interval is discarded, which guarantees the freshness and relevance of the collected information and also suits the requirements of a memory-constrained node. This collected data is passed on to the data analyzer component, which extracts useful information from these control messages and checks for any deviation from normal behavior. The analyzer extracts important information about the following:

- The location of broken links in route error messages
- The address of a node that was unable to deliver a packet in an ICMP destination unreachable (DU) message and the destination of that packet
- The address of a node that dropped a packet whose TTL had expired from an ICMP time-exceeded message and the destination of the original packet
- The destination of a TCP packet that timed out
- The time that each message was received or each event occurred

Finally, the data analyzer processes this information to determine if any malicious activity is taking place. In the case of undesirable activity, the node is alerted so that it can take appropriate action.

The main function of the "data collection" component is to record all the route errors received on a per flow basis at the source node. The collection component waits until it receives any route error message. Then it extracts pertinent information from the packet and records the occurrence on a per flow basis. If any route error messages older than the "detection interval" arrive, it purges those messages to maintain the freshness and relevance of the information. The information gathered by the data collection component is given to the data analyzer component whenever it detects a TCP time-out at that source. *Algorithm 1* illustrates the process implemented by the data collection component.

Algorithm 1 Data Collection (*detection interval*)
loop
  if *route error message received* **then**
    fid ← *Flow on which the route error was received; store the received route error in the store corresponding to fid;*
    current time ← *get current time;*
    if *there are messages older than (current time – detection interval)* **then**
  purge those messages from store;
    end if
  end if
end loop

The "data analyzer" component waits for the occurrence of a TCP time-out at the source node. Whenever a time-out occurs, it obtains the corresponding flow on which this time-out occurred and obtains the route error information from the data collection component. It then tries to correlate the time-out with any of the route error messages recorded by the data collection component. If it fails to find a route error message in the given "detection interval," it raises a flag informing the source node of possible malicious behavior in the corresponding flow. The source node can then use this information to take corrective action for the detected malicious behavior. *Algorithm 2* summarizes the process used by the data analyzer.

Algorithm 2 Data Analyzer (*detection interval*)
loop
  if *TCP time-out occurred* ***then***
    fid ← *Flow on which time-out occurred;*
    current time ← *get current time;*
    if *any route error messages received for fid* **then**
      if *no route error messages in (current time – detection interval, current time)*
      then
        raise a flag indicating malicious activity;
      end if
    else
      raise a flag indicating malicious activity;
    end if
  end if
end loop

The main assumptions being made in the unobtrusive monitoring technique are as follows:

- The ad hoc network has dynamic source routing (DSR [19]), ad hoc on-demand distance vector routing (AODV [24]), or some source routing protocol as the underlying routing protocol, and TCP is the underlying transport layer protocol.
- A node chosen to behave maliciously does so starting at some random time and from that time onward, it drops all the packets it receives.
- All packet drops are due to either malicious packet dropping or broken link(s) at some intermediate node in the source route.
- Nodes have enough memory to store information about the route error messages received within the detection interval.

Example Scenario

The following example illustrates how the unobtrusive monitoring technique uses information from different network levels to detect malicious behavior in the network. In this example, we have used DSR routing protocol. The technique works similarly with other source routing algorithms such as AODV.

In the DSR routing protocol, a node sends a DSR route error message back to the source if it is unable to deliver the packet to its next-hop neighbor as indicated in the source route. This could be due to node movement—which could result from a new network topology, which means the old route no longer exists—or a temporary broken link between the node and its next hop, which means the node is unable to find an alternate route to that destination using the routes it already discovered. A TCP time-out normally occurs when a sender does not receive an acknowledgment within a specific interval. This may happen because the packet was dropped at an intermediate node due to congestion at that node, the packets' TTL value has expired, some malicious node was trying to disrupt the network, or there was a broken link at some intermediate node in the source route. The distinction between packet dropping due to congestion and packet dropping due to malicious behavior can be handled at the transport layer. We developed a transport protocol,

TCP–manet—capable of detecting among congestion, wireless errors, and malicious packet drop—which is discussed in [8]. This TCP–manet can be used to distinguish between congestion-related packet drops and malicious packet drops in place of TCP without extensions. The unobtrusive monitoring technique at the source node uses such control information passed between the source and the destination to detect the presence of malicious nodes in that route.

In our example, as shown in *Figure 1*, node *S* is the source, *D* is the destination, and a TCP connection is enabled from node *S* to node *D*. The path from *S* to *D* with *M* as an intermediate node was found by the routing protocol during the discovery phase. The source node *S* is equipped with the unobtrusive monitoring technique. Initially there is no malicious activity (i.e., where malicious nodes are dropping packets) in the network. The data collection component will still collect the control data that falls within the detection interval, and the data analyzer component will extract useful information out of that data. Since there is no malicious activity in the network, it will only silently monitor the status of the network. Now assume that malicious node *M* drops packets originating from node *S*, which are destined for node *D* at some random time. The source node keeps sending out data until its congestion window is not full and the acknowledgment timer has not timed out. Once the acknowledgment timer times out, the analyzer component in the node *S* starts looking for any DSR route error messages received within the detection interval. Detection interval is the duration within which the data collection component gathers control information from different levels of the network stack (which is finally fed to the data analyzer component) and purges any old information that falls outside the interval. If there are any route error messages, then the time-out is attributed to a broken link rather than congestion or malicious behavior. If there are no route error messages within the detection interval, the data analyzer component alerts the node of a potential malicious activity in the source route from the source to the destination.

In case of packet dropping when a TCP time-out occurs, the monitoring daemon first checks if there are any route error messages indicating broken links. When the monitoring daemon receives an ICMP message with no accompanying route error message indicating a broken link, it raises a flag indicating malicious activity. Packet dropping due to broken links must be distinguished from those due to malicious behavior.

Performance Evaluation

To study the performance evaluation, we used the network simulator *ns-2* to simulate an ad hoc wireless network running the DSR routing protocol. We identify the performance metrics used to evaluate the unobtrusive monitoring technique, describe our experimental setup, and present our simulation results.

Performance Metrics

The objective of this performance analysis is to determine how effective the technique is in identifying malicious behavior. Hence, the primary metrics for evaluating the proposed unobtrusive monitoring technique are detection effectiveness and false positives. For these metrics, we investigate scalability in terms of the fraction of malicious nodes and mobility. We also study the effect of protocol parameters such as detection interval so as to identify the ideal range of detection interval.

Detection Effectiveness

Detection effectiveness measures how well the technique detects malicious events. For example, if the node detects every instance of a malicious packet drop, then the detection

FIGURE 1

Example Scenario for Packet Dropping

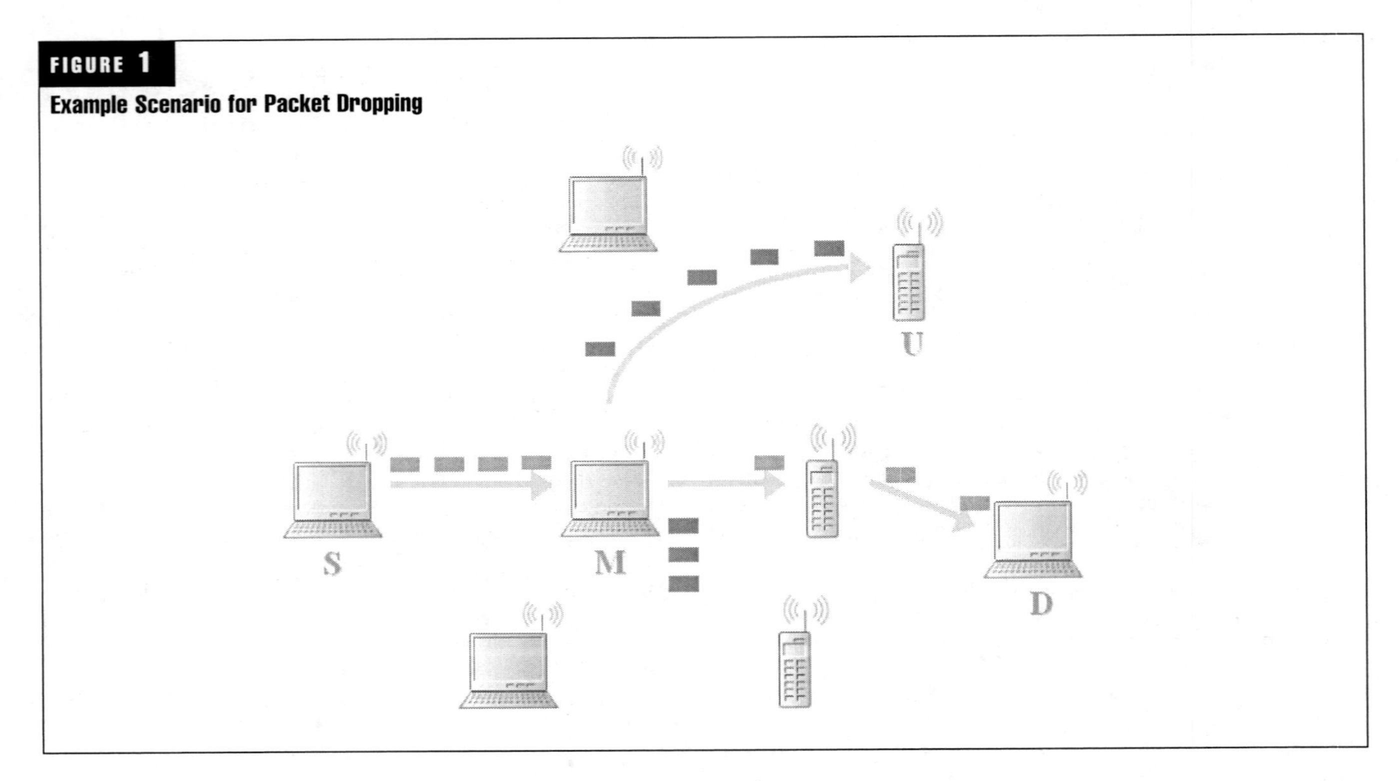

effectiveness is 100 percent, but if it does not detect any misbehavior, the detection effectiveness is 0 percent.

False Positives

Reporting malicious behavior when none occurred is a "false positive." To permit comparison between different scenarios, the number of false positives is normalized by the number of reported events. A perfect system will have 100 percent detection effectiveness and 0 percent false positives, but this is not always possible [14]. Techniques such as setting alert thresholds can reduce the number of false positives, but this would also reduce detection effectiveness.

Mobility Models

In our simulations, the following mobility models were used to evaluate the unobtrusive monitoring technique:

- *Random waypoint model*: In this model, a node moves from its current location to a new location by randomly choosing a direction and speed in which to travel and pauses between changes in direction and/or speed. This is a memory-less mobility pattern, meaning it retains no knowledge of its past locations and speed values. The current speed and direction of a node is independent of its past speed and direction. This characteristic can generate unrealistic movements such as sudden stops and sharp turns [7].

- *Gauss-Markov model*: This model was designed to adapt to different levels of randomness. Initially each node is assigned a current speed and direction. At fixed intervals, movement occurs by updating the speed and direction of each node. Specifically, the value of speed and direction at the nth instance is calculated based upon the value of speed and direction at the $(n-1)^{th}$ instance. The main advantages of this model are that it eliminates the sudden stops and sharp turns present in the random waypoint model and is close to being realistic.

- *Reference point group mobility (RPGM) model*: This model represents the random motion of a group of mobile nodes as well as the random motion of each individual node within the group. Group movements are based on the path traveled by a logical center for the group. The motion of the group center completely characterizes the movement of its corresponding group of mobile nodes, including their direction and speed. Individual nodes randomly move about their own pre-defined reference points, the movements of which depend on the group movement.

- *City section mobility model*: In this model, the simulation area is a street network that represents a section of a city where the ad hoc network exists. The streets and speed limits on the streets are based on the type of city being simulated. Each mobile node begins the simulation at a defined point on some street. A mobile node then randomly chooses a destination, also represented by a point on a street. The movement algorithm from the current destination to the new destination determines the shortest travel time between the two points; in addition, safe driving characteristics such as a speed limit and a minimum distance allowed between any two mobile nodes exist. Upon reaching the destination, the mobile node pauses for a specified time, then randomly chooses another destination (i.e., a point on some street) and repeats the process.

For each of these simulations, a network with 50 nodes in flat 670m by 670m topography was created; this is very common in current ad hoc network research and in the simulation examples provided with ns-2. Each simulation was run for 200 seconds. In case of random waypoint networks, the maximum node speed was set to 20 m/s to simulate high-mobility networks and 5 m/s to simulate medium-mobility networks. Also, the pause times were set to five seconds and 30 seconds to simulate high- and medium-mobility networks, respectively. Fixed-size packets of 512 bytes were continuously injected.

Communication Patterns

Networks with random waypoint mobility were generated using the CMU's transmission control protocol (TCP)/constant bit rate (CBR) traffic-scenario generator script [15], and networks with other mobility patterns listed earlier were generated using University of Bonn's BonnMotion [12], a mobility scenario generator and analysis tool. A communication pattern specifies the source, destination, and other parameters associated with the connection. In each experiment, 20 traffic connections were used for the nodes to communicate. The results were taken as the average of different communication patterns.

Misbehaving Nodes

The ns-2 simulator was modified to enable particular node(s) to be configured as malicious. The configuration also takes in a time parameter that specifies the time that node starts behaving maliciously. Beginning from that time, the node drops all the packets received at that node until the end of the simulation. Each network is designed to contain 20 malicious nodes, reflecting misbehavior of 40 percent of the nodes as suggested in [21]. The number and placement of the malicious nodes ensures that they will be along active paths in the network. To determine the effectiveness of our approach, the percentage of misbehaving nodes was varied from 0 percent to 40 percent in 5 percent increments. A staggered approach was introduced to ensure that at least one node was misbehaving for most of the simulation. The tool command language's (TCL's) built-in pseudo-random generator is used to designate misbehaving nodes randomly.

Detection Interval

Detection interval specifies how long a source node keeps track of all control messages (i.e., route error messages) it receives. The choice of the detection interval determines the detection effectiveness and false positive rate of the unobtrusive monitoring technique. Increasing the detection interval might allow unrelated route error messages to be associated with any TCP time-out at the source node and hence hamper the technique's ability to detect malicious activity. Having a very small detection interval also has a negative effect on the false positive rate of the technique. With a smaller detection interval, we might quickly jump to a conclusion about a TCP time-out at the source without waiting long enough to include any delayed route error message(s) for that flow and hence increasing the false positive rate of the technique. Increasing the detection interval implies that a node has to store information for a longer period of time. If the node is receiving a lot of messages, this can drastically increase the storage overhead, which burdens the memory-constrained mobile nodes. Therefore, choice of detection interval has a very significant impact on the technique's performance metrics and storage overhead.

In our simulations, we have experimented with detection intervals ranging from five to 50 seconds in five-second increments. We observed that at a lower detection interval, we have high detection effectiveness and a high false positive rate. For the detection intervals of 30, 35, and 40 seconds, our technique had good detection effectiveness with a lower false positive rate. To show how the detection effectiveness and false positive rate vary with the choice of detection interval, we present the detection effectiveness and false positive rate for high-mobility random waypoint networks for the following detection intervals: 20, 30, 35, 40, and 50 seconds. For the remaining models, we present the results only for detection intervals of 30, 35, and 40 seconds. The detection effectiveness and false positive rate corresponding to these intervals for different mobility models are presented in the next section.

Results and Discussion

The different configuration parameters used for generating the required mobility scenarios using BonnMotion [12] are given in *Table 1*.

Random Waypoint Model

The detection effectiveness for medium- and high-mobility networks is shown in *Figures 2* and *3*, respectively. From these figures, we can see that for both medium- and high-mobility networks, increasing the detection interval lowers the detection effectiveness. When the detection interval is larger, the probability of getting unrelated route error messages within this interval is much higher. Also, the detection effectiveness is better in medium-mobility networks than in high-mobility ones. Due to higher mobility, links get broken quite frequently and there are many route error messages sent out by the nodes in the network. This also increases the probability of receiving unrelated route error messages within the detection interval at a source node, thus leading the source node to correlate any TCP time-outs with the received route error message, which in turn decreases detection effectiveness.

TABLE 1

Configuration Parameters Used for Different Mobility Models

Mobility	*Speed in m/s*	*Pause time in seconds*
Random waypoint (medium mobility)	5 (max.)	30
Random waypoint (high mobility)	20 (max.)	5
RPGM (medium mobility)	5 (max.)	30
RPGM (high mobility)	20 (max.)	5
Gauss-Markov	20 (max.)	Default
Manhattan grid	20 (max.)	Default

FIGURE 2

Detection Effectiveness – Random Waypoint (Medium Mobility)

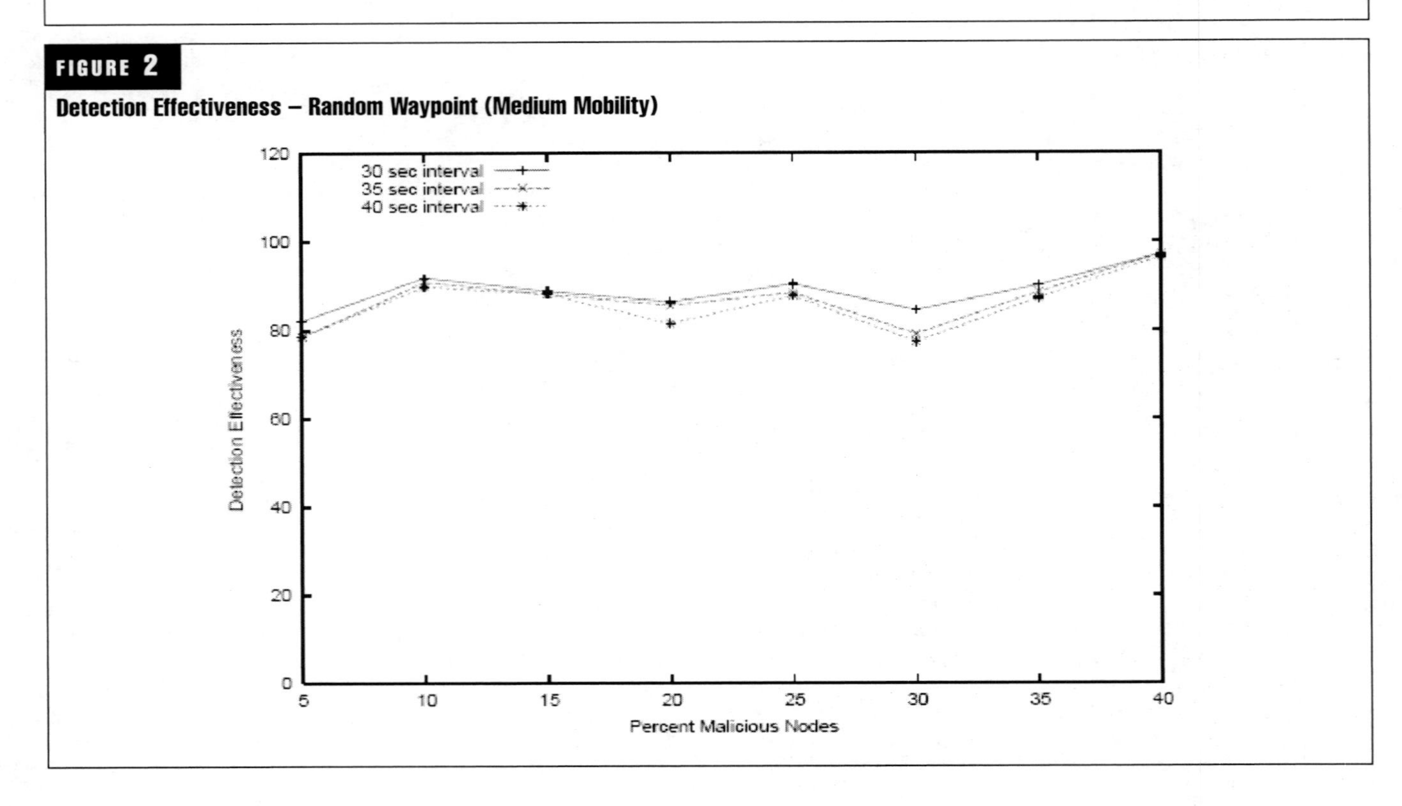

FIGURE 3

Detection Effectiveness – Random Waypoint (High Mobility)

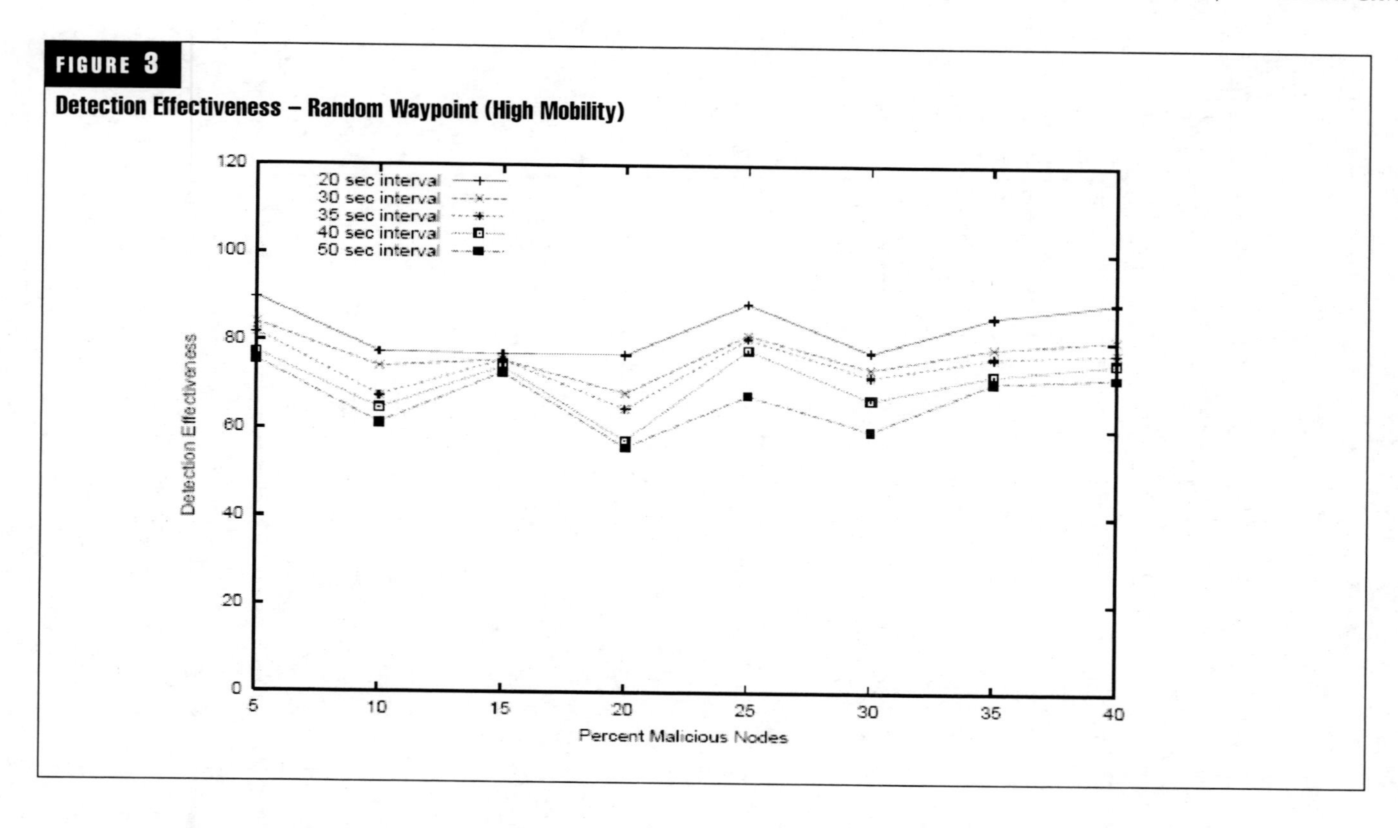

The false positive rate for medium- and high-mobility networks, as shown in *Figures 4* and *5*, respectively, decreases when the detection interval increases. With a larger detection interval, there is enough time to receive any delayed route error messages. When the nodes move faster, the links between the nodes get broken all the more. As a result, the route error messages sent out by intermediate nodes might get dropped before they reach the source node. When the source node times out, it will not find any related route error messages received by it within the detection interval and might conclude that the time-out is due to some malicious behavior in the source route. Therefore, higher mobility could result in a higher false positive rate due to the loss of route error messages.

FIGURE 4

False Positives – Random Waypoint (Medium Mobility)

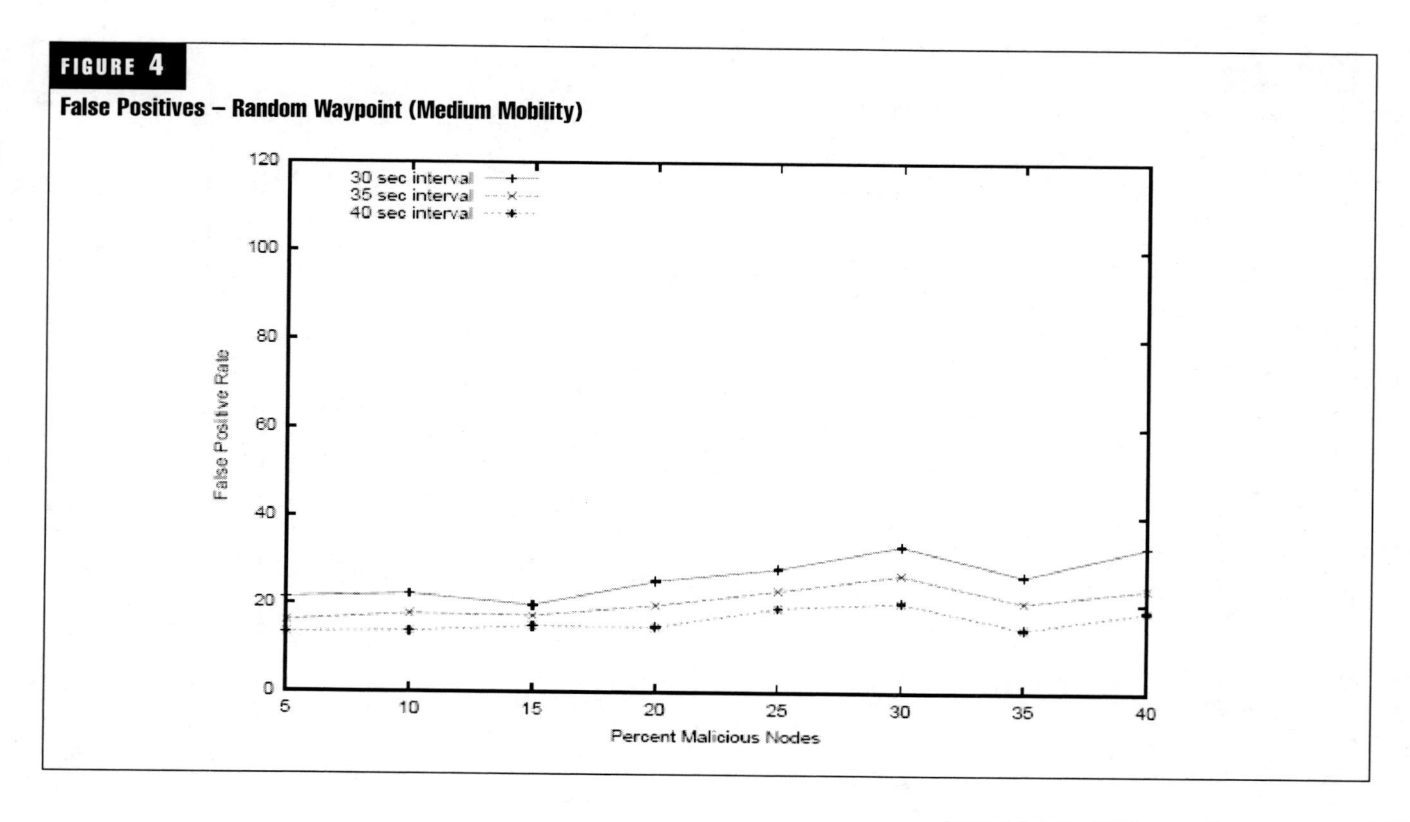

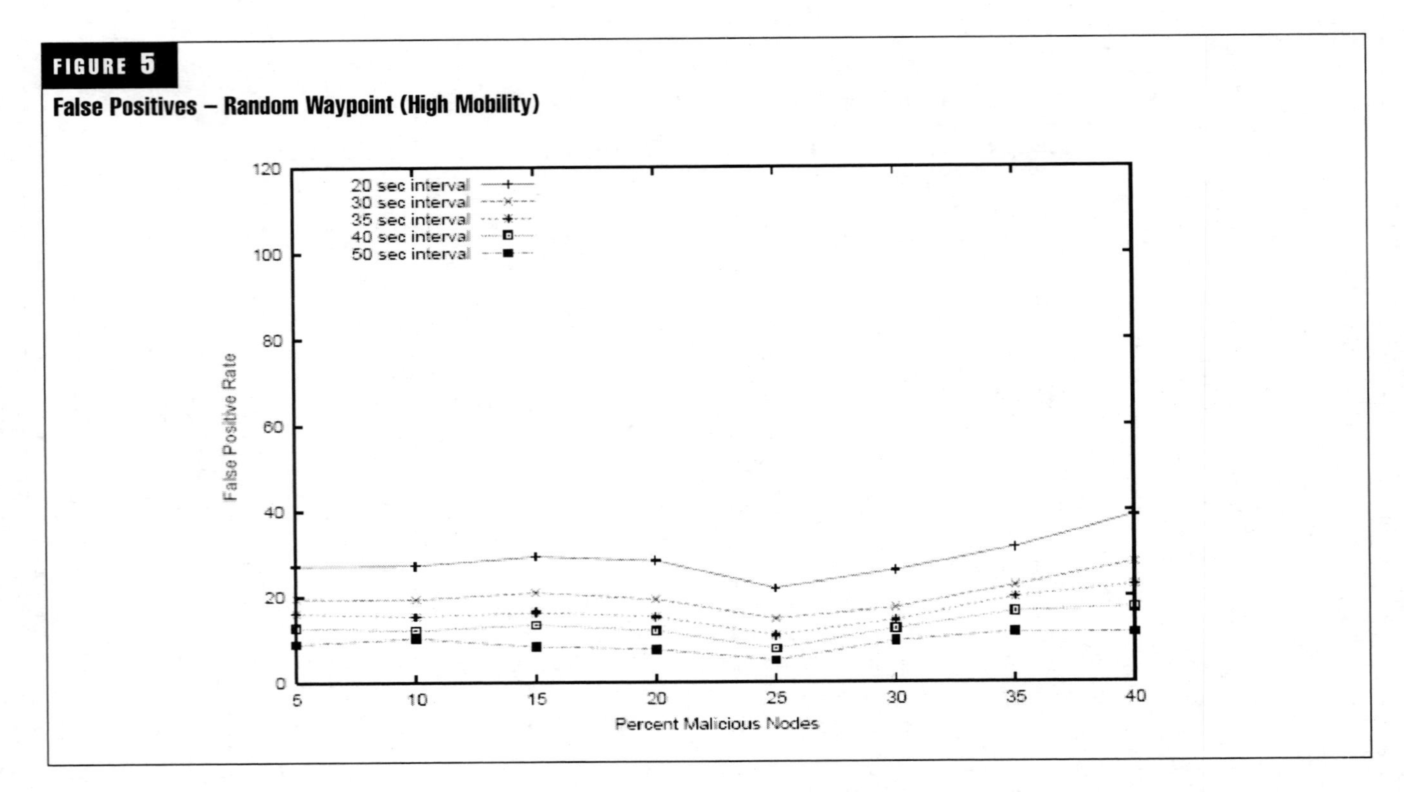

FIGURE 5

False Positives – Random Waypoint (High Mobility)

RPGM

Figures 6 and *7* show the detection effectiveness for RPGM for medium- and high-mobility scenarios, respectively. Increasing the detection interval decreases the detection effectiveness, as in the case of random waypoint mobility networks. This decrease could be attributed to the unrelated route error messages being considered for correlation with the TCP time-outs when using a higher interval. Also, there is a trend similar to random waypoint networks when mobility increases—an increase in mobility will lead to more broken links and more route error messages sent to the source nodes. The source node could miss some malicious drop events by correlating a TCP time-out with one of the many route error messages it receives due to increased mobility in the network. But since the nodes move in groups and with reference to a leader, there will not be as many bro-

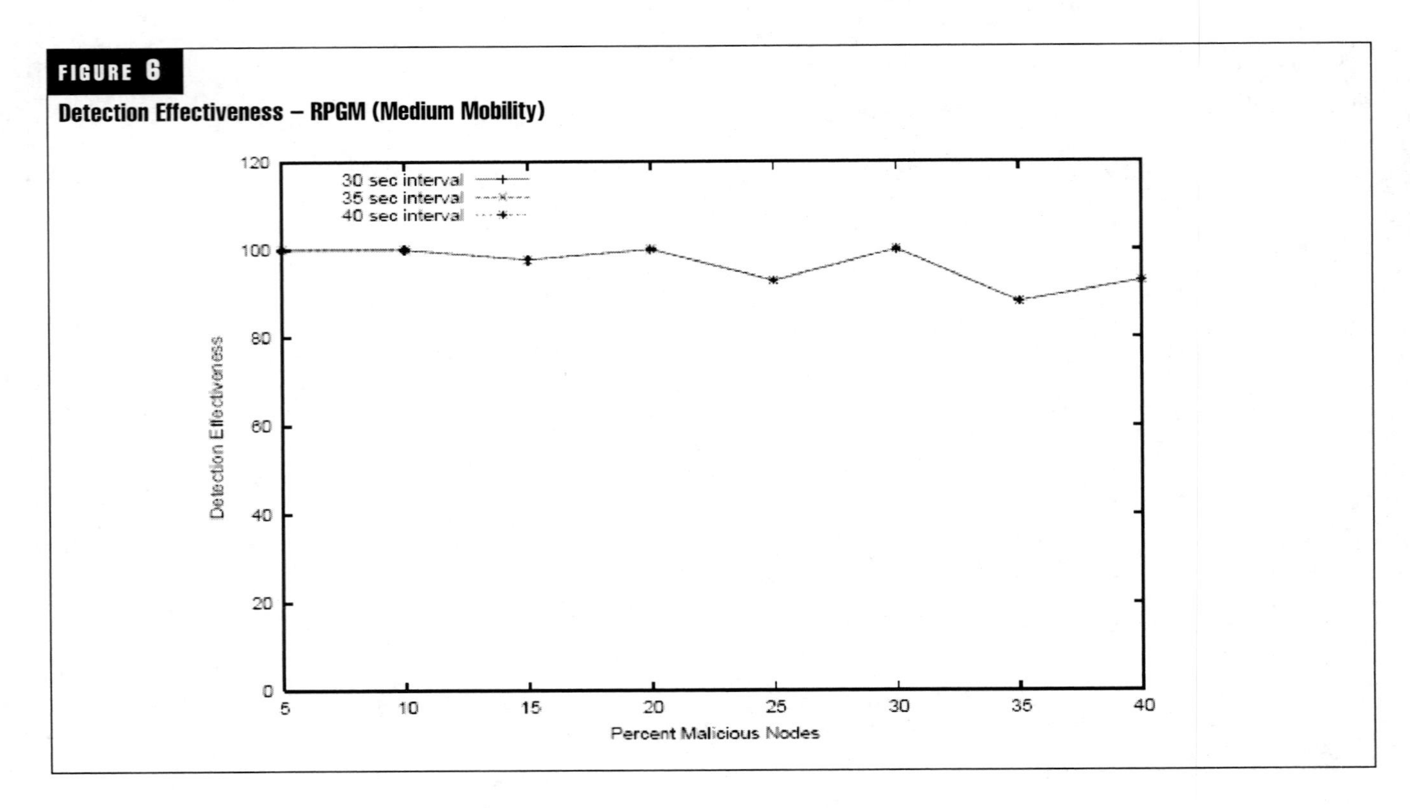

FIGURE 6

Detection Effectiveness – RPGM (Medium Mobility)

FIGURE 7

Detection Effectiveness – RPGM (Medium Mobility)

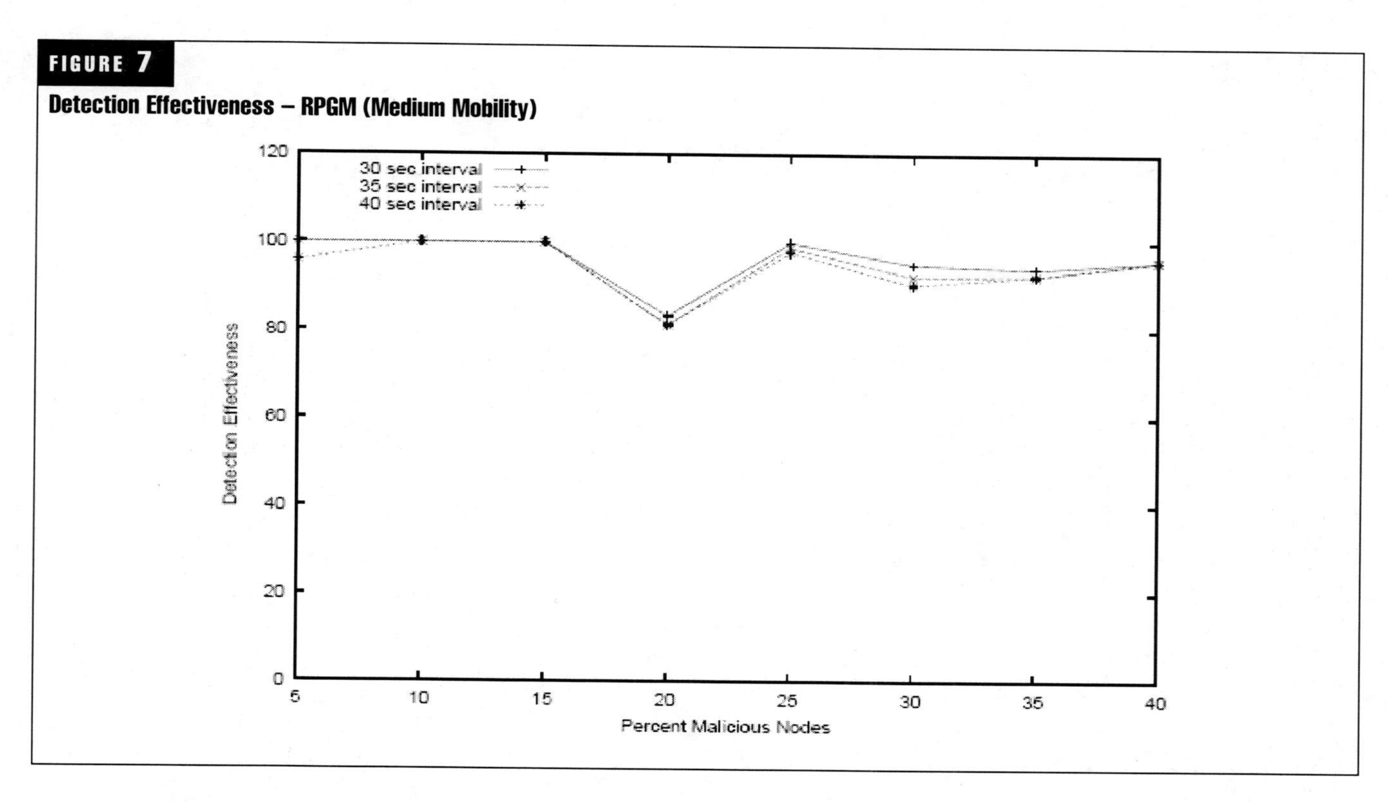

ken links as in the case of random waypoint mobility. The chance of getting a stray route error message within the detection interval is much lower with RPGM than with random waypoint. Hence detection efficiency is much higher when compared to random waypoint.

The false positive rate for medium- and high-mobility scenarios is shown in *Figures 8* and *9*, respectively. In this case also, the increase in mobility increases the false positive rate of the technique. It is more likely for the route error messages to be dropped by some intermediate nodes before they actually reach the intended source node, so the source node, which is unaware of the loss of route error messages, sees any genuine time-outs due to broken links as malicious. This leads to an increase in the false positive rate at higher mobility. Also, in RPGM, there will be fewer broken links

FIGURE 8

False Positives – Reference Point Group (Medium Mobility)

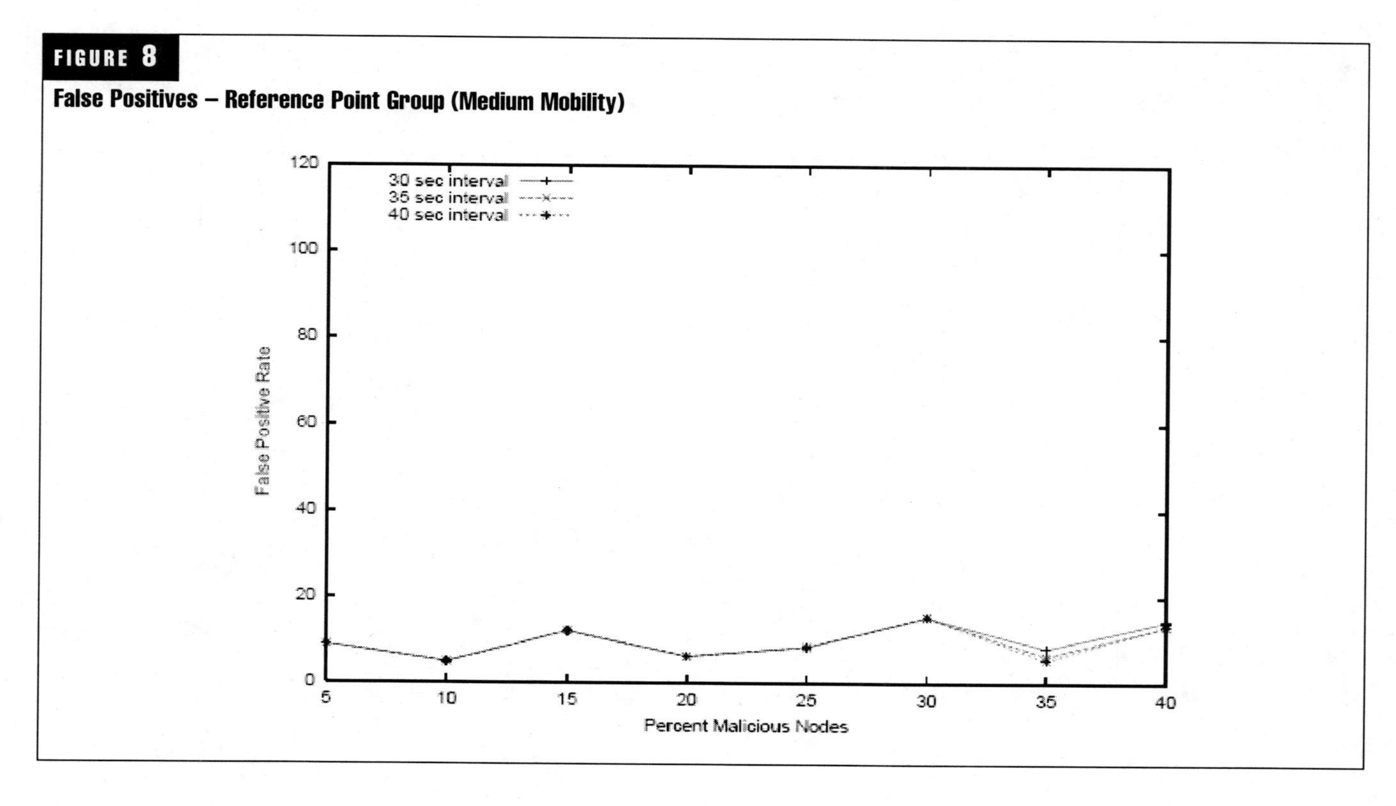

FIGURE 9

False Positives – Reference Point Group (High Mobility)

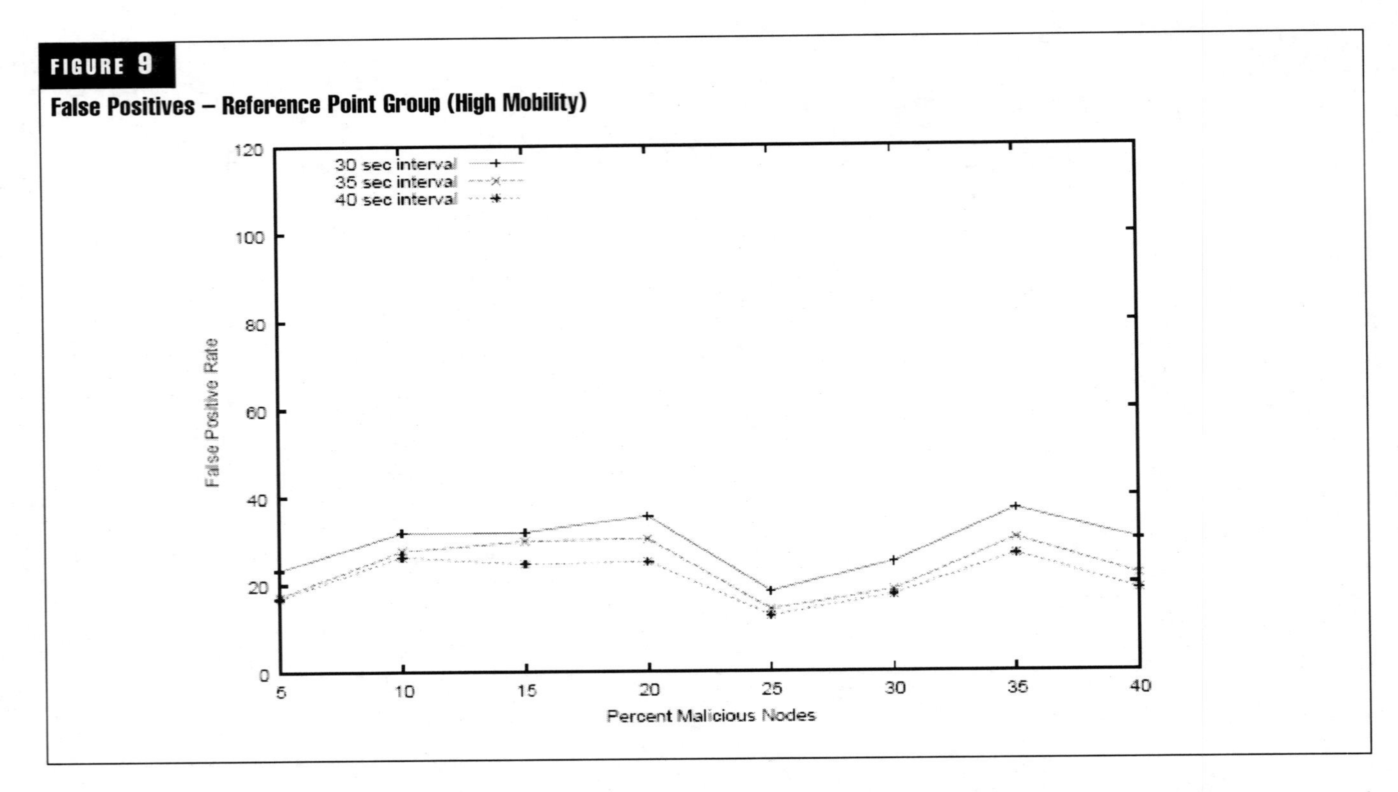

than in random waypoint mobility due to the nature of movement of the nodes. This means that wrongly classifying a single normal time-out as malicious may greatly affect the false positive rate.

Gauss-Markov Model

In this model, the nodes also move about at a maximum speed of 20 m/s, representing a high-mobility scenario. The detection interval affects the detection effectiveness and false positive rate as in the previous mobility models.

The detection effectiveness decreases as the detection interval increases, as shown in *Figure 10*. As mentioned earlier, this is a more realistic mobility model than the random waypoint model, where nodes choose a speed and a random destination and start moving toward the destination. So when

FIGURE 10

Detection Effectiveness – Gauss-Markov

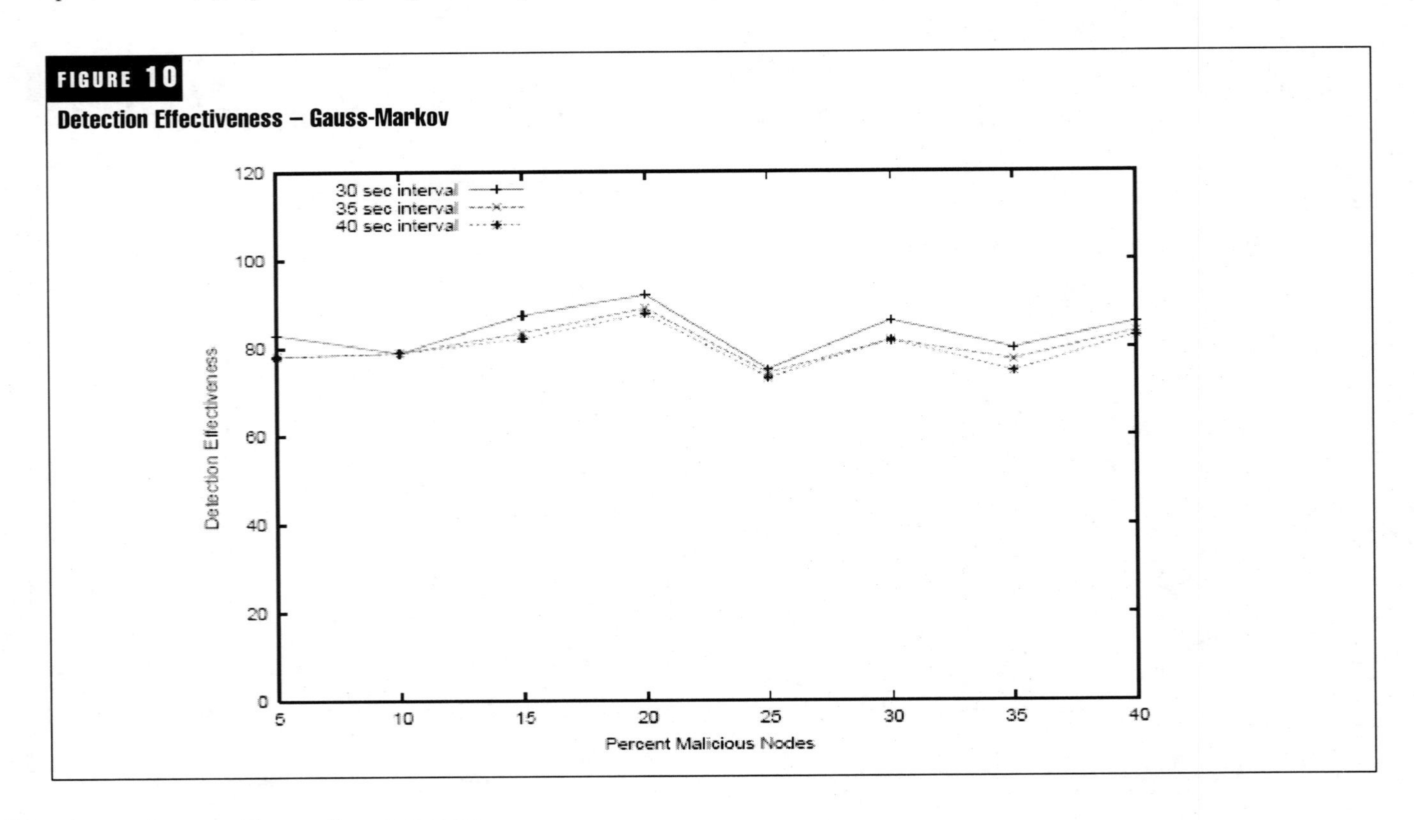

FIGURE 11

False Positives – Gauss-Markov

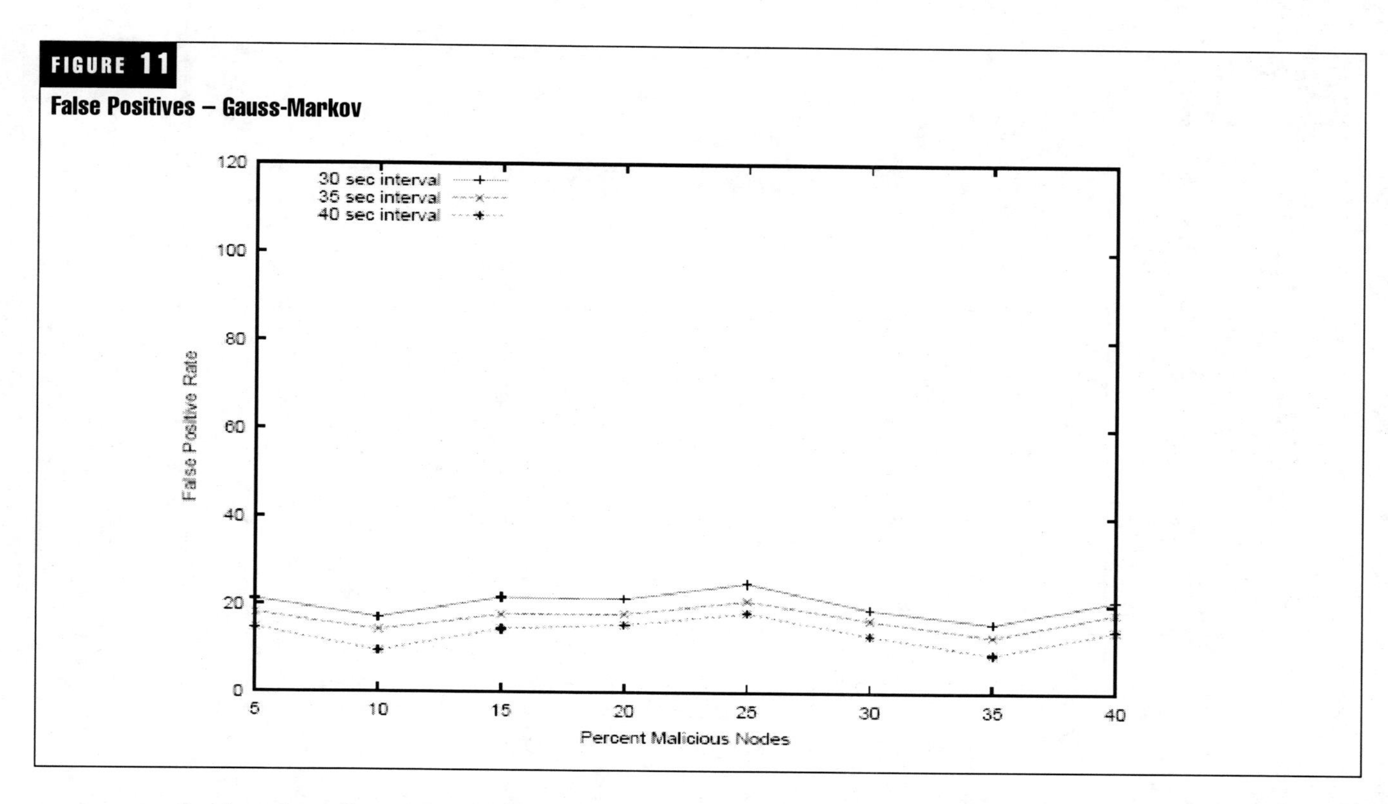

we compare the detection effectiveness of this model with the high-mobility random waypoint model, we can observe that the Gauss-Markov model exhibits slightly better detection effectiveness than the random waypoint model. Because of more realistic mobility in the model, the chances of getting unrelated route error messages is lower with the Gauss-Markov model than with the random waypoint mobility model, which leads to better detection effectiveness.

Figure 11 shows the false positive rate of the technique for Gauss-Markov mobility model. The results are similar to the previous models where the false positive rate decreases when the detection interval increases. Compared to random waypoint mobility model, Gauss-Markov has a slightly lower false positive rate. This can again be attributed to more realistic motion in the Gauss-Markov model. We speculate that there are more dropped route error messages in the random waypoint mobility model than in the Gauss-Markov mobility model, which leads to a lower false positive rate for Gauss-Markov mobility model.

Manhattan Grid Model

Figure 12 shows the detection effectiveness of the technique for the Manhattan grid mobility model. Here also, the detection effectiveness decreases with increase in the detection interval, which is similar to the results of the other models. This model can be seen as a special case of random waypoint mobility model, with the nodes choosing a random destination either in the horizontal or vertical directions. This model demonstrates slightly better detection effectiveness than high-mobility random waypoint mobility networks. Similar to the Gauss-Markov model, this could be due to the more restricted movement of the nodes in the network, which leads to slightly fewer broken links and a subsequently lower chance of getting a stray route error message.

Figure 13 shows the false positive rate for the Manhattan grid mobility model. The false positive rate decreases when the detection interval increases, which is similar to the other models.

Discussion

In all the mobility models, we have observed that the detection interval plays an important role in determining the detection effectiveness and false positive rate of the technique, as anticipated. A higher detection interval leads to a much better false positive rate and much lower detection effectiveness. Similarly, lowering the detection interval increases the detection effectiveness and at the same time leads to a much higher false positive rate. Therefore it is essential that we choose an appropriate detection interval that enables the technique to exhibit good detection efficiency while keeping a reasonable false positive rate. We varied the detection interval from 5 to 50 seconds in five-second increments and observed that detection intervals in the range of 30 to 40 seconds provide good detection effectiveness while maintaining reasonable false positive rates.

Comparison with Other Techniques

Papers covering watchdog [21] and nodes bearing grudges [5, 6] have provided simulation results. However, results are only provided for CBR data sources with no reliability requirements, and not with reliable transport protocol such as TCP.

- *Throughput*: Papers covering these techniques use throughput to measure the performance of DSR when their techniques are active. We have measured throughput and found that our technique provides higher throughput. However, we believe that comparing our technique with these two techniques would be unfair because TCP provides reliable data transfer as opposed to unreliable CBR traffic.

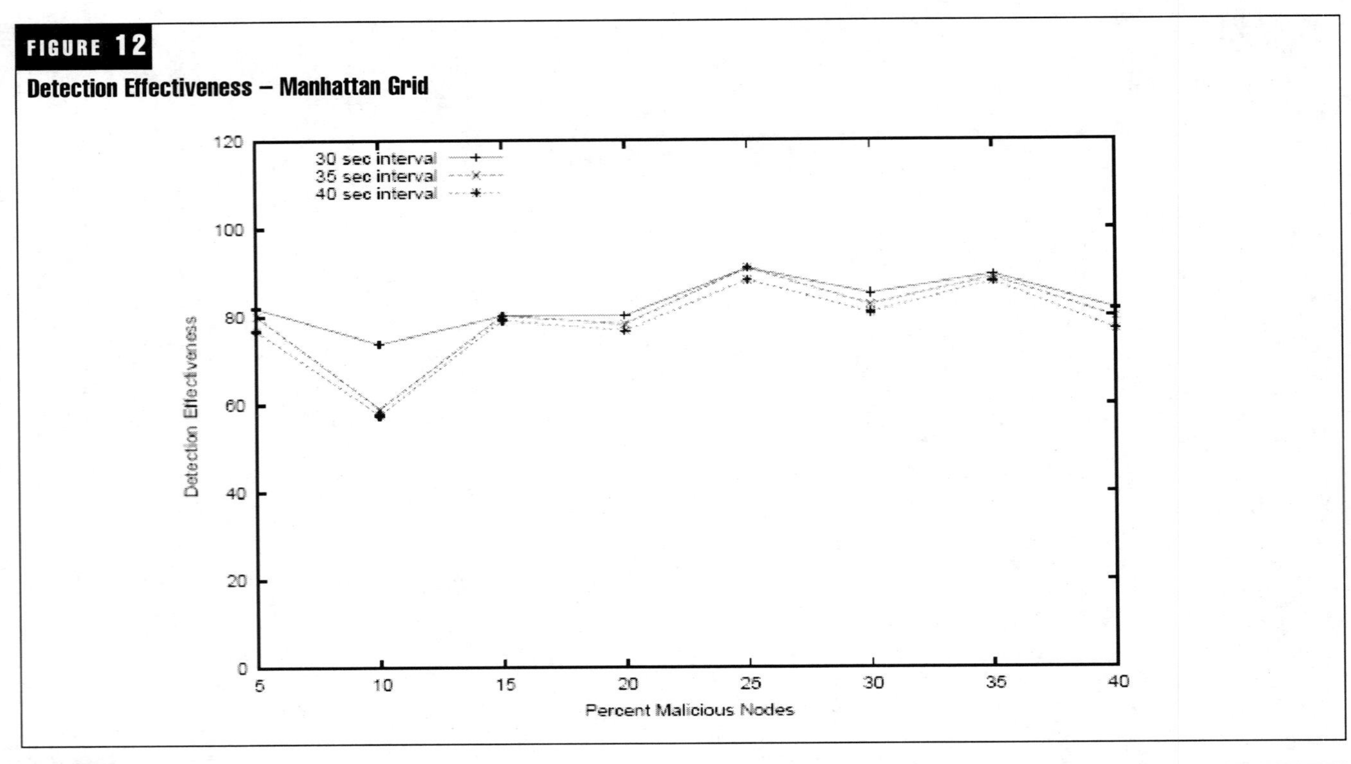

FIGURE 12

Detection Effectiveness – Manhattan Grid

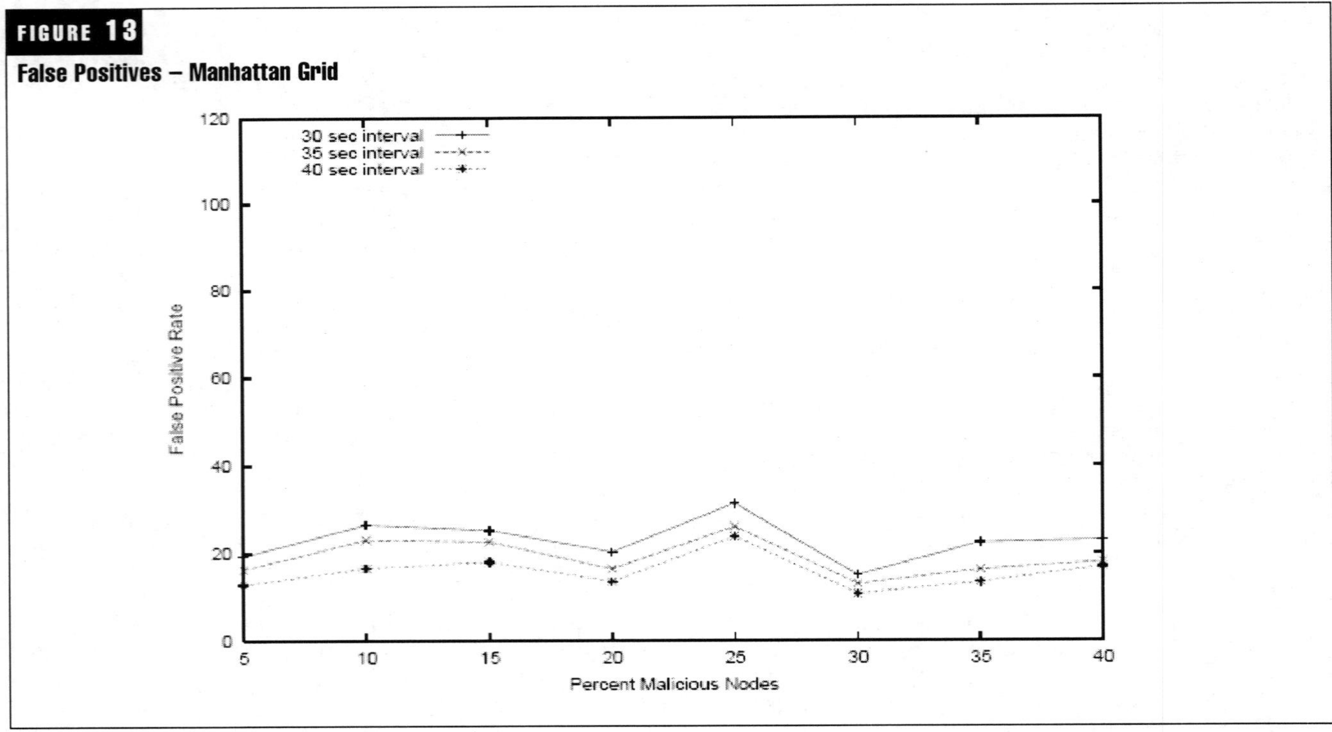

FIGURE 13

False Positives – Manhattan Grid

- *Overhead*: The second measure used in these papers is the computation of overhead. Watchdog [21] computes the routing-related overhead introduced by their technique; this overhead ranges from 11 to 33 percent. The CONFIDANT [6] implementation of nodes bearing grudges computes the ratio of alarm messages to total number of control messages transmitted, which is about 3 percent. Neither technique provides the overhead introduced by introducing security associations between the nodes. In our technique, we rely on readily available information in the network at different layers and do not introduce any additional messages. Our technique does not incur any overhead due to any additional messaging itself. The only overhead incurred is the storage overhead in the nodes, which is about 20 packets.

- *Utility*: Another measure used by nodes bearing grudges is to measure "utility," which is the effect of cooperation between nodes. In our approach, we do not rely on cooperation from other nodes.

- *Detection effectiveness*: None of the papers on watchdog or nodes bearing grudges measure the detection effectiveness of their mechanisms. One paper on nodes bearing grudges also fails to measure the false positive rate. In [21], the effect of false positives on throughput was studied, and it was reported that watchdog does not improve network throughput.

Conclusions and Future Research

Securing mobile ad hoc networks is an emerging research area with immediate practical applications. In this paper, we present a solution to identify malicious or faulty behavior such as dropped or misrouted packets. A key feature of our technique is that it can be deployed on a single node and can work without the cooperation of other nodes. Moreover, it works with existing protocols and conserves energy because it uses readily available information from various network levels to detect malicious behavior.

We implemented our technique using an ns-2 network simulator and conducted extensive experiments using mobility models such as random waypoint, Gauss-Markov, group reference point, and Manhattan grid models. Results show that this technique has high detection effectiveness in detecting malicious behavior with a low false positive rate. The detection effectiveness in the Gauss-Markov and group reference point models was higher than that of the random waypoint model because the random waypoint model tends to have more unrealistic movements than the others. The false positives in Gauss-Markov and group reference point models are lower than in the random waypoint model—as was expected—because the number of route error messages dropped is higher in the random waypoint model than in others. The detection effectiveness is higher and the false positives are lower in the Manhattan grid model—which can be considered a special kind of random waypoint model—than in random waypoint model because in Manhattan grid model, the movement is only along the x and y axes. The detection interval, which is the duration in which a source node keeps track of all the control messages (route error messages) received at that node, plays an important role in determining the detection effectiveness and false positive rate of the technique. Through our experiments, we observed that detection intervals in the range 30 to 40 seconds provide good detection effectiveness while maintaining a reasonable rate of false positives for all mobility models. Currently we are evaluating the effectiveness of our technique with additional routing protocols such as AODV and DSDV.

References

[1] B. S. Bakshi, P. Krishna, N. H. Vaidya, and D. K. Pradhan, "Improving Performance of TCP over Wireless Networks," Technical Report 96-014, Texas A&M University, 1989.

[2] C. Barakat, E. Altman, and W. Dabbous, "TCP Performance in Heterogeneous Networks: A Survey," *IEEE Communication*, 38:40–46, 2000.

[3] S. M. Bellovin, M. Leech, and T. Taylor, "ICMP traceback messages," IETF Internet Draft, draftietf-itrace-01.txt, October 2001.

[4] K. A. Bradley, S. Cheung, N. Puketza, B. Mukherjee, and R. A. Olsson, "Detecting Disruptive Routers: A Distributed Network Monitoring Approach," In *IEEE Symposium on Security and Privacy*, 115–124, 1998.

[5] S. Buchegger and J. Y. Le Boudec, "Nodes Bearing Grudges: Towards Routing Security, Fairness, and Robustness in Mobile Ad Hoc Networks," in *Proceedings of the Parallel, Distributed and Network-Based Processing*, 403–410, January 2002.

[6] S. Buchegger and J. Y. Le Boudec, "Performance Analysis of the CONFIDANT Protocol (Cooperation of Nodes: Fairness in Dynamic Ad-Hoc Networks)," in *ACM MOBIHOC*, 226–236, 2002.

[7] T. Camp, J. Boleng, and V. Davies, "A Survey of Mobility Models for Ad Hoc Network Research," *Wireless Communication and Mobile Computing*, 483–502, 2002.

[8] J. Ding and S. R. Medidi, "Distinguishing Congestion from Malicious Behavior in Mobile Ad-Hoc Networks," in *Proceedings of SPIE AeroSense Conference on Digital Wireless Communications*, volume 5400, 193–103, April 2004.

[9] J. Elliot, "Distributed Denial of Service Attack and the Zombie Ant Effect," *IT Professional*, 55–57, March/April 2000.

[10] P. Ferguson and D. Senie, "RFC 2267: Network Ingress Filtering: Defeating Denial of Service Attacks which Employ IP Source Address Spoofing," Internet standard, January 1998.

[11] L. Garber, "Denial-of-Service Attacks Rip the Internet," *Computer*, 12–17, April 2000.

[12] M. Gerharz and C. de Waal, "Bonn Motion," University of Bonn Report, 2003.

[13] R. L. Griswold, "Malicious Node Detection in Wireless Ad Hoc Networks," Master's thesis, Washington State University, School of Electrical Engineering and Computer Science, May 2003.

[14] R. L. Griswold and S. R. Medidi, "Malicious Node Detection in Ad-Hoc Wireless Networks," in *Proceedings SPIE AeroSense Conference on Digital Wireless Communications*, volume 5100, 40–49, April 2003.

[15] X. Hong, Taek J. Kwon, M. Gerla, D. L. Gu, and G. Pei, "A Mobility Framework for Ad Hoc Wireless Networks," *Lecture Notes in Computer Science*, 1987:185–196, 2001.

[16] J. R. Hughes, T. Aura, and M. Bishop, "Using Conservation of Flow as a Security Mechanism in Network Protocols," in *IEEE Symposium on Security and Privacy*, 132–141, 2000.

[17] Van Jacobson, "Traceroute," computer software. Available from ftp://ftp.ee.lbl.gov/traceroute.tar.gz.

[18] R. Jain and K. K. Ramakrishnan, "Congestion Avoidance in Computer Networks with a Connectionless Network Layer," Concepts, Goals and Methodology, in *Proceedings of the IEEE Computer Networking Symposium*, 134–143, 1988.

[19] D. B. Johnson and D. A. Maltz, "Dynamic source routing in ad hoc wireless networks," in *Mobile Computing*, volume 353, Kluwer Academic Publishers, 1996.

[20] C. Ko, M. Ruschitzka, and K. Levitt, "Execution Monitoring of Security-Critical Programs in Distributed Systems: A Specification-Based Approach," in *IEEE Symposium on Security and Privacy*, 175–187, 1997.

[21] S. Marti, T. J. Guili, K. Lai, and M. Baker, "Mitigating Routing Misbehavior in Mobile Ad Hoc Networks," in *Proceedings of ACM SIGCOMM*, 255–265, 2001.

[22] NightAxis and Rain Forest Puppy, "Purgatory 101: Learning to Cope with the SYNs of the Internet," www.packetstrom.securify.com/papers/contest/RFP.doc, 2000.

[23] K. Park and H. Lee, "On the Effectiveness of Route-Based Packet Filtering for Distributed DoS Prevention in Power-Law Internets," in *Proceedings of ACM SIGCOMM*, 15–26, 2001.

[24] Charles E. Perkins, Elizabeth M. Belding-Royer, and Samir R. Das, "Ad Hoc on-Demand Distance Vector (AODV) Routing," Internet draft, February 2003. draft-ietf-manet aodv-13.txt.

[25] J. B. Postel, "RFC 768: User Datagram Protocol," Internet standard, August 1980.

[26] R. Rao and G. Kesidis, "Detecting Malicious Packet Dropping Using Statistically Regular Traffic Patterns in Multihop Wireless Networks That Are Not Bandwidth Limited," in *Proceedings of IEEE GLOBECOM*, 2957–2961, 2003.

[27] Y. Rebahi and D. Sisalem, "Malicious Packet Dropping with Bottleneck Consideration in Ad Hoc Networks," in *Proceedings of International Workshop in Wireless Security Technologies*, 77–81, 2005.

[28] G. Sager, "Security Fun with Ocxmon and Cflow," presentation at the Internet 2 Working Group, November 1998.

[29] S. Savage, D. Wetherall, A. Karlin, and T. Anderson, "Practical Network Support for IP Traceback," in *Proceedings of ACM SIGCOMM*, 295–306, August 2000.

[30] A. C. Snoeren, C. Partridge, L.A. Sanchez, C. E. Jones, F. Tachakountio, B. Schwartz, S. T. Kent, and T. Strayer, "Single-Packet IP Traceback," *IEEE/ACM Transactions on Networking*, 10(6):721– 734, 2002.

[31] D. X. Song and A. Perrig, "Advanced and Authenticated Marking Schemes for IP Traceback," in *Proceedings of IEEE INFOCOM*, 878–886, 2001.

[32] F. Stejano and R. Anderson, "The Resurrecting Duckling: Security Issues for Ad-Hoc Wireless Networks," in *Proceedings of the International Workshop on Security Protocols*, 172–194, April 1999.

[33] T. Verwoerd and R. Hunt, "Intrusion Detection Techniques and Approaches," *Computer Communications*, 25(15):1356–1365, September 2002.

[34] X. Zhang, S. Wu, Z. Fu, and T. L. Wu, "Malicious Packet Dropping: How It Might Impact the TCP Performance and How We Can Detect It," in *Proceedings International Conference on Network Protocols*, 263–272, 2000.

[35] Y. Zhang and W. Lee, "Intrusion Detection in Wireless Ad-Hoc Networks," in *Mobile Computing and Networking*, 275–283, August 2000.

Performance of Transport Protocols in Wireless Networks

Sirisha Medidi
Assistant Professor, School of Electrical Engineering and Computer Science
Washington State University

Muralidhar Medidi
Associate Professor, School of Electrical Engineering and Computer Science
Washington State University

Jin Ding
Staff Scientist
Broadcom Corporation

Abstract

The increasing popularity of wireless networks indicates that wireless links will play an important role in future internetworks. Transport protocol as an essential part of the whole protocol hierarchy provides end-to-end communication between two or more communication hosts. In theory, transport protocol should be independent of the technology of the underlying network layer. However, in practice, transmission control protocol (TCP), as a standard transport protocol for Internet, is tuned for wired networks and performs poorly in wireless networks. This paper provides a survey of research on transport protocols for wireless networks. It first summarizes the proposals that enhance TCP performance in wireless networks and then provides a classification of transport protocols in ad hoc networks.

Introduction

The increasing popularity of wireless networks indicates that wireless links will play an important role in future internetworks [1]. The transport layer, as an essential part of the whole protocol hierarchy, provides reliable, cost-effective data transport from the source machine to the destination machine [2]. In theory, transport protocols should be independent of the technology of the underlying protocols. However, in practice, TCP, as one of the main standard protocols in the transport layer, is tuned for wired networks, with wired links and stationary hosts. Typically, congestion control mechanism is triggered upon a packet loss, since TCP assumes all the packet drops are due to congestion, and lower layers can handle all other errors in the network. This mechanism works well in wired networks due to the low packet error rate but performs poorly when used over wireless links. Several researchers have developed new transport protocols to enhance TCP performance over wireless communication networks.

This paper reviews the research that has been proposed on transport layer protocols in wireless ad hoc and sensor networks. Section 3 reviews the techniques that were proposed to improve TCP performance in wireless networks. Then the research in enhancing the TCP protocol design for ad hoc networks is reviewed in section 4, and a classification is provided. A summary of the paper is provided in section 5.
TCP in Wireless Networks

Jacobson initially studied TCP behavior over wired networks, where congestion is a regular cause for packet loss [3]. When wireless networks first came into existence, it usually meant the cellular networks or satellite networks, which we refer to as heterogeneous wired-wireless networks. In these networks, there is a base station in between the wireless network and wired network. Base station controls the communication within the cell and acts as an interface between the mobile node and Internet, as shown in *Figure 1*. Research results [1, 2, 4] have shown that TCP throughput (i.e., sending rate) is degraded in the presence of random and burst errors along with long propagation delays. To alleviate this problem, researchers focused on the development of architectures (e.g., wireless proxies) that assist the protocol's operation over wireless networks in order to introduce minor changes to the protocol itself. In these proposals, most commonly used approach for improving TCP performance between mobile nodes and fixed nodes in the Internet is proxy-based solution [11]. In this mechanism, there is a proxy implemented at the base station at the wired-wireless boundary to hide the effect of wireless errors from a wired network, avoid time-outs or fast retransmission at the sender, and avoid the exponential backoff of the time-out value.

The proxy-based solution usually buffers data segments at the proxy and retransmits them over the local wireless link if they get lost due to transmission error. Duplicated

FIGURE 1

Wireless Cellular and Satellite Networks

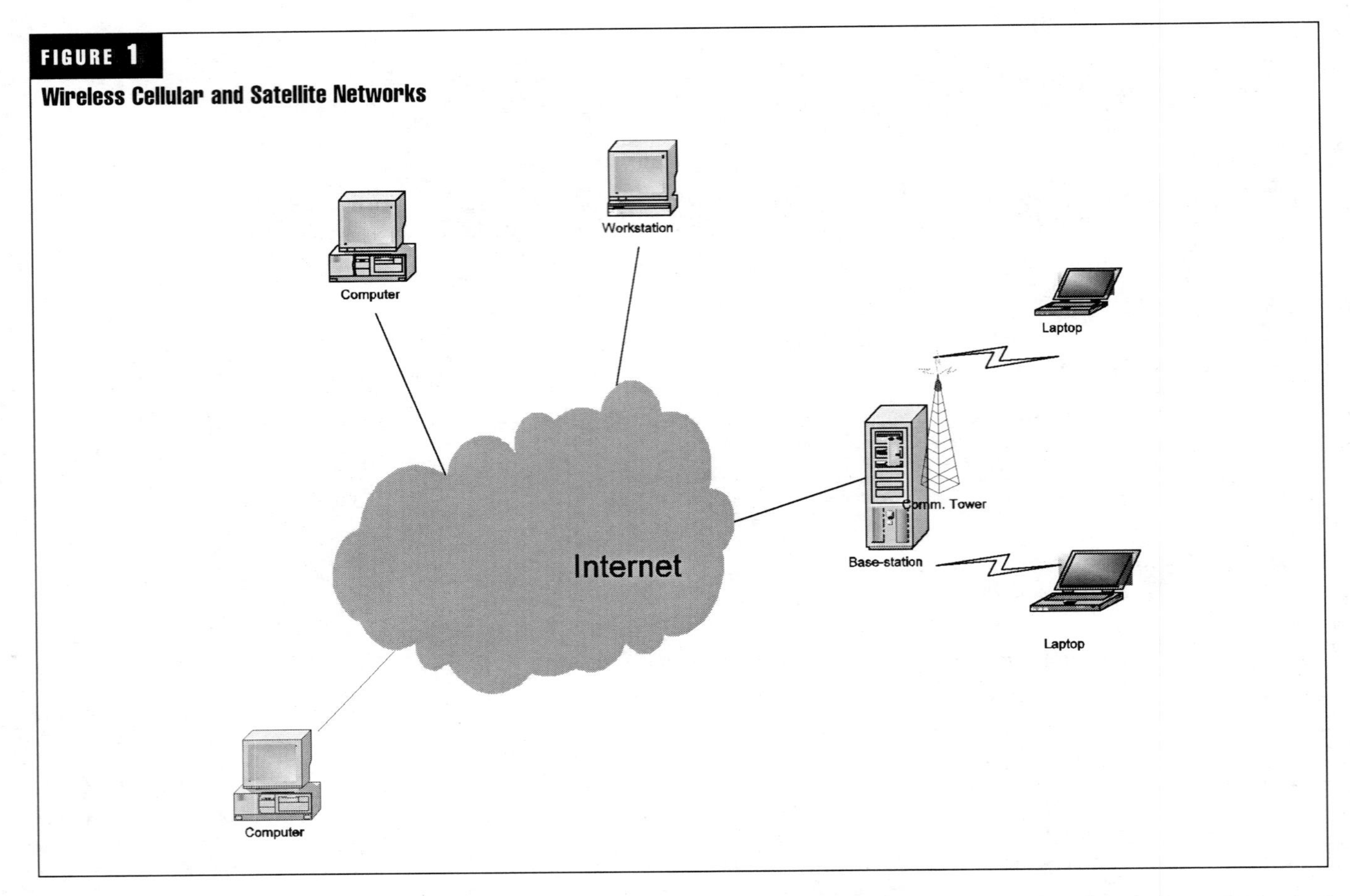

acknowledgments (ACKs) resulting from wireless losses are dropped to prevent triggering the fast retransmission, therefore avoiding window shrinkage at the TCP sender. For some explicit approaches, explicit congestion notifications (ECNs) are sent to the sender to shield the effect of wireless losses on the retransmission time-out maintained at the sender. Then the sender reacts to the congestion quickly and allows the proxy a chance to locally repair wireless losses.

The indirect TCP (ITCP) [5] splits the TCP connection into two connections. The first connection goes from the sender to the base station. The second one goes from the base station to the receiver. Hence, the base station maintains two TCP connections—one over the fixed network and another over the wireless link. The base station copies packets between the connections in both directions to hide the wireless link errors from the wired network. ITCP does not maintain end-to-end TCP semantics. It relies on the application layer to ensure reliability. That means that if the application has the ability to provide reliability, it uses ITCP; otherwise, it chooses TCP. This implies that the mobile hosts in the network must be aware of an application's ability to provide reliability for choosing a proper protocol for the transport layer. The base station should be informed of the mobile host's selection. The advantage of ITCP is that both connections are now homogeneous. Parameters can be tuned separately for different connections. The disadvantage of this scheme is that it violates the semantics of TCP. Since each part of the connection is a full TCP connection, the receipt of an ACK does not mean that the receiver got the segment, only that the base station got it.

MTCP [6] is similar to ITCP, except that the last TCP byte of the data is acknowledged to the source only after the mobile host receives it. If the sender does not receive the ACK for the last byte, it will resend all the data, including those that may have already been received by the mobile host. The base station sends zero window adjustment (ZWA) to freeze the source during handoffs so the window and time-out are not affected.

Explicit bad state notification (EBSN) [7] uses local retransmission from the base station to shield wireless link errors and improve the throughput. The sender time-out may be avoided by using an explicit feedback mechanism while the wireless link is in a bad state. In EBSN approach, the base station sends an EBSN to the source for every retransmission of a segment to the mobile host, and the sender will reinitialize the timer upon the received EBSN. The main disadvantage of this approach is that it requires TCP code modification at the source to be able to interpret the EBSN.

Wireless TCP (WTCP) [8] is also similar to ITCP with the exception being that the base station acknowledges a TCP segment to the sender only after that segment is acknowledged by the mobile host preserving the TCP end-to-end semantics. It hides the time spent by a TCP segment in the base station buffer to avoid affecting the round-trip time (RTT) estimates and time-out maintained at the sender. This is achieved by modifying the time-stamp field in the ACK instead of using an explicit feedback message. For the reliable connection from the base station to the mobile host, the base station reduces its window size to one segment in case of a time-out. It assumes that a typical burst loss will follow, rapidly reducing the window size to avoid the wasteful retransmissions and interference with other channels. Upon each ACK, the WTCP sender will assume that an ACK indicates that the wireless link is in a good state and will set its window size to the window size advertised by the receiver (mobile host). For duplicate ACKs, WTCP does not alter the wireless transmission window, assuming that the reception of the duplicate ACK is an indication that the wireless link is in good state, and immediately retransmits the lost segment. The base station can decide the number of duplicated ACKs to send the receiver to cease the retransmission at the sender side, therefore improving the utilization of the wireless channel.

Snoop [9] is similar to WTCP except that it is implemented at the link layer of the base station. The base station sniffs the link interface for any TCP segments destined for the mobile host and buffers them if buffer space is available. The base station does not forward the retransmitted segments that have already been acknowledged by a mobile host. The base station also sniffs the ACKs from the mobile host. It detects the loss by duplicated ACKs. If it is a wireless link error and the segment is buffered, the base station retransmits the lost segment and starts a timer. Although Snoop does not break the semantics of TCP, it makes several small modifications to the network layer code in the base station. One of the changes is the addition of a snooping agent that observes and caches TCP segments going out to the mobile host and ACKs coming back from it. When the snooping agent sees a TCP segment going out but does not see an ACK coming back before its (relatively short) timer goes off, it just retransmits that segment without informing the source that it is doing so. It also retransmits when it sees duplicate ACKs from the mobile. Duplicate ACKs are discarded to avoid having the source misinterpret them as congestion.

Fast retransmission [10] reduces the effect of mobile host handoff by delayed ACKs for controlling the sender's transmission rate at the receiver. During a mobile host handoff from one base station to another, TCP segments can be lost or delayed and the source can time out. Because of the typical coarse granularity of the TCP clock, the time-out period is much longer than handoff time and the mobile host has to unnecessarily wait for a long duration to receive a retransmission of a lost TCP segment from the source. In the fast-retransmit approach, immediately after completing the handoff, the IP in the mobile host triggers TCP to generate a certain number of duplicate ACKs. Since most TCP implementations now have fast retransmit, these duplicate ACKs cause the source to retransmit the lost segment without waiting for the time-out period to expire.

Several proxy-based mechanisms have been proposed, and a comparison of some of these mechanisms is given in [11]. For a good-quality wireless link, the throughput of ITCP is better. When the wireless link quality degrades, WTCP yields better throughput mainly because of its aggressive retransmission policy over the wireless link. WTCP achieves throughput values that are four to eight times higher than TCP–Tahoe. However, the aforementioned policy slightly degrades the utilization of the wireless link. In addition, though Snoop achieves throughput values comparable to ITCP and WTCP at low-loss situations, the throughput of Snoop is poor when the wireless link is very bursty (i.e., in a bad error state) for long durations. The reason is that Snoop triggers retransmission only after the base station receives a duplicate ACK. When the wireless link is in a bad state, ACKs might be lost.

TCP in Ad Hoc Networks

A mobile ad hoc network (MANET) is a self-configuring network made up of exclusively of mobile hosts connected by wireless links to form an arbitrary topology. The network has no access points (APs) and operates in peer-to-peer operating mode, as shown in *Figure 2*. The mobile hosts are free to move randomly and organize themselves arbitrarily;

FIGURE 2

Ad Hoc Network

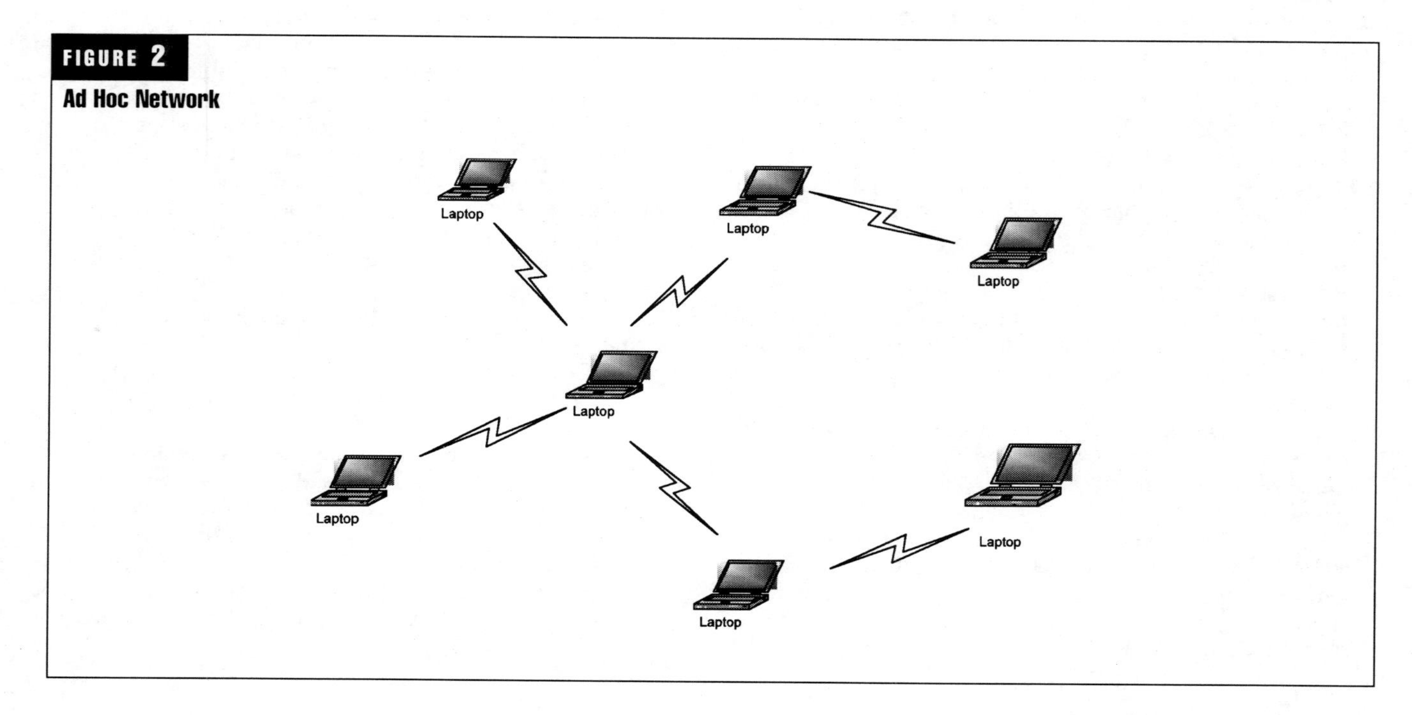

thus the network's wireless topology may change rapidly and unpredictably.

Ad hoc networks are different from cellular and satellite networks. Since they have no infrastructure, there is no base station in the network. In addition, since the ad hoc network is highly dynamic and all the links are wireless, it is not feasible to assign a proxy in the network to take charge of the data communications.

In addition, ad hoc networks could operate in a stand-alone fashion or be connected to the larger Internet. The mobile hosts in ad hoc networks usually have the entire protocol stack just as the fixed hosts in wired networks do to provide interoperability and compatibility with the Internet. Hence, it is desirable to develop new strategies to improve the traditional TCP performance instead of developing new transport protocols.

Several solutions have been proposed to improve TCP performance in ad hoc networks and can be classified into two categories: cross-layered design and layered design. Cross-layered design usually involves more than one open-systems interconnection layer. In these proposals, the lower layer can give the sender explicit information about the nature of the error or attempt to hide the error from the sender. Layered-design detection mechanism does not need information from other layers or intermediate nodes. It tries to impose minimal demands, if any, on any host other than the sender or the receiver. In this approach, intermediate nodes in the network need not declare the reason for packet losses; therefore, the intermediate node in the network need not be modified. Additionally, this approach does not incur overhead at the end host and the network, whereas intermediate node assistance mechanisms incur overhead due to explicit notification.

Cross-Layered Design

The ECN scheme proposed in [12] adds ECN to the IP to trigger TCP congestion control. The intermediate nodes in the network send ECN to the sender to notify the sender that its limited buffer space has been full. Upon receiving the ECN message, the TCP sender triggers the congestion control algorithm to avoid congestion collapse. The ECN message could be lost; not receiving an explicit notification does not mean a detected drop was not caused due to congestion. This is not suitable for ad hoc networks, since it cannot be guaranteed that all the intermediate nodes will be ECN–capable.

TCP–feedback [13] tries to handle the effect of the node mobility in ad hoc networks. It is a feedback-based scheme in which the TCP sender can distinguish between route failures and network congestion by receiving route failure notifications (RFNs) from intermediate nodes. When an RFN is received, TCP–feedback will push the TCP sender into a snooze state where TCP stops sending packets and freezes all its variables such as timers and congestion window size. When the new route to the destination is re-established, the sender will receive a route-re-establishment notification (RRN). The sender can leave the frozen state and resume transmission using the same variable values prior to the interruption. In addition, a route failure timer is used to prevent infinite wait for RRNs. It is triggered whenever an RFN is received, and in case it expires, the frozen timers are reset, allowing the TCP congestion control to be invoked normally. TCP–feedback showed gains over standard TCP in conditions where the route re-establishment delays are high. It also performs better in scenarios with high rates. However, since RGNs and RRNs should be carried by the routing protocol, no such protocol was considered in the evaluation.

An explicit link failure notification (ELFN)–based approach [3] provides meaningful enhancements over standard TCP by using an ELFN message to interact with the routing protocol in order to detect route failure and take appropriate actions when it is detected. When a wireless link error happens, the node detecting the failure will send the ELFN back to the sender. The ELFNs contain sender and receiver addresses and ports, as well as the TCP sequence number. The TCP–ELFN is able to distinguish losses caused by congestion from the route failure. When the TCP sender receives an ELFN, it enters a standby mode, which implies that its timers are disabled and probe packets are sent regularly toward the destination in order to detect the route restoration. Upon receiving an ACK packet, the sender leaves the standby mode and resumes transmission using its previous timer values normally. This scheme was evaluated for the DSR routing protocol where the stale route problem was found to be crucial for the performance of this modified TCP as well. Additionally, the length of the interval between probe packets and the choice of which type of packet to send as a probe were also evaluated. Only the former turned out to be really relevant. It suggests that a varying interval based on RTT values could perform better than the fixed probe interval used in this algorithm.

Ad hoc TCP (ATCP) [4] does not impose changes to the standard TCP itself. It implements an intermediate layer between network and transport layers. In particular, this approach relies on the Internet control message protocol (ICMP) and ECN scheme to detect network partition and congestion, respectively. In this manner, the intermediate layer keeps track of the packets to and from the transport layer so that the TCP congestion control is not invoked when it is not required. When three duplicate ACKs are detected, indicating a high-loss channel, ATCP puts TCP in persist mode and quickly retransmits the lost packet from the TCP buffer; after receiving the next ACK, the normal state is resumed. In case an ICMP destination unreachable message arrives, pointing out a network partition, ATCP also puts the TCP in persist mode, which only ends when the connection is re-established. Finally, when the receipt of an ECN detects network congestion, the ATCP does nothing but forward the packets to TCP so that it can invoke its congestion control normally.

In [14, 15], fuzzy logic was used to assist the TCP error detection mechanism in ad hoc networks. RTT measurement and variance as input data were used to infer the status of the network. It is an end-to-end scheme, since it requires only end-node cooperation. However, because it uses a time to live (TTL) field within the IP header to identify the number of hops from the sender to the receiver to estimate the RTT, it needs help from the network layer. In addition, it does not provide recovery strategy after the error detection.

Split TCP [16] splits long TCP connections into shorter localized segments to improve the throughput and fairness of TCP in wireless ad hoc networks. In split TCP, certain nodes along the route take up the role of being proxies for the TCP connection. The proxies buffer packets upon receipt and administer rate control. When a proxy receives packets, it buffers them and acknowledges their receipt to the source or previous proxy by sending a local ACK (LACK), and then it takes over the responsibility of delivering the packet, at an appropriate rate, to the next local segment. Upon receiving a LACK, a proxy will purge the packet from its buffer. Split TCP does not change the end-to-end ACK system of TCP—that is, the source will not clear a packet from its buffer unless it is acknowledged by a cumulative ACK from the destination. If there is a need for a packet to be retransmitted, the buffering at proxy can enable dropped packets to be recovered from the most recent proxy, thus avoiding having to go all the way back to the source. The main contribution of this work is that, by splitting the TCP connection, it localizes the congestion control, flow control, and error recovery and achieves better parallelism in the network.

Layered Design

Fixed retransmission time-out (RTO) [17] relies on the idea that routing failure recovery should be accomplished in a faster fashion by the routing algorithm. This approach is based on the observation that if disconnections were treated as transitory periods, exponential backoff would be unnecessary. It disables exponential backoff mechanism when two time-outs indicating the route failure occur and make the TCP sender retransmit at regular intervals instead of increasingly exponential ones. In fact, the TCP sender doubles the RTO once and if the missing packet does not arrive before the second RTO expires, the packet is retransmitted again and again, but the RTO is no longer increased. It remains fixed until the route is recovered and the retransmitted packet is acknowledged. The authors evaluated this proposal considering different routing protocols as well as the TCP selective and delayed ACKs options. They report that significant enhancements were achieved using fixed RTO with on-demand algorithms, and only marginal improvements were noticed regarding the TCP options mentioned.

TCP–door [18] focuses on the idea that out-of-order (OOO) packets can happen frequently in an ad hoc network environment as a result of node mobility, and it might be enough to indicate link failure inside the network. TCP–door detects OOO events and responds accordingly. Since data packets as well as ACK packets can experience OOO deliveries, TCP–door implements detection mechanisms at both sender and receiver. This is achieved by adding a 1-byte option for ACKs and a 2-byte option for data packets in TCP options. For every data packet, the sender increments its own stream sequence number inside the 2-byte option regardless of whether it is a retransmission or not; standard TCP does not increment the sequence number of each retransmitted packet. This enables the receiver to detect OOO data packets and notify the sender via a specific bit into an ACK packet. The receiver then increments its own ACK stream sequence number inside the 1-byte option for every retransmitted ACK so that the sender can distinguish the exact order of packets sent, even if they are retransmissions. These mechanisms provide the sender with reliable information about the order of the packet stream in both directions, allowing the TCP sender to act accordingly. The TCP–door sender can respond to OOO events by temporarily disabling congestion control and instant recovery during congestion avoidance. When the congestion control is disabled, the TCP sender keeps its state variables constant for a while (T1) after the OOO detection in order to avoid unnecessary congestion control invocation. In the instant recovery mechanism, when an OOO condition is detected, the TCP sender checks if the congestion control mechanism has been invoked in the recent past (T2). If so, the connection state prior to the congestion control invocation is restored, since such an invocation may have been caused by temporary disruption instead of by congestion itself. Performance results show that TCP–door improved TCP performance significantly—by 50 percent on average.

A simple receiver-based mechanism [19, 20] is implemented at the receiver to distinguish congestion losses from corruption losses. This works in only if the last hop to the receiver is a wireless link and has the smallest bandwidth among all links in the connection path. The receiver attempts to detect the real cause of the packet loss and inform the TCP sender to take the appropriate actions. Specifically, if the loss is a transmission error, the receiver can speed up the recovery and avoid shrinkage of the sender's congestion window.

TCP–probing [21] incorporates a probing mechanism into standard TCP. It uses a probe cycle that consists of a structured exchange of probe segments between the sender and receiver to monitor the network condition. When a packet loss is detected, the sender initiates a probe cycle during which data transmission is suspended and only probe segments (header without payload) are sent. A lost probe or ACK reinitiates the cycle, hence suspending data transmission for the duration of the error. When the probe cycle is completed, the sender compares the measured probe RTTs and determines the level of congestion. The protocol allows for three tactics in response to the nature of the error detected: slow start (for congestion detected by time-out), fast recovery (for moderated congestion detected by three ACKs), and immediate recovery (for a congestion-free path). A fourth tactic (i.e., conservative recovery due to frequent link errors) is not included in the recovery strategy. The reason is that a probing cycle can be extended when a probe or an ACK is missing. Hence, a probing device models two properties—it inspects the network load whenever an error is detected and rules on the cause of that error, and it suspends data transmission for as long as the error persists, thereby forcing the sender to adapt its data transmission rate to the actual conditions of the channel. Probing can be more effective than TCP–Reno and TCP–Tahoe when the sending window is not too small. The immediate recovery mechanism of TCP–probing avoids the slow start and/or the congestion avoidance phase of TCP–Tahoe and TCP–Reno. In this case, probing immediately adjusts the congestion window to the recorded value prior to the initiation of the probe cycle. When congestion is indicated, the protocol complies with the congestion control principles of standard TCP.

TCP–real [22] proposes modifications to enhance the real-time capabilities of the protocol over wired and wireless networks and tackle the problem of asymmetry by decoupling the size of the congestion window from the time-out. It is receiver-oriented and detects errors at the receiver using wave patterns. A wave is a fixed pattern of data exchange between sender and receiver that enables the receiver to measure the perceived level of congestion based on the time required for a wave to be delivered. In this mechanism, congestion window size is included in the TCP header for the receiver to communicate with the sender to direct the sender's congestion control. The receiver measures the number of successfully delivered segments within a wave and the wave delivery time respectively to estimate the level of loss and changes in current conditions. The wave delivery time is the time difference between the reception of the first and the last segment of the wave and it could be much smaller than the RTT. Congestion control in TCP–real has therefore two additional properties: it can avoid unnecessary congestion window adjustments due to path asymmetry, and it can determine the level of loss and jitter with better precision, since the size of the sending window is known to the receiver.

TCP–Westwood (TCPW) [23] is a simple modification of the TCP source that allows the source to estimate the available bandwidth. It uses the bandwidth estimation to recover faster and achieve higher throughput. TCPW exploits the end-to-end estimation of the available bandwidth and uses the estimate to set the slow start threshold and the congestion window. TCPW does not require any assistance from network layer or proxy agents. TCPW source continuously estimates the packet rate of the connection by averaging the rate of returning ACKs. The estimate is then used to compute the permissible congestion window and slow start threshold to be used after a congestion episode is detected, that is, after three duplicate ACKs or after a time-out. Unlike TCP–Reno, which simply halves the congestion window after three duplicate ACKs, TCPW attempts to make a more informed decision. It selects a slow start threshold and a congestion window size that is consistent with the effective connection rate at the time congestion is experienced. The key idea is to use the bandwidth estimate to drive the window instead of using it to compute the backlog. The rationale is that if a connection is achieving a given rate, it can safely use the window corresponding to that rate without causing congestion in the network.

Freeze–TCP [24] is another inactive method. It avoids time-outs at the sender during handoffs, since a time-out shrinks the sending window to a minimum in all TCP versions. To this end, freeze–TCP exploits the ability of the receiver to advertise a window of zero. It assumes all the errors are due to congestion. Error recovery schemes such as the congestion window adjustment, the ACK strategy, and the time-out mechanism and factors such as receiver advertised window and slow start threshold contribute to the efficiency of the recovery process. The idea is that if the protocol is able to distinguish the nature of the error, the recovery strategy can be more aggressive. The recovery strategy is to shrink the congestion window and extend the time-out period due to congestion or freeze the window and time-out upon drops due to non-congestion error.

Receiver-based rate estimation is used in [25] for the proactive detection of incipient congestion in rate-based protocols. It calculates the receiving rate on the receiver side instead of the sender side. The receiver infers the average of the sending-rate in a given interval. If the difference between the sending rate and the receiving rate is over a certain threshold, it is indicative of congestion. By matching the actual and expected receiving rate, congestion can often be detected before packet losses occur. Simulation results show that this modification can help avoid packet losses and stabilize the transmission rate quicker at session start-up. They point out that the same principle can be used to avoid packet losses in other situations as well, including the adaptation to rapidly changing link characteristics in wireless systems.

The self-tuning proportional and derivative (ST–PD) TCP [26] employs a ST–PD control based TCP congestion control scheme to decouple the congestion control and error control functionalities. It is control theory–based and uses a PD controller to change the TCP congestion window size to keep the buffer occupancy of the bottleneck node on the connection path at a desired operating level. The input to the PD controller is the control error, which is the difference between the designed buffer occupancy and the observed buffer occupancy. The output of the PD controller is the TCP congestion window adjustment. Because TCP operates over a variety of networking environments, the bandwidth-delay product of TCP connections can vary significantly. To handle this, control gains of the PD controller are tuned on-line by a fuzzy logic controller to estimate the bandwidth-delay product of the TCP connection. ST–PD TCP is an end-to-end design that uses an ACK from the receiver to probe available bandwidth without causing buffer overflows.

TCP–Santa Cruz [27] replaces the round-trip delay measurements of TCP with estimations of delay along the forward path and uses an operating point for the number of packets in the bottleneck.

Summary

TCP has an abysmal performance in wireless networks, since it assumes that all packet losses are only due to congestion. TCP does not react well to wireless losses and losses due to mobility and invokes congestion control mechanisms in these cases. This slows down the sending rate to reduce the network load to alleviate the perceived congestion, which in reality does not exist. In this paper we reviewed the research proposals that have attempted to improve the TCP performance over wireless networks. Then we have provided a classification for transport protocols in ad hoc networks. The proxy-based solutions split the TCP connection into wired and wireless parts to improve TCP performance. Layered approaches do not require any kind of explicit notification from intermediate nodes and also do not require any additional information from the network layer. Cross-layered designs, on the other hand, involve more than one protocol layer. These proposals try to improve TCP performance in wireless networks in a different way.

References

[1] H. Balakrishnan, V. N. Padmanabhan, S. Seshan, and R. H. Katz, "A comparison of mechanisms for improving TCP performance over wireless links," IEEE/ACM Transactions on Networking, vol. 5, no. 6, pp. 756–769, 1997.

[2] A. S. Tanenbaum, Computer Networks, fourth edition Prentice Hall PTR, 2003.

[3] G. Holland and N. Vaidya, "Analysis of TCP Performance over Mobile Ad Hoc Networks," Ph.D. thesis, August 1999.

[4] J. Liu and S. Singh, "ATCP: TCP for mobile ad hoc networks," IEEE Journal on Selected Areas in Communications, J–SAC, vol. 19, no. 7, pp. 1,300–1,315, 2001.

[5] A. Bakre and B. R. Badrinath, "ITCP: Indirect TCP for Mobile Hosts," in Proc. 15th Int'l Conf. on Distr. Computer Systems, IEEE, pp. 136–143, 1995.

[6] K. Brown and S. Singh, "M–TCP: TCP for mobile cellular networks," pp. 19–43, 1997.

[7] B. S. Bakshi, P. Krishna, N. H. Vaidya, and D. K. Pradhan, "Improving performance of TCP over wireless networks," Tech. Rep. Technical Report 96-014, Texas A&M University, 1996.

[8] K. Ratnam and I. Matta, "WTCP: An efficient mechanism for improving TCP performance over wireless links," in Proceedings of the Third IEEE Symposium on Computer and Communications (ISCC98), June 1998.

[9] H. Balakrishnan, S. Seshan, E. Amir, and R. Katz, "Improving TCP/IP performance over wireless networks," in Proceedings of the 1st ACM Int'l Conf. on Mobile Computing and Networking (Mobicom), November 1995.

[10] R. Caceres and L. Iftode, "Improving the performance of reliable transport protocol in mobile computing environment," IEEE Journal of Selected Areas in Communications, vol. 13, June 1995.

[11] V. Tsaoussidis and I. Matta, "Open issues on TCP for mobile computing," Wireless Communications and Mobile Computing—Special Issue on Reliable Transport Protocols for Mobile Computing, February 2002.

[12] K. Ramakrishnan and S. Floyd, "A proposal to add explicit congestion notification (ECN) to IP, RFC 2481," January 1999.

[13] K. Chandran, S. Raghunathan, S. Venkatesan, and R. Prakash, "A feedback-based scheme for improving TCP performance in ad hoc wireless networks," Personal Communications, IEEE [see also IEEE Wireless Communications], vol. 8, February 2001.

[14] R. Oliveira and T. Braun, "A delay-based approach using fuzzy logic to improve TCP error detection in ad hoc networks," in IEEE Wireless Communications and Networking Conference (WCNC 2004), March 2004.

[15] R. Oliveira and T. Braun, "A fuzzy logic engine to assist TCP error detection in wireless mobile ad hoc networks," in Next Generation Teletraffic and Wired/Wireless Advanced Networking (NEW2AN'04), February 2004.

[16] S. Kopparty, S. V. Krishnamurthy, M. Faloutsos, and S. Tripathi, "Split TCP for mobile ad hoc networks," in proceedings of IEEE GLOBECOM '02, November 2002.

[17] T. D. Dyer and R. V. Boppana, "A comparison of TCP performance over three routing protocols for mobile ad hoc networks," in Proceedings of the 2nd ACM Symposium on Mobile Ad Hoc Networking and Computing, pp. 55–66, October 2001.

[18] F. Wang and Y. Zhang, "Improving TCP performance over mobile ad hoc networks with out-of-order detection and response," in proceedings of MOBIHOC'02, pp. 217–225, June 2002.

[19] S. Biaz and N. H. Vaidya, "Discriminating congestion losses from wireless losses using inter-arrival times at the receiver," IEEE Symposium on Application—Specific Systems and Software Designing and Technology, ASSET'99, 1999.

[20] S. Cen, P. C. Cosman, and G. M. Voelker, "End-to-end differentiation of congestion and wireless losses," IEEE Trans. on Networking, pp. 703–717, 2003.

[21] V. Tsaoussidis, H. Badr, X. Ge, and K. Pentikousis, "Energy/Throughput Tradeoffs of TCP Error Control Strategies," in Proceedings of the 5th IEEE Symposium on Computers and Communications (ISCC), 2000.

[22] C. Zhang and V. Tsaoussidis, "TCP real: Improving real-time capabilities of TCP over heterogeneous networks," in Proc. of the 11th IEEE/ACM NOSSDA1, 2001.

[23] S. Mascolo, C. Casetti, M. Gerla, M. Y. Sanadidi, and R. Wang, "TCP–Westwood: Bandwidth estimation for enhanced transport over wireless links," in Mobile Computing and Networking, pp. 287–297, 2001.

[24] T. Goff, J. Moronski, and D. Phatak, "Freeze–TCP: A true end-to-end enhancement mechanism for mobile environments," in proceedings of the INFOCOM, 2000.

[25] S. Baucke, "Using receiver-based rate matching for congestion detection in rate-based protocols," in Wireless Communication and Networking, 2003, WCNC 2003. 2003 IEEE, vol. 3, pp. 1,784–1,789, 2003.

[26] W. Xu, A. Qureshi, and K. W. Sarkies, "Novel TCP congestion control scheme and its performance evaluation," in IEEE Proc. of Communications, vol. 149, pp. 217–222, August 2002.

[27] C. Parsa and G. Aceves, "Improving TCP congestion control over Internets with heterogeneous transmission media," in IEEE ICNP 99, 1999.

The Emergence of the Wireless VoIP Phone

Chin Yit Ooi
Staff Engineer, Sort Test Technology Department
Intel Microelectronics Malaysia

Just as voice over Internet protocol (VoIP) increased competition and lowered calling rates for wireline local and long-distance services, wireless VoIP brings changes to mobile phone networks and services. The main drivers behind this change are the rise of VoIP on landline networks, new technologies, and the fixed-mobile convergence strategies. From there comes the wireless VoIP phone.

Some call it a wireless VoIP phone. Others call it a voice over wireless local-area network (VoWLAN) phone. Still others call it a wireless fidelity (Wi-Fi) phone. As it provides cost-effectiveness calls through IP networks, enterprises are now considering how to extend those savings by enabling IP voice over WLANs. This leads to further awareness of wireless VoIP phones.

Usage Model

Inside enterprises use VoIP, which is cheaper to use than cellular phones. Today, cellular phone calls made within the same campus make up 30 to 50 percent of total enterprise cellular bills. Implementing wireless VoIP allows people to carry their desk phones with them, eliminating cellular phone calls and related per-minute costs.

In wireless VoIP, a caller's voice is converted into tiny bits of data that stream out from the phone's antenna onto Wi-Fi radio waves, which then feed into a small stationary antenna connected to the company data network. From there, those data bits reverse their steps and travel to another wireless VoIP receiver or funnel out onto the stationary VoIP desk-phone network. So regardless of where a user is, a colleague can reach the user simply by dialing the user's wireless VoIP phone extension. Another advantage that wireless VoIP phones have over cellular is reception—cellular phones do not always work inside a large office, manufacturing buildings, or basement offices, whereas VoIP phones do. The technology is well-suited for large offices, large retailers, hospitals, and for people who must travel from office to office. For those who must always have access to the phone, having a wireless VoIP phone clipped to their belts instead of at their desk would save time.

Also, phones that sense a Wi-Fi hot spot will switch to a Wi-Fi mode, which enables functions such as sending e-mail and downloading files. Large phone makers are working on this technology.

A dual-mode smartphone is a consolidated mobile computer and phone that supports the Global System for Mobile Communications (GSM), general packet radio service (GPRS), and Wi-Fi network access and is packaged in a personal digital assistant (PDA); the growing popularity of smartphones signals user acceptance of that format for receiving voice and data. Not as many users accepted personal computers (PCs) with an accompanying headset as a voice delivery device. As the design of dual-mode smartphones continues to improve and new power-efficient network access technologies and standards-based wireless VoIP–enabled applications are added, the line between phone and mobile computer will be blurred even further.

Wi-Fi has already been integrated onto laptops. Wi-Fi–enabled dual-mode smartphones are the next step, a fact that might be alarming to some wireless service providers. The access-selection feature built into the phone will enable the dual-mode smartphone to determine the best way for the user to communicate. In areas where Wi-Fi connectivity is prevalent, the phone will recognize the better quality of service and cost advantage available through Wi-Fi and connect the user via VoWLAN. And in cases where VoWLAN will not work, such as in a moving vehicle or where Wi-Fi coverage is poor, the phone's access-selection feature will recognize that the best quality of service (QoS) is available via the cellular network and will connect a call accordingly.

Current and Potential Market

According to In-Stat (www.in-stat.com), wireless VoIP is spreading throughout the business market. In a recent survey of more than 300 mid-size businesses and large enterprises, 23 percent of decision-maker respondents said that they had already deployed wireless VoIP in some manner, and another 30 percent said they were planning or evaluating to implement it in the next six to 12 months.

Wireless VoIP presents carriers with a lucrative new opportunity if they market the service smartly. Carriers should look at wireless VoIP as another way to provide seamless access to customers.

The high-tech research firm found the following:

- Survey respondents were most interested in having the ability to make phone calls from a laptop computer, allowing employees to make phone calls from a PDA, and unified messaging, which allows users to access e-mail messages from their voice-mail boxes.

- The number of cellular and WLAN subscribers is expected to exceed 256 million worldwide by 2009. By 2009, the numbers of subscribers using WLAN for voice is expected to exceed those using WLAN for data only.

- Overall, about 60 percent of decision-maker respondents believed that it would be beneficial to have a solution that integrates WWAN with WLAN.

From the Infonetics research, the worldwide Wi-Fi handset revenue will grow to $3.4 billion in 2009, which is equivalent to 36 million units in the potential market with average pricing at less than $100 per handset.

The market is getting off the ground with wireless VoIP solutions, particularly through integrated cellular and Wi-Fi offers. Widespread deployment of these solutions is expected for 2006, but not until 2008 on cellular networks. The success of wireless VoIP will depend mainly on the rates charged for voice calls on third-generation (3G) networks, the availability of unlimited data packages, the regulatory situation, and the availability, price, and format (e.g., Wi-Fi and dual-mode cellular/Wi-Fi) of the handsets.

Regulation and Standards Issues

Wireless VoIP Challenges

As anyone who has used a WLAN knows, typical WLANs offer highly variable data rates and quality of service (QoS). Actual bandwidth and connection quality depends largely on the user's distance from the nearest access point (AP), the number of other users connected to the same AP, the bandwidth demands of each user, and the amount of interference from nearby APs or other sources. Given these variables and the overhead associated with 802.11 packet communication, WLAN throughput can vary from a maximum of about half the specified capacity (5.5 Mbps out of 11 Mbps on an 802.11b network, for example) down to a fraction of that (1 Mbps or less on an 802.11b network). Unfortunately, IP voice devices cannot operate without basic guarantees of low delays, low jitter, and extremely low packet loss. As we have seen, ensuring these capabilities even on wired networks required some significant improvements in QoS mechanisms. Replacing the wires with radio frequency (RF) channels adds a whole other level of complexity.

Elements of a VoIP–Enabled WLAN

To truly make 802.11 wireless VoIP an extension of wired VoIP, corporate WLANs must abandon technology designed to support simple hot-spot access for data traffic in favor of technology used in a true wireless communications network such as the cellular phone system, where the network infrastructure controls all client interactions and base stations collaborate to avoid interference and balance subscriber load. The following are key technical requirements for such a WLAN:

- The network must be able to predictably prioritize and deliver packets regardless of the number of clients connected to its APs and how much other traffic is congesting the network.

- The APs must be designed to work as a coordinated system and take the existence of co-channel interference as a core design assumption.

- The architecture must support all existing Wi-Fi client devices.

The current cellular telephone networks present a time-tested model for how enterprise WLANs should work—they reliably control every transmission on the network all the way to the clients (cellular phones). In addition, cellular base stations are coordinated with each other to allow them to govern voice QoS without interfering with each other.

The difference between this approach and traditional 802.11 implementations is simple: infrastructure-controlled communication versus client-controlled communication. In a cellular network, the infrastructure controls the transmissions from origin to client, and base stations collaborate to assure efficient, non-interfering coverage (infrastructure-controlled communication). In traditional 802.11 networks, clients dictate timing of communication, and APs do not coordinate with one another (client-controlled communication). Instead, the clients choose when to connect and which AP to connect with, and APs choose when to respond to each client.

With an infrastructure-controlled approach, the WLAN can decide which AP or client transmits when and can guarantee packet delivery while dynamically reserving bandwidth over the air for VoIP communication.

In the coming years, VoIP technology will become increasingly important to corporations. Cellular and wireless vendors are already starting to collaborate to provide seamless interoperability between cellular and VoIP networks and, one day, wired telephone handsets may be relics of the past. Go-anywhere wireless VoIP and true converged communications require a new WLAN architecture.

By moving toward an infrastructure-controlled WLAN architecture, we can deliver the scalability and QoS that will allow enterprises to reap the benefits of wireless data applications as well as the additive benefits of wireless VoIP. Rather than continually patching a fundamentally flawed enterprise WLAN system built on collections of stand-alone APs driven by off-the-shelf technology, forward-looking network architects will begin thinking about a new WLAN that manages wireless clients as well as the cellular network manages callers.

VoWLAN in the enterprise market could arguably require the use of 802.11a, not 802.11b or 802.11g, because with 802.11a the range is shorter, which results in less interference and means that it is possible to deploy APs more densely and support a larger concentration of VoIP users. However, it could be a significant problem if companies are required to install 802.11a networks to support VoWLAN, as most have deployed the more prevalent 802.11g/b standards, which cannot provide the level of service needed by VoIP. With only three channels that do not overlap, 802.11g/b is able to support only six to eight calls at a time. In comparison, 802.11a boasts 21 non-overlapping channels able to handle 25 VoIP calls from a single WLAN access point. For converged networks, voice should be deployed over 802.11a while data is deployed over 802.11g. Security is essential for VoWLAN, and wireless security is finally improving with the newly ratified 802.11i standard, but this has not been fully adopted into the VoWLAN equipment yet. 802.11r is a brand-new Institute of Electrical and Electronics Engineers (IEEE) task force created specifically to address VoWLAN security issues.

Business Models That Companies Can Adopt

Two business models for wireless VoIP could emerge. The first, developing initially with businesses with independent Wi-Fi access networks as their driving force, and later mobile operators, will resemble the Internet model. A key reason for the significant growth projection for Wi-Fi phones is the widespread carrier transition to packet infrastructure for many communication networks; carriers are looking for operational expenditures (OPEX) savings, network consolidation, and new services via packet networks, so the way they connect to customers will change. VoWLAN is likely to be a key application in driving widespread adoption of WLANs throughout the enterprise, and will certainly have a marked positive impact on adoption of Wi-Fi phones through our forecast period. As voice applications make WLANs more desirable and mobility makes VoIP more valuable, it is natural that the two will converge to form a powerful mobile voice solution for the enterprise.

For the past two years, as businesses of all sizes move to WLAN and VoIP; vendors of WLAN and VoIP equipment have been educating the corporate market on the virtues, benefits, and paybacks of VoWLAN so companies are ready to embrace what no longer looks like newfangled technology. Vendors such as Vocera and SpectraLink launched Wi-Fi–enabled IP voice devices in 2002 (with Symbol Technologies having an 802.11 voice handset available back in 2000), and other enterprise-orientated vendors such as Cisco and Intermec have now joined them, launching wireless VoIP handsets in 2004.

Major companies' handsets with VoWLAN as their only voice-access capability include the following:

- Alcatel: Mobile IP Touch 300
- Alcatel: Mobile IP Touch 600
- Cisco: 7920
- Hitachi: Mobile Communicator NPD-10JWL*
- Intermec: 760 Mobile Computer*
- Motorola: VP1000
- SpectraLink: NetLink i640
- SpectraLink: NetLink e340
- Symbol Technologies: NetVision
- Viper Networks: WiFi vPhone
- Vocera Communications: Badge
- ZyXEL: Prestige 2000W

The second business model for wireless VoIP is likely to involve wireless VoIP integrated by mobile manufacturers and incorporated into their product plans. Competition is increasing in the Wi-Fi phones market as WLAN technology and services continue to gain acceptance; 2005 saw WLANs emerge from being a niche technology to being a part of the IT mainstream, complementing other networking and wireless technologies such as GPRS and 3G. The fact that major cellular vendors such as Nokia and Motorola sought to incorporate Wi-Fi into their infrastructure products shows that they are taking a more active stance in this space and looking to serve operators building out hot spots and/or develop dual-mode Wi-Fi/cellular handset solutions.

The extension of VoWLAN solutions to cellular services, blurring the lines between GSM/code division multiplex access (CDMA) and 802.11 networks, was heralded by the partnership between Avaya, Proxim, and Motorola announced in 2003. Motorola developed a dual-mode CDMA/GSM and 802.11 phone and mobility manager that allows handoffs between Wi-Fi and mobile operator networks in mid-conversation. Proxim developed Wi-Fi access points that support 802.11e (QoS) and 802.11i (Wi-Fi security protocol, which incorporates temporal key integrity protocol [TKIP] and advanced encryption standard [AES]), allowing IP telephony calls to flow over public and private Wi-Fi networks. In addition, Avaya extended its MultiVantage IP telephony features over Wi-Fi networks, adding SIP to endpoints, and enabling presence-based messaging. These products are expected to be available in 1H05 and as a combined solution could immediately begin to have a marked impact on the enterprise market for VoWLAN and Wi-Fi handsets.

Major companies' dual-mode Wi-Fi/cellular handsets include the following:

- Fujitsu-Siemens: PocketLOOX 610 BT/WLAN*
- HP: iPAQ h6315*
- HP: iPAQ h6340*
- i-mate: PDA2K GSM/GPRS Pocket PC*
- Motorola: MPx
- Motorola: CN620
- NEC: N900iL
- Nokia: 9500 Communicator*
- Toshiba: e750*
- UTStarcom: F-1000

VoIP Announcements in 2005

1. VoIP provider Vonage announced that it formed a partnership with vendor UTStarcom to produce a Wi-Fi phone capable of connecting to Vonage's VoIP network from remote locations. Vonage said it had not determined how it will price the product and that it was considering offering it for free to new customers. Vonage has more than 200,000 customers in the United States so far, and such a partnership could produce a significant rise in Wi-Fi phone shipments.

2. Chip manufacturer SyChip announced a drop-in 802.11g Wi-Fi module for use in a variety of mobile devices, including smartphones, cell phones, PDAs, handheld game consoles, and media players. Nearly all PDAs and smartphones that integrate Wi-Fi do so through the 802.11b standard. The module is optimal for handhelds because of its advanced power-saving functions, security features (802.11i), and

small form. Mobile device vendors can mount the module to a device's circuit board using industry-standard mounting technology for high-volume production. While 802.11g does not necessarily enhance the performance of VoWLAN, the faster data speeds offered by 802.11g are likely to make wireless devices, including Wi-Fi VoIP handsets, more popular and could also enable cellular-like applications over IP such as photo attachments to text messages.

3. Texas Instruments announced the integration of VoWLAN with a processing architecture it uses in many cellular phones: The TNETV1600 platform provides the cost savings, flexibility, and power management to allow developers to manufacture a WLAN IP phone with talk and standby time comparable to cell phones. The new chipset couples its WLAN silicon with the company's Telogy Software for VoIP. TI's WLAN chip will support 802.11g rather than just 802.11b as in earlier versions.

Future of Wireless VoIP Phones

While the year-on-year massive triple-digit annual unit and revenue growth through 2009 looks spectacular, it must be noted that this growth is from tiny base figures; nonetheless, actual units shipped become significant in global terms from 2007 onward as annual shipments pass the 2 million mark.

The forecast of 30 million units by 2009 represents a fairly conservative estimate of 5 percent of the projected total of 618 million cellular phones shipped in 2009 being Wi-Fi–enabled. Wi-Fi will gradually become a regular feature of cellular phones in the same way that it is becoming a standard embedded feature already in most new laptops and PDAs. However, it should be noted that it will take longer for Wi-Fi to become a standard cellular phone feature as QoS for voice over Wi-Fi (VoWi-Fi) is yet to be resolved and the inclusion of Wi-Fi on a cellular phone requires an additional chip, which drains the battery of the device. As this category becomes an increasingly significant proportion of the massive global cellular phone market, it is likely to continue to dramatically grow beyond our forecast period. Mobile operators will subsidize dual-mode Wi-Fi/cellular handsets for subscribers, as they do with all other types of cellular phones, making them far more affordable for consumers.

Advances in wireless VoIP use may still be a few years away, but for now, companies that develop and sell wireless VoIP equipment and services will need to look beyond the traditional "cost efficiencies" selling point. Wireless VoIP has other value propositions, including better QoS and greater mobility. Focusing on these additional values will help increase consumer interest in wireless VoIP.

Acknowledgments

1. Sean Kent, *VoIP in the Palm of Your Hand*, 2005.
2. Martha McKay, *George Gonzales Explains New Wireless VoIP Phone*, March 7, 2004.
3. Joel Vincent, *Maximizing Cost Savings with Wireless VoIP*, October 14, 2004.
4. Richard Webb, *Infonetics: Wi-Fi Phones Annual Worldwide Market Size and Forecasts*, July 2005.
5. Market research from In-Stat, *Wireless VoIP Gaining Traction in Business Market*, February 8, 2005.

Effect of Interference and Control Error on Cellular Mobile Communication Networks

Shamimul Qamar
Assistant Professor, Department of Electronics and Communication Engineering
College of Engineering (COE), Roorkee

Dr. S. C. Sharma
Associate Professor, Department of Paper Technology
Indian Institute of Technology, Roorkee

Dr. Mohan Lal
Assistant Professor, Institute Computer Centre
Indian Institute of Technology, Roorkee

Abstract

Multiple access schemes are used to share resources for establishing calls in mobile communication system. In this paper, we review the various multiple access schemes such as FDMA, TDMA, and CDMA; compute the radio capacity of CDMA as the number of users increases, and discuss the effect of other user interference, including power control error, on the radio capacity of CDMA cellular communication systems. It is observed that as interference increases, the number of users decreases. The effect of power control error on the system capacity is also discussed.

Introduction

Multiple access is a technique whereby many users or local stations can use communication channels at the same time, despite the fact that their individual transmissions may originate from widely different locations. It is desirable that in a multiple access system, the sharing of resources of the channel be accomplished without causing serious interference between users of the system. There are five multiple access schemes. In frequency division multiple access (FDMA), disjoint sub-bands of frequencies are allocated to the different users on a continual basis. In time division multiple access (TDMA), each user is allocated the full spectral occupancy of the channel, but only for a short duration of time called a time slot. Polarization division multiple access (PDMA) serves the calls with different polarization. PDMA is not applied to mobile radio. In space division multiple access (SDMA), resource allocation is achieved by exploiting the spatial separation of the individual users. In code division multiple access (CDMA), CDMA serves the calls with different code sequences.

A traditional narrowband system based on FDMA or TDMA is a dimension-limited system. The number of dimensions is determined by the number of non-overlapping frequencies for FDMA or by the number of time slots for TDMA. In a TDMA system, once all the slots are assigned, no additional users can be added. Thus, it is not possible to increase the number of users beyond the dimension limit without causing an intolerable amount of interference to a mobile station's reaction at the cell-site receiver. PDMA and SDMA are not used in cellular mobile communication networks.

Spread spectrum (wideband) systems can tolerate some interference. Each mobile station introduces a unique level of interference. So the introduction of each additional active radio increases the overall level of interference to the cell-site receivers receiving CDMA signals from mobile station transmitters. The number of CDMA channels in the network depends on the level of total interference that can be tolerated in the system. Thus, the CDMA system is limited by interference, and the quality of system design plays an important role in its overall capacity. The paper is organized as follows: The next section describes the key elements in designing the cellular communication system. The following section describes the capacity of multiple access schemes.

Key Elements in Designing Cellular Communications Systems

The frequency reuse concept is used in designing cellular communication systems [8].

Co-Channel Interference Reduction Factor (CIRF)

The minimum separation between two co-channel cells, D_S, is based on a co-channel interference reduction factor. The co-channel interference reduction factor (q) is expressed as q = D_S/R, where R is the cell radius. The value of q is different for each system. For analog cellular systems, q = 4.6 is based on the channel bend width B_C = 30 kHz and the carrier-to-interference ratio equal to 18 dB. This factor is used in frequency management.

Handoff

The handoff is a unique feature of cellular communication systems. When a mobile unit moves from one cell area to another cell area, the handoff of the ongoing call takes place between one base station and another base station. It switches a call to a new base station without either interrupting the call or altering the user. Unnecessary handoffs should be limited. The main feature of the CDMA access technique is the soft handoff. The soft handoff uses a "make before break" strategy. Only necessary handoffs should be made. The soft handoff occurs in CDMA communication systems.

Reverse-Link Power Control

The reverse-link power control is for reducing near-end–to–far-end interference. The interference occurs when a mobile unit close to the cell site can mask the received signal at the cell site so that the signal from a far-end mobile unit is unable to be received by the cell site at the same time. The CDMA system is an interference-limited system in which link performance depends on the ability of the receiver to detect a signal in the presence of interference. For a CDMA link to perform satisfactorily, we must specify a frame error rate (FER). The key issue in CDMA network design is the minimization of multiple-access interference. The power control is used to reduce multiple-access interference.

Forward-Link Power Control

In the forward direction (BS to MS), a pilot signal is used by the mobile demodulator to provide a coherent reference. The forward-link power control is used to reduce the necessary interference outside its own cell boundary.

Capacity Enhancement

The capacity of cellular systems can be increased by handling q in the following two conditions:

Within Standard Cellular Equipment

The value of q given by:

$$q = \frac{D_s}{R}$$

remains a constant. Reduce the cell radius R, and D_S reduces. For a smaller D_S, the same frequency can be used more often in the same geometrical area, which is why we are trying to use small cells (sometimes called microcells or picocells).

Chosen from Different Cellular Systems

We look for those cellular systems that can provide smaller values of q. when q is smaller, D_S can be less, even if the cell radius remains unchanged. By choosing a smaller q in a new system, we can increase the amount of capacity without reducing the size of the cell based on the old q of an old system. That is why we are choosing a new digital system to replace the old analog system. If we reduce the cell size in a system, the number of cells will increase, and that is always costly. Therefore, the development of a digital cellular system is the right choice.

Voice Activity Factor

One advantage of CDMA is that it can readily exploit the nature of human conversation to increase system capacity. Most of the time, there is a silent period, i.e., a time when the speakers are silent. CDMA can suppress the transmission from user to user when there is no voice present—in other words, transmission is activated only when the voice is present. Most existing digital vocoders can monitor user voice activity, and extensive studies show that typical speech is active only 35 to 40 percent of the time. Exploitation of this silent period introduces less interference to the system, since it reduces the average transmit power during periods of silence and hence increases system capacity. Typical values for the voice activity factor range from 0.35 to 0.4.

Capacity of Multiple Access Schemes

The multiple access schemes are used to share resources for establishing calls. The capacity is defined as the maximum number of channels per cell. The number of channels in a particular all depend on the $\left(\frac{C}{I}\right)$ and. $\left(\frac{E_b}{I_o}\right)$

Carrier-to-Interference Ratio

These multiple access schemes, FDMA, TDMA, and CDMA can be applied in digital system. The carrier-to-interference ratio $\left(\frac{C}{I}\right)$ is related to $\left(\frac{E_b}{I_o}\right)$ at the base band as follows:

$$\frac{C}{I} = \left(\frac{E_b}{I_o}\right)\left(\frac{R_b}{B_c}\right) \quad (1)$$

where is the energy per bit; I_o is the interference power per Hertz, R_b is the number of bits per second, and B_c is the radio channel bandwidth in Hertz. In digital FDMA or TDMA, there are dedicated channels or time slots for calls. Thus, $R_b = B_c$ and $\left(\frac{C}{I}\right) = \left(\frac{E_b}{I_o}\right)$ for FDMA and TDMA, while in CDMA, all the coded sequences share only one radio channel; thus B_c is much greater than R_b. The notion B_c is replaced by B_{SS} which is spread-spectrum channel bandwidth.

The radio capacity in terms of $\left(\frac{C}{I}\right)$ is given by the following:

$$m = \frac{B_t}{B_c\sqrt{\frac{2}{3}\left(\frac{C}{I}\right)}} \quad (2)$$

where B_t is total bandwidth, B_c is channel bandwidth, and $\left(\frac{C}{I}\right)$ is the minimum required carrier-to-interference ration per channel.

C/I for the Forward Link in Uniform Cell Scenario

Let the cells are uniform, i.e., they are similar in size. We are considering forward link, i.e., from base station to the mobile location. The (C/I) ratio is given as follows: by [9]

$$\left(\frac{C}{I}\right)_{Req} = \frac{v_f S_c}{(N-1)v_f S_c + I_{oc}} \quad (3)$$

This equation explains that the number of channels in a particular cell decreases because of the interference of other cell mobile users. The maximum number of channels can be found out by assuming no adjacent cell interference $I_{oC} = 0$ (No adjacent cell interference)

We have

$$N = \frac{1}{(C/I)} + 1 \text{ ------------------(4)}$$

$$\text{If } (C/I) = -17\text{dB } (50), N = 50 + 1 = 51,$$

The maximum voice channel in a cell is 51.

Thus, the capacity of the code division multiple access scheme is limited by the interference of the self-cell or other cell. The capacity in CDMA can be increased by sectorization, voice activity. In a cellular CDMA system a strong signal received at a BS receiver from a nearer MS will mask the weak signal received from another MS. In general the nearer MS is the MS which is closer to the receiver. This interference is known as the near-far effect.

To combat the near-far effect, a power control scheme is applied. With this scheme the power transmitted from the MS is adjusted according to its received power in the BS. In a perfect power control situation, all the signals from the MS reach the BS at equal power. If there are M simultaneous users in a given BS receiver and each of them produce signals with equal power P at a BS then the Signal to Interference (Noise) Ratio can be written as

$$SNR = \frac{UserPower}{InterferencePower} = \frac{P}{(M-1)P} \quad (5)$$

For a system with a data bit rate = R, and bandwidth = W, we may rewrite equation (5)in terms of bit Energy (Eb) to Noise density (No) ratio to obtain,

$$\frac{E_b}{N_0} = \frac{P/R}{(M-1)P/W} = \frac{W/R}{M-1} \quad (6)$$

If the required Eb/No is given then the maximum number of users can be easily calculated. From equation (6) the total number of simultaneous active users in a system for a given Eb/No is,

$$M \leq \frac{W/R}{E_b/N_0} + 1 \quad (7)$$

This is the case of neglecting the interference.

In equation (7), W/R represents a parameter known as the spread spectrum processing gain. For the Qualcomm CDMA system (IS-95) the processing gain is 128. Eb/No represents the operating energy per bit to noise ratio requirement and is related to the BER performance of the system. In our analysis we will use the limiting value of BER to BER =0.01 as in previous sections.In order to find the interference distribution we need to model our system.

The topology of the mobile cellular system to be analyzed is assumed to be a regular array of hexagonal cells each having a BS in the centre. Now assume other cell interference, To reflect the total received power including interference, the maximum number of mobile is given by-

$$M_{\max} \approx G_p \left[\frac{\eta_c}{(E_b/N_o)v_f(1+f)}\right] \quad (8)$$

where f is the interference effect from the neighboring cell. Table no. 1 shows the effect of the power control factor Ë and other cell interference factor effect. Assume processing gain=128 and Ë=0.8

As the interference from the other cell increases the number of maximum user decreases. For the lower value of signal to noise ratio, the number of user get increased in the cellular communication system.

TABLE 1

Effect of Other Cell Intereference

f	E_b/N_o	M_{max}
0.50	7 dB	27
0.55	7	26
0.60	7	25
0.75	7	23

TABLE 2

Effect of Power Control Accuracy factor

η	$E_b/N_{o\ dB}$	M_{max}
0.10	7	3
0.25	7	8
0.255	7	9

Table 2 shows the effect power control accuracy in CDMA cellular communication system.

As the power control accuracy factor increases, the number of users in CDMA cellular communication system increases. This is because the CDMA system needs the perfect power control scheme.

Conclusion

In this paper we have analyzed the reverse link of a CDMA under uniform cell loading. Such a capacity analysis is crucial for the adequate provisioning of network resources at the cell site. This analysis shows that the no. of user decreases as the number of interfering users increases in a cell or nearby cells. It is shown that the capacity of CDMA system depends on the voice activity factor and the other user interference. The effect of imperfect power control has been discussed. It is concluded that the capacity of CDMA system is interference limited.

References

1. W. C. Y. Lee, "Mobile cellular telecommunication system," New York: Mc-Graw Hill, 1989.
2. W. C. Y. Lee, "Mobile communications engineering," New York: Mc-Graw Hill, 1982.
3. J. G. Proakis, "Adaptive equalization for a TDMA digital mobile radio," IEEE Transition Veh. Technology, pp 333–341, 1991.
4. S. N. Crozier, "Short-block equalization techniques employing channel estimation for fading time-dispersive channel," in proc. IEEE Veh. Technology Conf., 1989.
5. W. C. Y. Lee, "Spectrum efficiency in cellular," IEEE Trans Veh. Tech. 1989.
6. K. S. Gilhousen, "On the capacity of cellular CDMA system," IEEE Trans. Veh. Technology 1991.
7. R. L. Pickhotz, L. B. Milstein, and D.L. S. Chiling, "Spread spectrum for mobile communications," IEEE Trans. Veh. Technology 1989.
8. W. C. Y. Lee, "Overview of cellular CDMA," IEEE Transaction on Vehicular technology, Vol. 40, 1991.
9. Kyoung II Kim, "CDMA Cellular Engineering Issues," IEEE Transaction on Vehicular Technology, Vol. 42, 1993.
10. M.A. J. Vieterbi, "Erlang Capacity of Power Controlled CDMA System," IEEE JSAC Vol. 11, 1993.
11. R. L. Pickholtz et al., "Spread spectrum for mobile communication," IEEE VT, Vol. 40, 1991.

Seamless Mobility

A Continuity of Experiences across Domains, Devices, and Networks

Raghu Rau

Senior Vice President of Marketing and Strategy
Motorola

A New Era Emerges

A Revolutionary Convergence

Every decade or so, technologies converge to create a generational leap. Today, we are in the midst of a revolution that will shift the center of gravity in the business of communications. Technologies and disparate information systems are morphing to allow for a seamless experience. Already, the digitization of everything is creating a more natural communications experience. Boundaries separating spatial domains and communication nodes are dissolving. So tomorrow, our car, our mobile phone, our home security system, our office, and all the systems that surround us will automatically communicate with each other to fill our environment with our preferences, our desires, our music collections, and everything we need or want to feel connected anywhere, anytime.

The Next Level of Connectivity: Seamless Mobility

Work and lifestyles are merging in new ways. From the local-area networks (LANs) to hot spots we are all so familiar with, even to emerging wireless access technologies, wireless broadband will soon become readily available to the person walking down the street, driving down the highway, sitting in the park, or even on a boat in the middle of a lake. As a result, we will refuse to settle for anything less than a life that is constantly fun, exciting, productive, and connected. The world of seamless mobility envisions easy, universal, uninterrupted access to information, entertainment, communication, monitoring, and control.

Creating the Seamless Mobility Infrastructure: A Tremendous Opportunity

Achieving the vision of seamless mobility requires technical innovation on many fronts, along with a strong orientation to the user experience. It requires understanding the issues, architectures, and needs of various wireline, wireless, cable, and full-service providers. This paper touches on some of the drivers that are shaping the industry and making seamless mobility a reality. It then proposes the framework and architectures necessary to achieve seamless mobility. Finally, it provides some concluding thoughts on the commitment needed by the organizations in this ecosystem to capture the true promises of seamless mobility.

The Vision of Seamless Mobility

The vision of seamless mobility mostly involves maximum mobility with minimal effort for the user. Seamless mobility is a set of solutions that will give the experience of being connected anywhere, anytime, to anything, with any service.

A number of forces are converging to create this possibility of seamless mobility. Computers have evolved from standalone devices to computer networks to the Internet, where information, content, and services ride upon a network of networked computers. The telephone has evolved from a fixed voice-centric service through mobile telephony to the mobile access of information and simple services on mobile Internet. The next natural step is to continue the trends toward service and content in the computer world and voice to data in the mobile world to one where the experience is delivered in a consistent and intuitive manner regardless of the device, service, network, or location from which it is delivered.

Instead of experiencing a disconnect as we move between devices, environments, and networks, seamless mobility will deliver fluid, end-to-end experiences that span the home, vehicle, office, and beyond. For example, a video teleconference will move transparently from a wireless LAN (WLAN) at the office to the traditional cellular environment for the drive home and to the fixed network when you arrive. A premium entertainment experience will start out on a digital television at home, shift effortlessly to a digital infotainment system in the car and conclude on the device formerly known as a cell phone.

Seamless mobility will call forth a cycle of innovation that may well exceed that seen in the information technology industry in the 1990s. Content developers will reach broader audiences, inspiring new forms and styles of content. Application developers will use the new mobility to create an explosion of voice, imaging, and other non-voice applications that will transform existing applications and spur new ones. In addition, both cable industry and cellular operators will find fresh ways to deepen customer loyalty, reduce churn, and realize new sources of revenue.

Framework for Seamless Mobility

The vision of seamless mobility focuses on continuity of experience across the traditional seams that can occur when devices, locations, networks, services, or domains change. It envisions interoperable devices and networks, with intelligence to manage the boundaries or seams and give the user a consistent experience.

The architecture incorporates a continuum of wide-area networks (WANs), including existing networks such as code division multiple access (CDMA), global system for mobile communications (GSM), and third generation (3G), as well as emerging 802.16 broadband wireless and fourth-generation (4G) networks. It also encompasses shorter-range networks such as 802.11 and ultra wideband (UWB) wireless, which may be deployed in homes, vehicles, enterprises, and hot spots. All are connected to a common IP core network either directly or through communication gateways (see *Figure 1*).

Within each home, vehicle, or other space is a local wired or wireless network and a local communications gateway that supports mobility and assists with seamless transitions between spaces. Devices running client software connect via gateways and/or directly to networks that find their way through other networks to a converged core, and ultimately to common user services. Intelligence is thus distributed across all architectural elements: devices, networks, servers, and gateways. Following best practices, the software foundation includes application services, a services-oriented software architecture, and network-aware middleware. Together, these elements support the rapid development of network-agnostic applications and mediate between those applications and the heterogeneous networks upon which the applications ride.

Seamless Mobility Functional Architecture

A primary technical challenge for seamless mobility is preserving key user-centric capabilities across the seams or boundaries among networks, services, devices, and environments to deliver information, data, and multimedia content. To accomplish this in complex, real-world environments, many types of technologies must collaborate and interoperate so devices can access multiple networks and network controllers can pass session control securely between disparate networks and redirect streams from one device to another.

Figure 2 depicts a functional architecture to support seamless mobility. This architecture addresses both application and network requirements. An experienced architecture guides the provisioning of capabilities for seamless mobility at the application level. Areas of focus include the following:

- *Intelligent interaction*: An intuitive, personal user experience that crosses devices, environments, and tasks and enables users to efficiently achieve their goals
- *Content handling*: Easy access to content anywhere, with the ability to share and communicate with protection and privacy
- *Real-time communications*: Enhanced experiences and sharing through the merger of content and communications

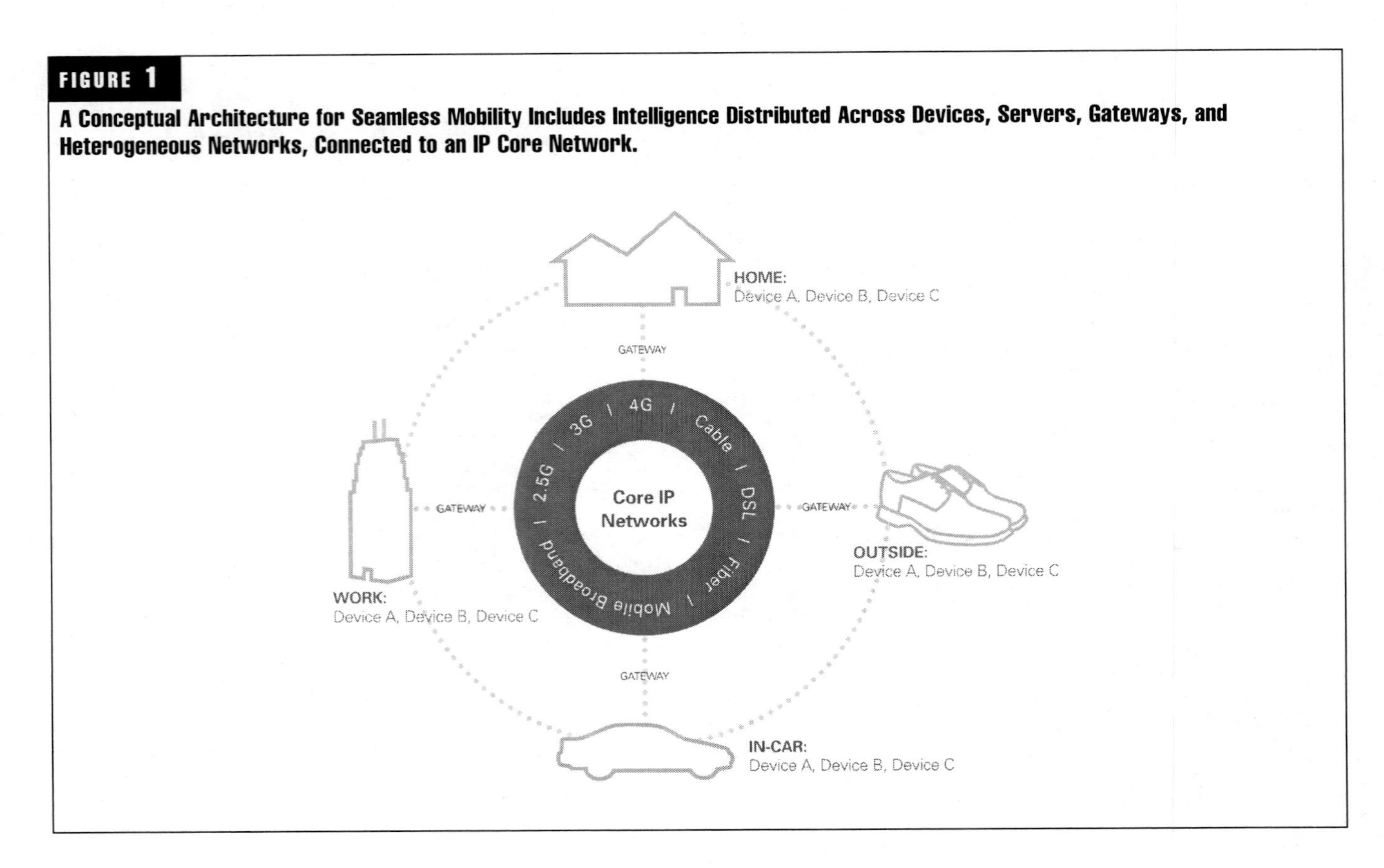

FIGURE 1

A Conceptual Architecture for Seamless Mobility Includes Intelligence Distributed Across Devices, Servers, Gateways, and Heterogeneous Networks, Connected to an IP Core Network.

FIGURE 2

Seamless Mobility Functional Architecture Encompasses an Experience or Application Architecture that Enables Application and User Interface Capabilities and a Connectivity Architecture that Manages Network and Session Details.

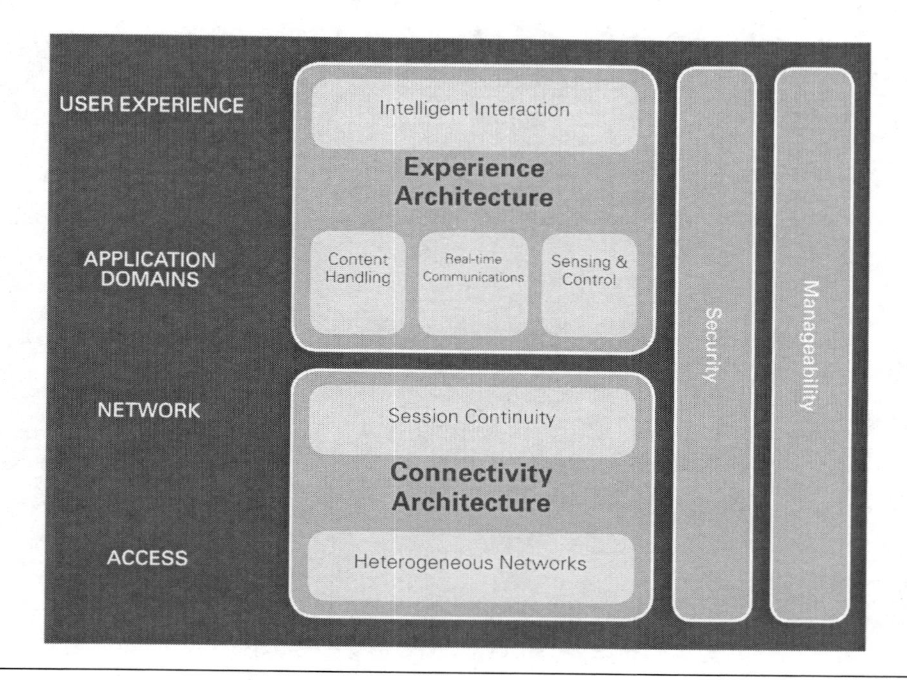

- *Sensing and control*: Giving "things" the means to talk, discover, monitor, report, and connect

A connectivity architecture resolves the technical issues necessary to provide seamless handover between sessions, including heterogeneous network access. The major elements are as follows:

- *Session continuity*: Universal accessibility of people, applications, and network features across devices and networks

- *Heterogeneous networks*: Ubiquitous availability of disparate networks

The connectivity architecture ties these technologies together and provides system-level features such as authentication, authorization, security, accounting, and quality of service (QoS). Security and manageability issues must be addressed at all levels of the functional architecture to prevent unauthorized access to information and services and to ensure that networks can be configured, operated, managed, and healed in real time based on business and performance objectives.

Intelligent Interaction

Seamless mobility solutions will enable context-dependent, personalized interactions and services across devices, networks, spaces, and user tasks. The key challenge is making complex capabilities more intelligent and intuitive so users do not have to set parameters. It requires personalizing content and applications so they can learn user preferences and make intelligent decisions based on the environment, context, and device capabilities. Solutions will have to provide a consistent and context-sensitive interface across devices and environments and the ability to adapt content and applications depending on user preferences.

Research topics include input and output modalities, context interpretation, user modeling, interaction management, and goal determination.

Content Handling

Seamless mobility solutions will promote sharing and experiencing content together by providing the ability to find, buy, use, share, and protect fresh, relevant, personalized content when and how you want it while putting the appropriate safeguards in place to preserve privacy and established digital property rights. It will require optimizing content according to device and network characteristics. The key challenges are digital rights management, storage management, intelligence to adapt content to the network and device characteristics, transparent synchronization of information access across devices, and content scaling depending on network bandwidth and device capabilities.

Areas of research include collaborative filtering and content intelligence, synchronization of media in different formats across user devices, searches for media across multiple devices, access to metadata about content, management of user profiles and preferences, and highest-definition device experience while minimizing battery and weight impacts.

Real-Time Communications

Seamless mobility solutions will merge real-time communications and content, making new levels of business and per-

FIGURE 3

Seamless Mobility Envisions Intelligent Interaction That Provides Greater Transparency and Freedom of Movement between Devices and Environments.

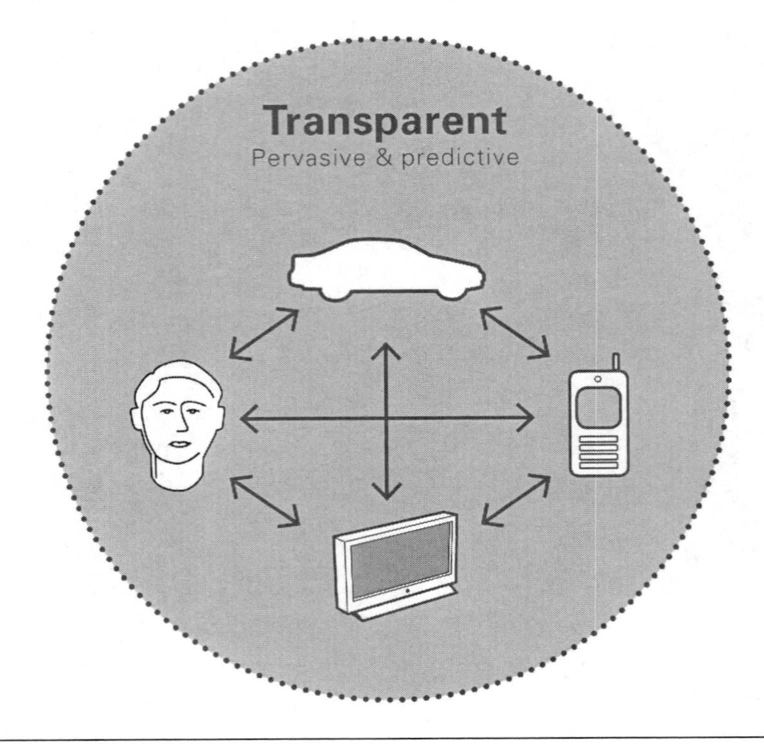

FIGURE 4

Seamless Mobility Envisions Content Handling Capabilities That Provide Content Portability, Allowing Users to Access, Interact with and Share Voice, Data and Video Content and Communications—Regardless of Device or Network.

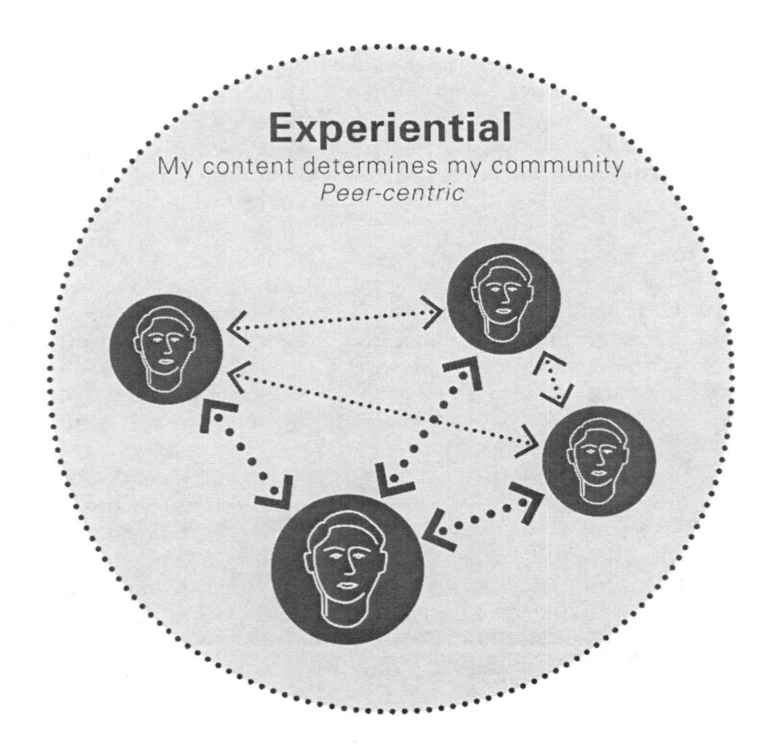

sonal collaboration possible and enabling people to share experiences from different locations. The key challenges are coordination of content and communications, the ability to set up or tear down a content session during a real-time communication, linking communications history with content and social contacts, and using presence and locations to enhance and personalize communications experiences.

Research includes the sharing of content—such as video, Web, and music—the control of annotations and real-time markup, location mapping, and the communications of media in different formats based on user devices. Push-to-x, including push-to-talk over cellular and push-to-view, and information management system (IMS)–based applications are areas of particular focus.

Sensing and Control
Seamless mobility solutions will bridge the physical and computing worlds, enabling consumer convenience and control in the home, in the auto, and on the go, as well as in enterprise control of manufacturing and other processes. As a result, users' lives will be simpler and more productive. The key challenges are implementing machine-to-machine communications, managing multiple network connections to multiple, low-power sensing devices, collecting and aggregating volumes of sensor data, extracting actionable information, enabling communication and collaboration, and driving down energy consumption and footprint while adding intelligence to and enhancing security of sensor devices.

Areas of research include things-to-things communications and self-organizing, self-optimizing, and self-healing networks. This requires further development of low-maintenance and low-power technologies, advanced architectures, standard sensor interfaces, and global profiles.

Session Continuity
Seamless mobility solutions will enable multimedia sessions to be established, maintained, and optimized across networks, devices, and environments. If the user moves out of one network's range, the session will be transparently linked to another network, thus ensuring a continuous and uncomplicated experience. It will require network operators to provide authentication, billing, and provisioning that is straightforward and transparent to the user. The key challenges are to establish, maintain, and optimize multimedia sessions across networks, devices, and environments, and to acquire and manage the network and QoS information needed by other layers of the architecture.

Areas of research include session setup and control based on identities, roles, and policies; software-defined, multiband radio; and IMS– and single inline package (SIP)–based solutions for network authentication, billing, and provisioning.

Heterogeneous Network Access
Seamless mobility solutions will promote continuous communications by offering transparent interoperability across disparate networks, allowing the user to choose the fastest and/or least expensive connection available. The key challenge in delivering heterogeneous access is to manage inter-

FIGURE 5

Seamless Mobility Envisions Real-Time Communications that Enable Shared Group Experiences through Rich Media Exchange, Enhancing Business Productivity and Social Interaction

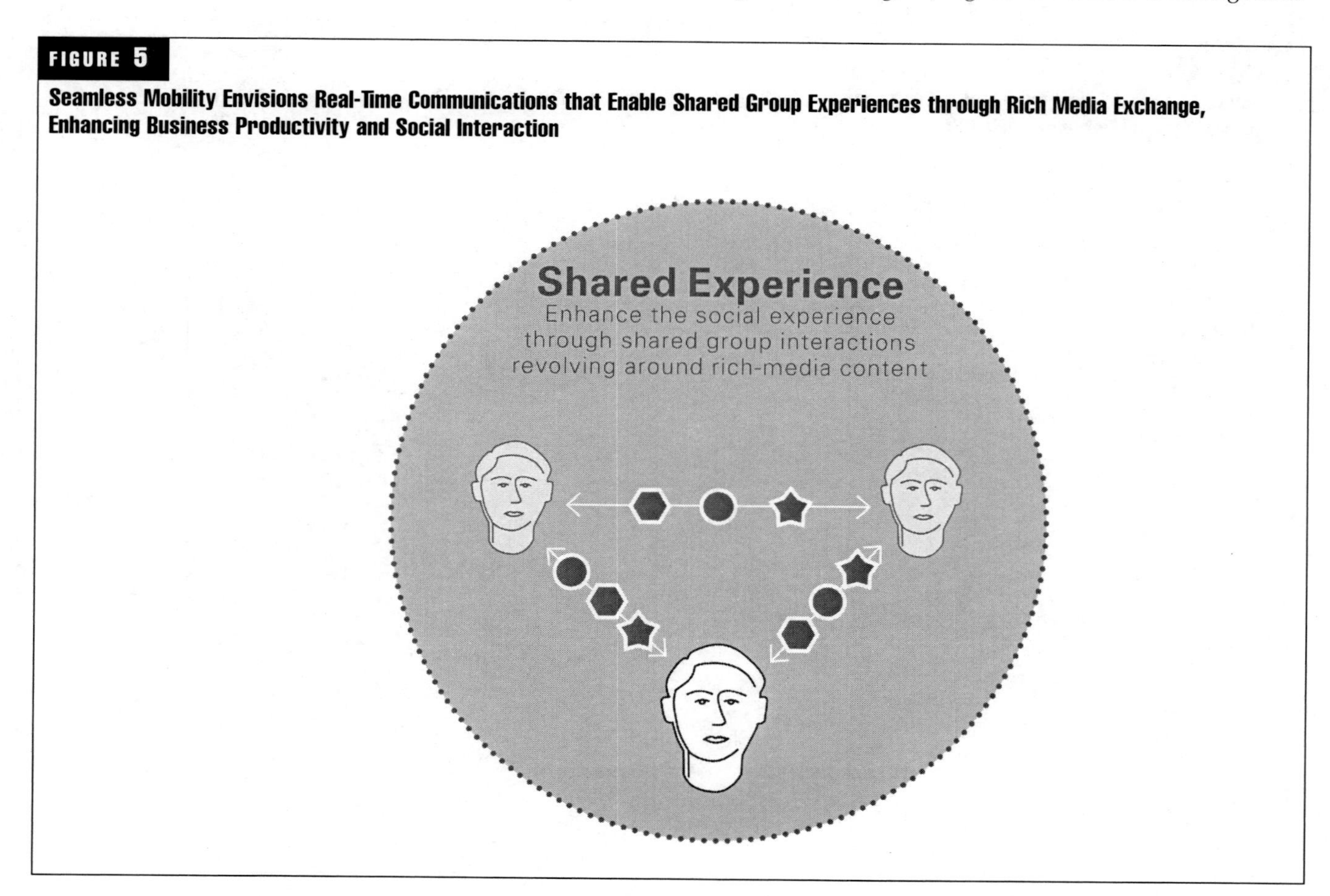

FIGURE 6

Seamless Mobility Envisions Sensing and Control Capabilities in which Near-Time Information and Control Will Be Transparently Delivered to Intelligent Applications via Wireless Networks that Connect Billions of Sensor Enabled Devices.

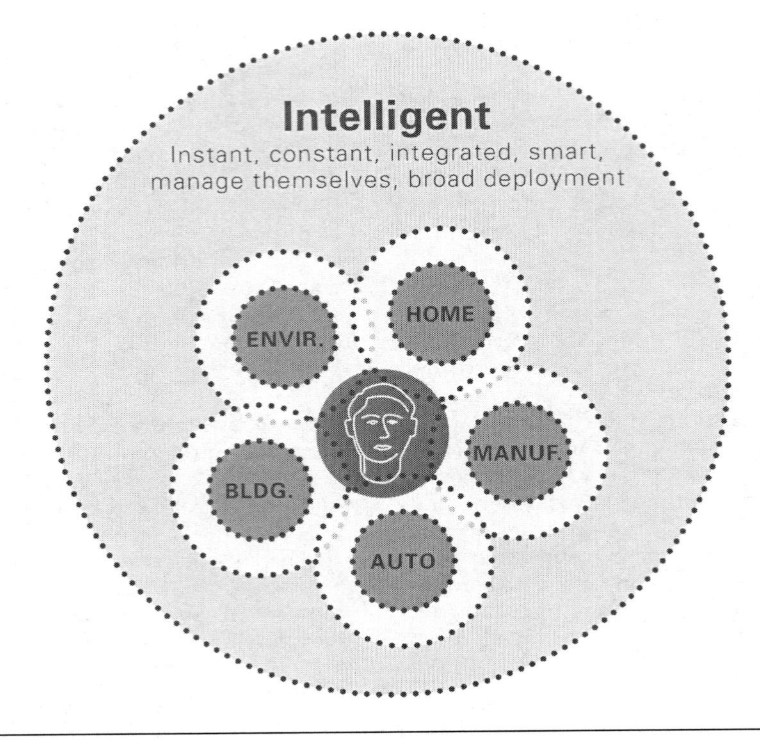

FIGURE 7

Seamless Mobility Envisions Session Continuity in which Multiple Operators Collaborate, Allowing Devices to Always Connect to the Most Optimal Network Available.

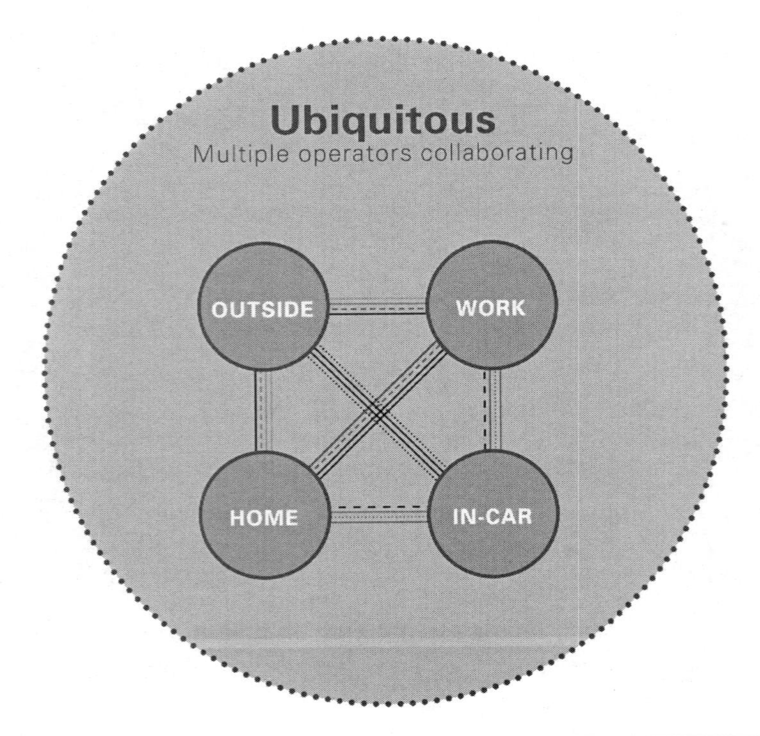

network interfaces to span the seams between networks through network discovery, link quality measurement, location information, power management, and device implementation. This requires a variety of technologies to enable devices to discover types of networks, measure their link quality and speed, provide authentication for network access, and access multiple wired and wireless interfaces, all while conserving and managing power efficiently.

Wireless access technologies required include orthogonal frequency division multiplexing (OFDM), mesh networking, antenna arrays, and channel coding to enhance the wireless broadband user experience, especially in buildings and for wide-area mobility. This includes UWB, Bluetooth, ZigBee, and IEEE 802.11n for short-range communications, as well as many new features in IEEE 802.16e and the radio access network evolutions in the third-generation partnership project (3GPP) and 3GPP2. In addition, key technologies for cognitive radios and software-definable radios are required to assist the industry in such standards as IEEE 802.22.

Security
Seamless mobility solutions will protect the information assets of users and content and network providers while maintaining a personalized user experience. Security will need to encompass individual content, peer-to-peer content, and purchased content. Challenges include authenticating the user to the device as well as providing secure, efficient storage and management of multiple types of valuable content within a device. The network and the applications that run over it must be secure from intrusion and abuse.

Areas of research include identity management capabilities in end-user devices and networks to enable single sign-on across networks and services. Biometric solutions are enabling secure personal identifications. Collaboration within the industry is required to enable seamless digital rights management of purchased content. Security architectures for trusted platforms will ensure that information can be stored securely and platforms can be trusted to administer the defined security and privacy policies. Engineers are exploring ways to fuse inputs from multiple authentication mechanisms to maintain confidence in the user identity.

Manageability
Seamless mobility will require intelligent, autonomic self-management by client devices supported by wired and wireless infrastructures that are context- and task-aware. It will require network systems that proactively share information, allocate resources, and coordinate service delivery based on real-time awareness of changes in user needs and the operating environment while in compliance with relevant business policy. Challenges include enabling seamless billing systems and businesses to drive the services and resources that the network provides, all while enabling a continuum of information to be securely and efficiently exchanged among diverse systems at the resource, service, and business layers.

Research includes a new generation of intelligent management software that will integrate key elements of autonomics, policy-based management, machine learning and reasoning, sensor network interaction, and end-to-end security enforcement.

FIGURE 8

Seamless Mobility Envisions Heterogeneous Networks that Allow Users to Move from Device to Device and from Network to Network without Re-establishing Connections or Dealing with a New User Interface.

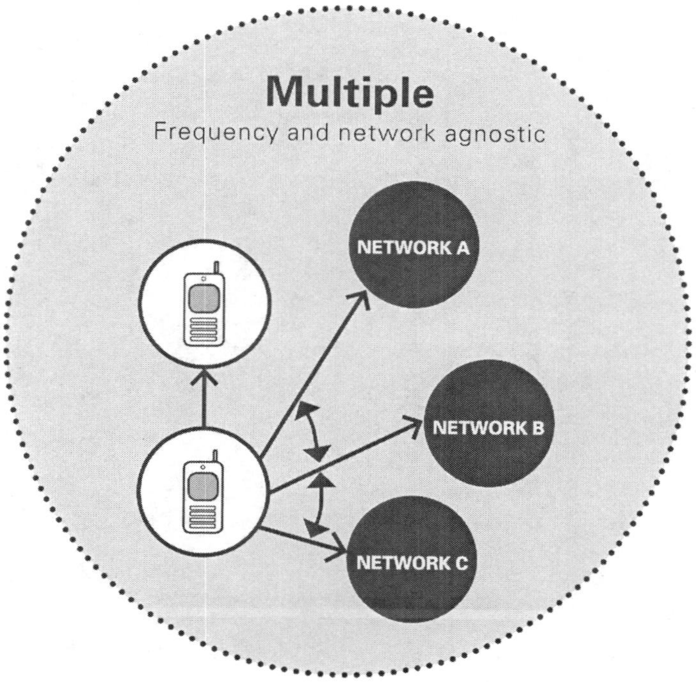

FIGURE 9

Seamless Mobility Envisions Mobile Security that Gives Users, Operators, and Content Providers Full Confidence in the Security and Confidentiality of Their Content and Communications.

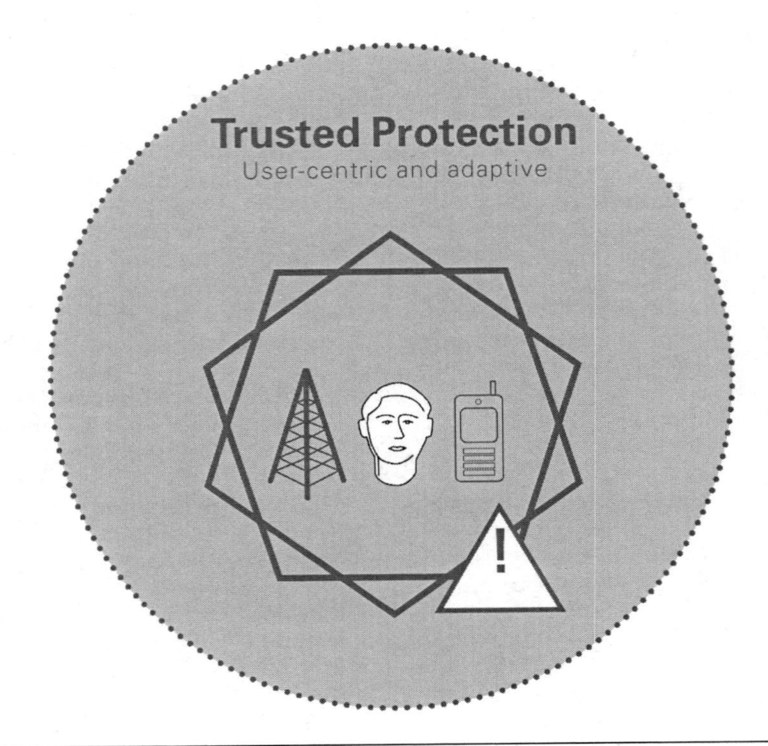

FIGURE 10

Seamless Mobility Envisions Manageability which Is Autonomic, User Centric Management, that Will Increase Integration Between Vendors and Reduce Manual Intervention by Users and Skilled Administrators.

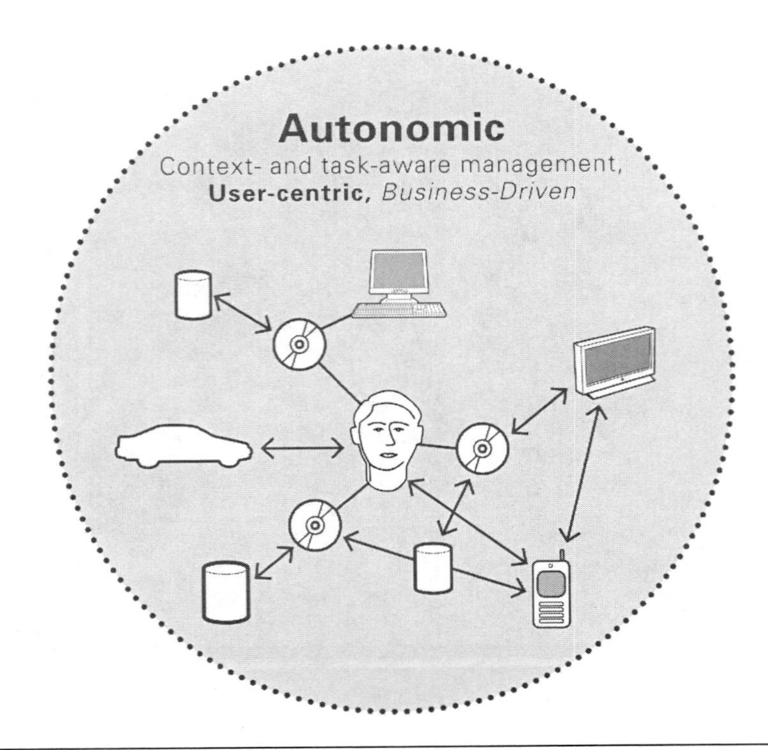

Software Implementation Architecture

Seamless mobility is all about experiences. Solutions and services will be built around products that ensure the consumer's requirements are met while focusing on ease of use, simplicity, and usability. Software plays a critical role in enabling the functionality in devices, networks, infrastructure, and everything in between to deliver seamlessly mobile experiences.

Whereas the seamless mobility functional architecture defines what needs to be considered, the implementation architecture defines how it will be addressed. To ensure that specific functionality requirements are addressed, the defined functionality within the functional architecture is mapped to different nodes in the implementation architecture—in devices, in servers and in various option sets—while taking into consideration customer needs, existing and emerging standards, integration points, and deployment requirements.

The implementation architecture answers questions such as: How does a mobile user request a service on demand? What kind of intelligence is implemented into the client's software, server software, and policy enforcement points in the devices? How does each device and service achieve true plug-and-play compatibility? How are content and services seamlessly delivered across multiple network domains via different devices? The architecture also addresses the end-to-end technology gaps such as security and management across various product lines and partner offerings.

The implementation architecture follows a services-oriented architecture model to allow for rapid deployment and platform-independent operations. It enables software agents to provide loose and dynamic binding between interfaces to build targeted end-to-end solutions rapidly. Applications and services can use the framework to provide consistent user experiences, optimized for each device and network. Applications and services can be tailored to each user's preferences, heightening the user's emotional bond with the experience and enhancing brand loyalty. Powerful application programming interfaces (APIs) and software development kits (SDKs) help developers use underlying architectural features to ensure software interoperability within and across network elements and applications and to maximize component reuse. This implementation architecture is geared at achieving higher interoperability, accelerated time to market and reduced development costs as it creates end-to-end solutions, services, and value-added products focused on user experience.

Moving Forward

Seamless mobility will provide users with new forms of entertainment and new levels of enjoyment, productivity, and connectedness. It is opening the door to a robust innovation cycle for business and government users, application and content developers, and operators. It is also offering tremendous potential to spur growth and innovation, enhance competitiveness, and promote inclusion.

Much work must be done to achieve seamless mobility's full potential. Network technologies must continue to advance toward a future all–Internet protocol (IP) network with easily configurable metrics such as QoS and security to deliver high-value, real-time data, voice, streaming video, and multimedia. Innovation must advance on many fronts, including nanotechnology, privacy, and security. Networking and software standards must continue to advance. The legislative and regulatory environment must embrace innovation and provide a level playing field for emerging and existing technologies.

Seamless mobility will enrich our lives, foster inclusion, drive economic expansion, and affect broad segments of our society. To capture the true promises of seamless mobility, commitment, and cooperation required by the organizations within this new and expanding ecosystem, contributors must do the following:

- *Network providers*: Embrace seamless mobility and use the opportunities it presents to increase efficiency and

FIGURE 11

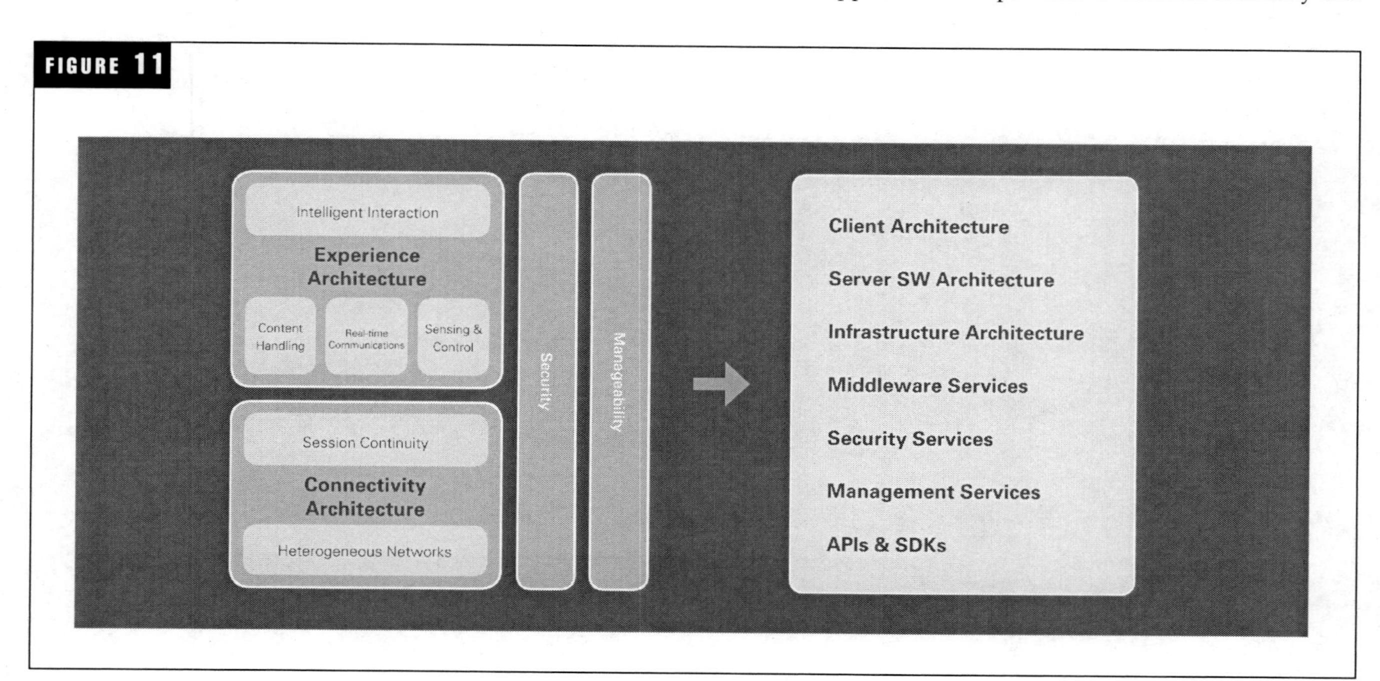

productivity in the workplace; deliver new services to consumers; reduce churn; enhance loyalty; increase revenue per user; and compete effectively in a rapidly changing marketplace

- *Content providers*: Work collaboratively to explore new business models for intellectual property monetization and protection while extending content to new media
- *Application developers*: Consider the transformative advantages for your customer base and design next-generation applications to exploit those advantages
- *Legislative and regulatory bodies*: Create a climate in which innovation and economic competitiveness flourish
- *Businesses and government agencies*: Take advantage of advances in mobile communications, computing, and entertainment to transform business and agency productivity; monitor and control processes more effectively; develop breakthrough products and services; build out mobile infrastructure; and proactively adopt new mobile usage paradigms to gain the advantages of mobility.

Optimizing Video over Wireless Using Performance and Architecture Modeling

Deepak Shankar
President and Chief Executive Officer
Mirabilis Design, Inc.

Wireless local-area network (WLAN) technologies such as 802.11 (Wi-Fi), Bluetooth, and ultra wideband (UWB) got their beginnings in providing a wire replacement for Internet access and digital data communications. The transfer rate robustness, high performance using embedded microprocessors, and dropping prices are making these viable for audio and video applications such as audio streaming, videoconferencing, and gaming. To accommodate the data rate characteristics as a function of distance on top of a wireless network, the issues associated with wireless contention from other same access point users and the added functionality of audio or video processing must be evaluated. Audio processing extends from voice-level to CD–level audio quality (pulse code modulation [PCM] or MP3, 64Kbps to 1.5Mbps). Video processing must support gaming, real-time video, or TV/VCR formats that have differing latency (10 ms to 40 ms), data rate (MPEG–2 or MPEG–4, 1.5 Mbps to 20 Mbps), and frame loss requirements (10 to 20 percent max).

What does quality of service (QoS) mean with a 15 percent frame loss? Discrete event modeling has the ability to evaluate the performance of these added applications over the basic wireless protocols, especially the contention of other users for the wireless airspace and internal contention of client hardware resources.

Some of the issues to be considered by system architects include the following:

- How should frame errors that cause retries, or contention from other wireless LAN users, be accounted for when the application is based on a constant bit rate (CBR) algorithm?
- How big should the application buffer be to handle CBR interruptions?
- What are the resource tradeoffs in implementing new applications and features in custom hardware or software?
- What is the buffer tradeoff versus end-to-end frame latency?
- How many users can the local coffee shop Wi-Fi network support before information loss renders the audio or video application unusable?

Modeling Options

The complexity of these technical decisions and the non-deterministic nature of the traffic profile require accurate modeling efforts to predict the resource requirements and the effective maximum throughput for a Wi-Fi network. Excel is useful for calculating the average throughput and latency with many assumptions along the way. This approach has historically had difficulty predicting the peak system and network performance. Designing a new system, based on the average calculations plus some design margin, has proven to be a trial-and-error approach. Ultimately, one or two prototype cycles are required to determine if the design margin was just enough, too much, or, in the worst case, not enough. These additional design "spins" can be very costly for a new chip or board re-layout. It is possible to create more complex models using Excel macros, Visual Basic and .NET to analyze the non-deterministic traffic distribution. However, these tools are very general and do not necessarily provide the needed infrastructure for a focused solution, such as discrete event modeling. In addition, custom development is required for the media access control (MAC) and physical wireless layers to define the functionality of the application layer.

Another approach is to use a graphical discrete event modeling tool that provides the Wi-Fi/Bluetooth protocols in terms of traffic modules, hardware and software resource modules, and a networking capability. VisualSim from Mirabilis Design, Inc., is a product that provides a Wi-Fi library and hardware and software architecture modules to enable engineers to conduct performance/architectural tradeoffs. Users can model applications on top of the existing wireless nodes by adding additional discrete event blocks.

The discrete event modeling modules can interface with current Excel spreadsheets and accept current C/C++/Java functions with limited modifications. The environment must be able to randomize the test vectors in a manner that is application-specific and analyze outputs for expected results. These random techniques may contain statistical (periodic, uniform, exponential, and Gaussian), empirical (CDF and self-similar), or application-specific traffic profiles.

Model Construction

The discrete event simulation model in VisualSim consists of traffic generators, nodes, access points, arbiters, and statistics generation. To reduce model building time, the network nodes have been modeled with both the transmitter and receiver. The arbiter was separated from the nodes to improve simulation performance and to experiment with the arbitration schemes. VisualSim's model of 802.11 transmits frames from multiple client nodes to an access point. Nodes are contending for service based on the standard 801.11 backoff procedure. The overall frame latency is measured for each packet exiting the central access point. Application level throughput is impacted by the availability of hardware resources. The application can request hardware resources defined as an ASIC or software resource from an embedded processor.

Each wireless node is defined as a state machine with five states: IDLE, RDY, CTS, RTS, and NEXT. The signaling and arbitration can accommodate single-frame, burst, and fragmentation modes of operation. The separate arbiter block provides arbitration between the node devices, performs backoff, and helps with slot retention/improvement for the losing node in the arbitration process.

The input frame is supplied with a source, destination, frame size, and timestamp. The model generates frames at periodic rates but can be easily modified to accommodate a network trace for more accurate modeling. The frame size and fragmentation limit are maintained as parameters that can be modified to test protocol functionality and system throughput. The network frequency and individual node timing can be varied from 1 MHz to 11.0 or 54.0 MHz. The dispersion compensation fiber (DCF) interframe spacing and short interframe spacing delays are also parameters of the model.

The protocol data (frame) and control frames (RTS, CTS, and ACK) are based on the 802.11 specification. Network access vectors (NAV_RTS and NAV_CTS) are signals to discourage other stations from accessing.

Some protocol standards have been not been modeled, especially those that have secondary impact to system throughput and latency. However, the model has been constructed with expandability in mind. Provision is also provided for mobile position, direction, and velocity in the input definition.

FIGURE 1

This Shows the Topology of a Wi-Fi Network in VisualSim, a Graphical System-Modeling Software. Nodes and Access Points Can Be Arbitrarily Added or Removed. This Model Contains Two Nodes, an Access Point, an Arbiter, and a Statistics Block. The Traffic Enters the System at Each Node and Leaves the Network at the Access Point.

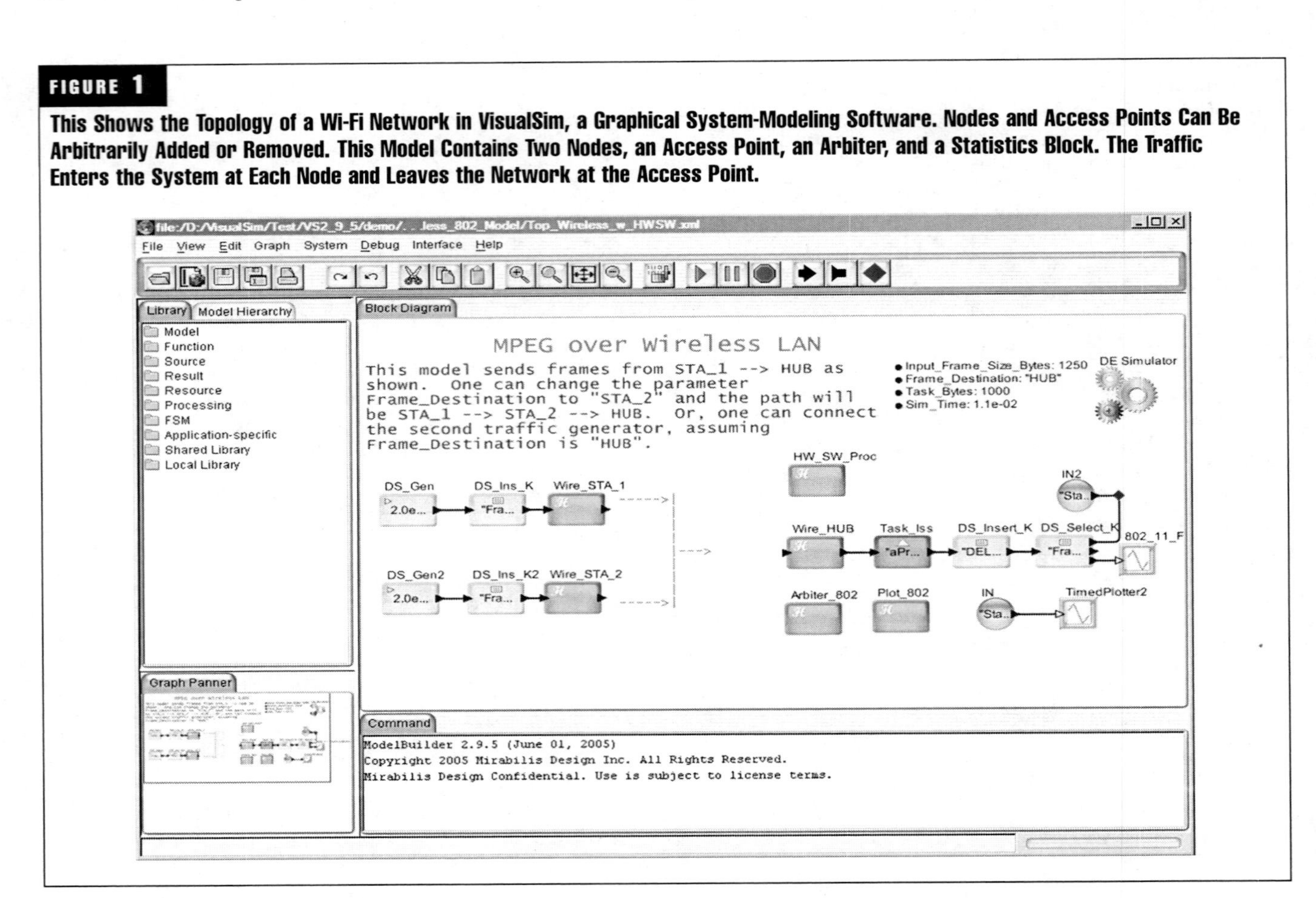

FIGURE 2

VisualSim Timeline Plot Contains all the 802.11 Signals: RTS, CTS, ACK, NAV_RTS, NAV_CTS, and FRAME. The Plot Shows Fragmentation Frames Being Transmitted. Contention is Shown in the First Sequence Where There Are Two RTS Frames.

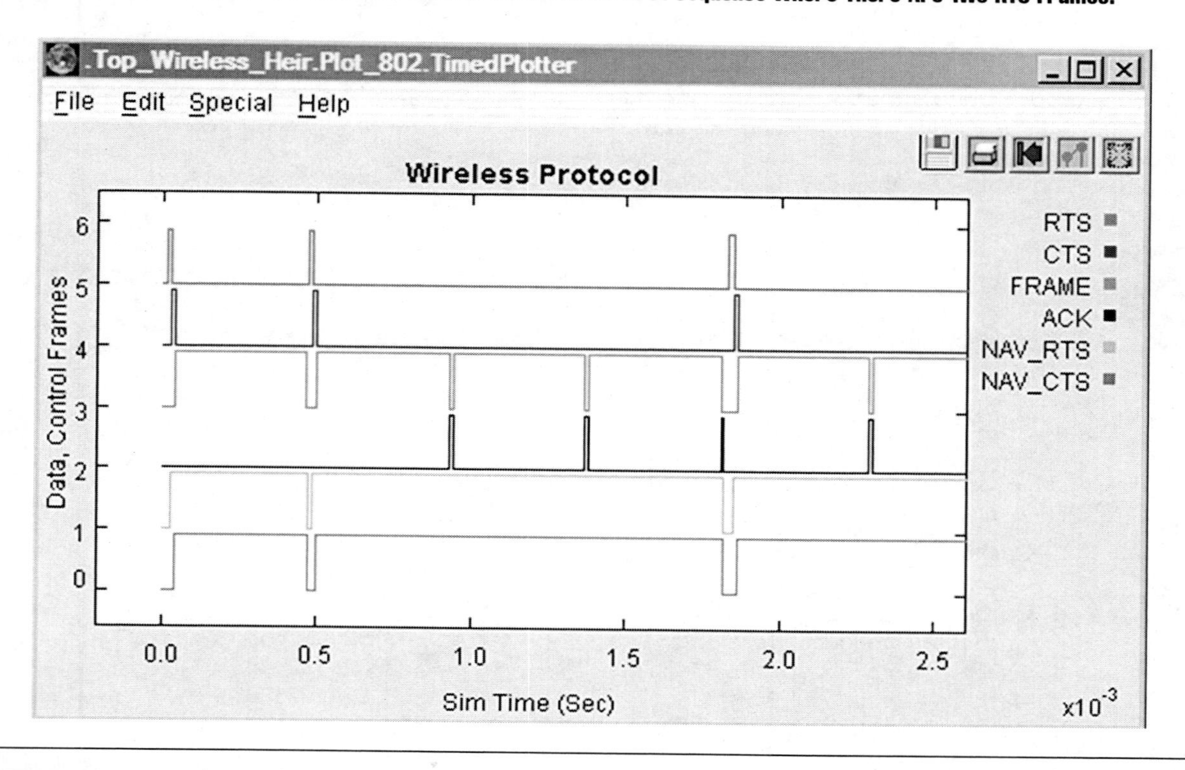

FIGURE 3

This Plots the Transmission Latency of the MPEG Frame from the Wireless Hub to the Individual Nodes.

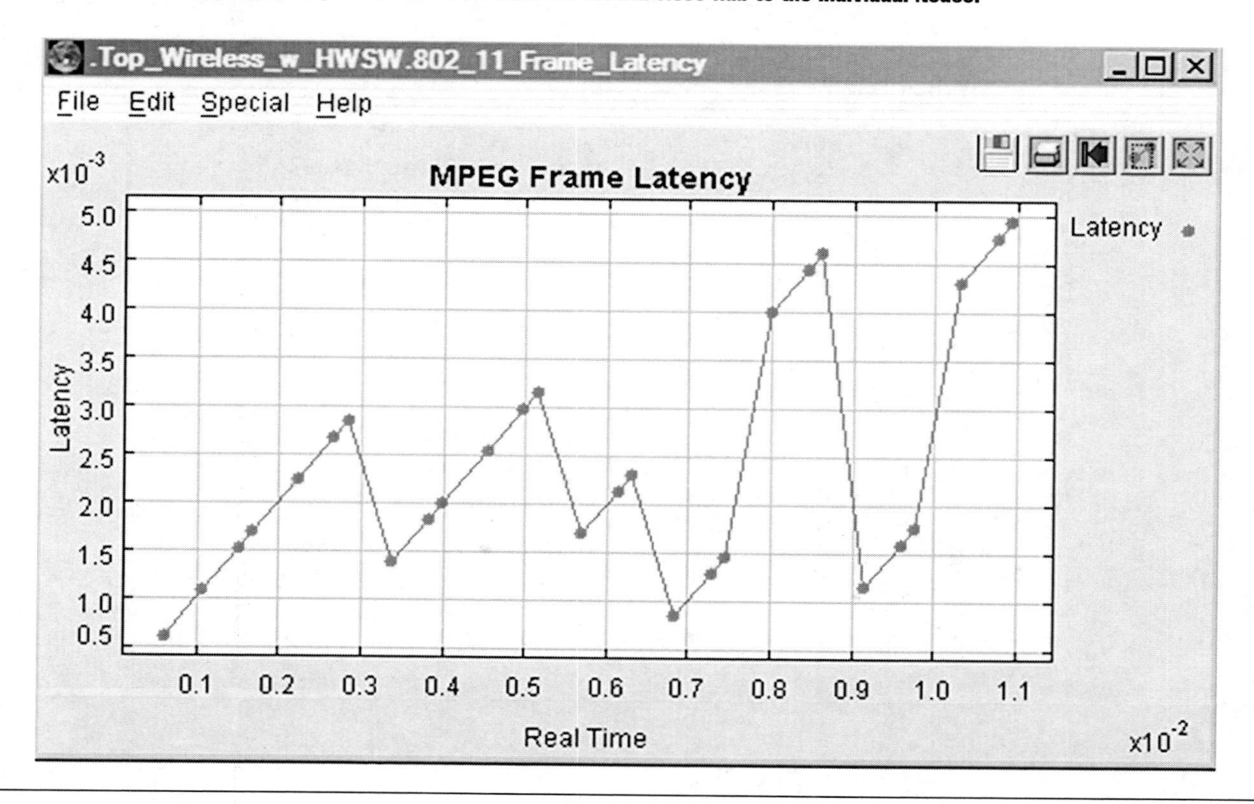

Analysis and Results

Studies on the following issues have been conducted:

- Performance of the arbitration contention algorithms when multiple users in the same network are contending for bandwidth.
- Verification of timing by testing for any frames not transmitted by a waiting station, when the current transaction completes
- The best- and worst-case throughput and latency of time-critical frames based on different rates of arbitration contention for a given topology of users.
- The impact on overall performance by management frames and other higher-priority activities on the hardware resources.
- Expected performance for recovery timing of a failed transmission by a competitor's 802.11 implementation of the protocol.
- Determination of the "peak" performance drivers and their frequency of occurrence.

FIGURE 4

This Summarizes the Statistics of the MPEG Software Operation on the Proposed Hardware Architecture.

```
Cache_1_Delay_Time_Max                       = 2.0E-9,
Cache_1_Delay_Time_Mean                      = 2.0E-9,
Cache_1_Delay_Time_Min                       = 2.0E-9,
Cache_1_Delay_Time_StDev                     = 0.0,
Cache_1_Hit_Ratio_Max                        = 100.0,
Cache_1_Hit_Ratio_Mean                       = 100.0,
Cache_1_Hit_Ratio_Min                        = 100.0,
Cache_1_H it_Ratio_StDev                     = 0.0,
Cache_1_Throughput_MBs_Max                   = 0.0398211928886,
Cache_1_Throughput_MBs_Mean                  = 0.0398211928886,
Cache_1_Throughput_MBs_Min                   = 0.0398211928886,
Cache_1_Throughput_MBs_StDev                 = 0.0,

PPC_7410_1_Bus_Delay_Max                     = 1.8000000000012E -8,
PPC_7410_1_Bus_Delay_Mean                    = 8.9999999999986E -9,
PPC_7410_1_Bus_Delay_Min                     = 3.9999999999546E -9,
PPC_7410_1_Bus_Delay_StDev                   = 6.4031242374381E -9,
PPC_7410_1_Bus_Throughput_MBs_Max            = 0.7167814719954,
PPC_7410_1_Bus_Throughput_MBs_Mean           = 0.7167814 719954,
PPC_7410_1_Bus_Throughput_MBs_Min            = 0.7167814719954,
PPC_7410_1_Bus_Throughput_MBs_StDev          = 0.0,
PPC_7410_1_Context_Switch_Time_Pct_Max       = 0.2001015440671,
PPC_7410_1_Context_Switch_Time_Pct_Mean      = 0.2001015440671,
PPC_7410_1_Context_Switch_Time    _Pct_Min   = 0.2001015440671,
PPC_7410_1_Context_Switch_Time_Pct_StDev     = 0.0,
PPC_7410_1_D_1_Hit_Ratio_Max                 = 93.75,
PPC_7410_1_D_1_Hit_Ratio_Mean                = 93.75,
PPC_7410_1_D_1_Hit_Ratio_Min                 = 93.75,
PPC_7410_1_D_1_Hit_Ratio_StDev               = 0.0,

PPC_7410_2_FP_1_ Throughput_MIPs_Max         = 1.1678052435651,
PPC_7410_2_FP_1_Throughput_MIPs_Mean         = 1.1678052435651,
PPC_7410_2_FP_1_Throughput_MIPs_Min          = 1.1678052435651,
PPC_7410_2_FP_1_Throughput_MIPs_StDev        = 0.0,
PPC_7410_2_Stall_Time_Pct_Max                = 0.8230436272908,
PPC_74 10_2_Stall_Time_Pct_Mean              = 0.8230436272908,
PPC_7410_2_Stall_Time_Pct_Min                = 0.8230436272908,
PPC_7410_2_Stall_Time_Pct_StDev              = 0.0,
```

Internetworking of Next-Generation IPv6–Based Mobile Wireless Networks and MPLS/GMPLS–Based Multiservice Backbone Networks

Mallik Tatipamula
Senior Project Manager, Service Provider Router Technology Group
Cisco Systems, Inc.

Hiroshi Esaki
Professor, Graduate School of Information Science and Technology
The University of Tokyo

Abstract

The new end-user services including Internet protocol (IP) multimedia applications and a common transport technology are the key drivers for the integration of mobile wireless and IP technologies. Support for IP multimedia applications in next-generation (NG) mobile core networks is being discussed by the third-generation partnership project (3GPP) as a part of Release 5 (R5) specifications. 3GPP R5 has mandated IP version 6 (IPv6) for IP multimedia applications in the IP multimedia subsystem (IMS) domain of 3G mobile operator architecture. The IMS domain has also been adopted by 3GPP2 for code division multiplex access 2000 (CDMA2000) architecture as an IP core network harmonization effort between the two standard bodies.

At the same time, many service providers have embraced multiprotocol label switching (MPLS) as the enabler for the required multiservice capabilities of their NG packet core/backbone network. Also, the widespread deployment of dense wavelength division multiplex (DWDM)–based optical transport systems in the core network to satisfy the tremendous need and increase in capacity demand, has led network planners to reconsider traditional approaches to "provisioning" and "network restoration" and to plan the integration of the optical layer into the MPLS infrastructure according to the emerging generalized MPLS (GMPLS) technology. It becomes critical for IPv6–based wideband CDMA (WCDMA)/general packet radio service (GPRS) mobile network to coexist and integrate with MPLS/GMPLS–based multiservice backbone network.

This paper presents an end-to-end network architecture for NG mobile wireless networks based on WCDMA/GPRS technologies. It discusses the role of IPv6 in WCDMA/GPRS networks. It presents various internetworking techniques of IPv6–based WCDMA/GPRS networks with service provider multiservice MPLS/GMPLS backbone networks. Finally, we provide a comparison of these internetworking mechanisms for NG wireless networks to coexist with NG wireline networks.

Introduction

There are three technological trends being discussed among the telecom and networking communities today for NG network architectures. They are as follows:

- Evolution of IPv6–based mobile wireless networks
- Evolution toward IP/MPLS–based multiservice backbone networks for supporting various access mechanisms, including mobile wireless, cable, xDSL, etc.
- Integration of IP/optical networking for intelligent multiservice IP/optical backbone networks, based on GMPLS

It is critical to understand the end-to-end network architecture, including access networks and multiservice carrier backbone networks, while looking at internetworking techniques based on above technical trends. In end-to-end network architecture, mobile operators' network extends IPv6 connectivity to the gateway GPRS support node (GGSN)/customer edge router, and carrier backbone network is used to carry all traffic between mobile operator networks. This paper looks in detail at an end-to-end network architecture and IPv6 traffic flow, and it defines internetworking techniques for this end-to-end network architecture.

Internetworking/transition to IPv6 in GPRS and WCDMA networks is being discussed in [1, 2, 3] with an assumption of only IPv6–over–IPv4 (non–MPLS/GMPLS) backbone networks. The paper [4] summarizes internetworking/transition mechanisms defined by Internet Engineering Task Force (IETF), including IPv6–over–IPv4 tunneling and dual-stack mechanisms. As mentioned above, since NG backbone networks are evolving toward multiservice backbone networks based on MPLS/GMPLS, it becomes critical to look at internetworking techniques for IPv6–based mobile networks with MPLS/GMPLS carrier backbone networks. In this paper, we define one of the important aspects of IPv6 co-existence and integration with MPLS/GMPLS–based networks, which has not been discussed in [1, 2, 3, 4].

This paper is organized as follows. Section 2 presents NG networking trends and defines a view of NG end-to-end network architecture. The end-to-end network architecture provides a two-tier network architecture consisting of mobile operator network and carrier backbone network for connecting various mobile operator network sites. Section 3 presents the role of IPv6 in WCDMA/GPRS mobile networks. Section 4 looks at some of the scenarios for internetworking between mobile wireless networks and carrier backbone wireline networks. There are two important concepts described in this paper for these internetworking scenarios. Section 5 discusses the first concept that describes IPv6 traffic flow within the mobile network. Section 6 describes the second concept of transport of IPv6 traffic from mobile network over MPLS/GMPLS–based carrier multiservice backbone network. In this section, we define various techniques for internetworking of WCDMA/GPRS with NG MPLS/GMPLS–based multiservice backbone networks. Finally, in section 7, we present the comparison of these techniques for internetworking of IPv6–based wireless and MPLS/GMPLS–based wireline networks. Since MPLS is considered part of GMPLS [16], we use GMPLS terminology to imply both MPLS and GMPLS, as the internetworking techniques in this paper do not differ between the two.

Next-Generation Networking Trends

This section briefly discusses some of the emerging trends in NG networks (NGNs). A view of end-to-end NGN architecture is defined.

Evolution of IPv6–Based Mobile Wireless Networks

With 3GPP R5, one more domain, the IMS is being added to the mobile core network. The introduction of the IMS is driven by the demand to offer more and enhanced services to end users. At the heart of IMS is an IP–based transport mechanism for both real-time and non-real-time services, plus the introduction of multimedia call control. 3GPP has selected session initiation protocol (SIP) as the only call control protocol between terminals (or user equipment [UE]) and the mobile network. 3GPP also decided to use IPv6 as the only IP–version protocol for the IMS components. IPv6 provides the end-to-end addressing required by the new multimedia environments for mobile phones and residential voice over IP (VoIP) gateways. IPv6 provides the services, including integrated auto-configuration, quality of service (QoS), security, and direct-path mobile IP, also required by IMS system environments.

As mentioned above, standards bodies for the wireless data services [5, 6, 7] and multimedia subsystem [8] are preparing for the future, and IPv6 provides the end-to-end addressing required by these new environments for mobile phones and residential VoIP gateways. IPv6 provides the necessary addressing space to enable peer-to-peer, always-on services. It can also improve the services, thanks to integrated auto-configuration, QoS and security, and direct-path mobile IP also required by these environments.

Evolution toward Multiservice Carrier Backbone Networks

At the same time, traditional telecom backbone network architecture is moving from a world of many separate backbone networks for voice (e.g., public switched telephone network [PSTN]), data (Internet, frame relay, asynchronous transfer mode [ATM]), and radio and cable into a world of a single high-bandwidth multiservice packet backbone network. The packet-switched backbone that can handle and scale a wide variety of services is a key component in any multiservice backbone network. It should be able to carry types of traffic that have diverse characteristics, e.g., real-time traffic, Internet traffic, imaging, and streaming voice and video.

Such backbone networks must deliver true, carrier-class resilience and serviceability, and support powerful IP border routing for peering with other IP networks. This type of multiservice packet backbone consists of routers running the MPLS protocol suite.

GMPLS–Based IP/Optical Carrier Backbone Networks

Finally, the widespread deployment of DWDM–based optical transport systems in the core network to satisfy the corresponding increase in capacity demand has led network designers to reconsider traditional approaches to "provisioning" and "network restoration." Consequently, there is a consensus to push the infrastructure integration paradigm one step further by integrating the optical layer into the MPLS control plane based on the emerging GMPLS technology [16]. GMPLS has emerged as an NG networking protocol for providing intelligent optical control plane for optical network. GMPLS is considered to be superset of MPLS, as it extends forwarding plane and control plane to include not only packet-based and cell-based, but also synchronous optical network (SONET)–, DWDM– and fiber-based network elements. In this paper, we use the term GMPLS to cover both existing MPLS multiservice backbone network and NG GMPLS multiservice backbone network, as the technique discussed here apply to both.

A View of Next-Generation End-to-End Network Architecture

In this section, we examine an end-to-end IP network architecture. As shown in *Figure 1*, the IP infrastructure is made up of the following main tiers:

- A backbone tier that is used to carry all traffic between mobile operator networks or sites or between mobile operator networks and corporate networks identified by access point name (APN) identifiers [6].

FIGURE 1

A View of NG End-to-End Network Architecture

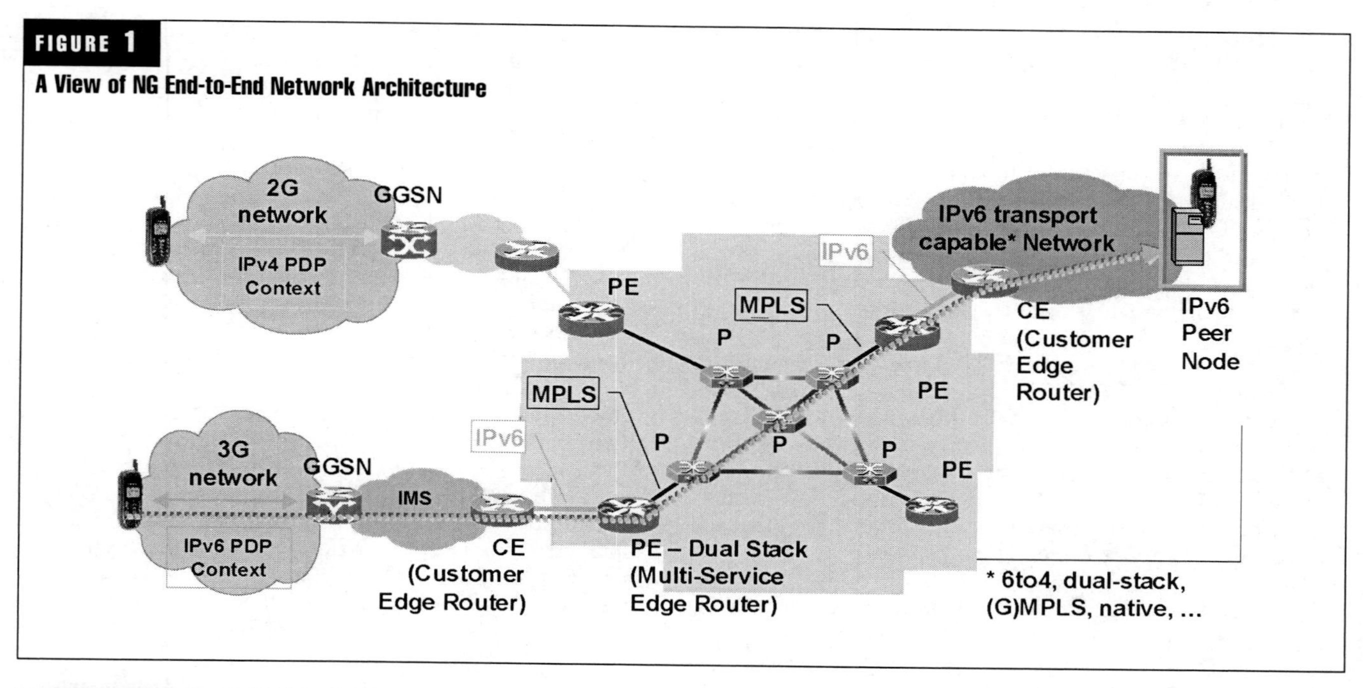

FIGURE 2

End-to-End Network Protocol Stack for Figure 1

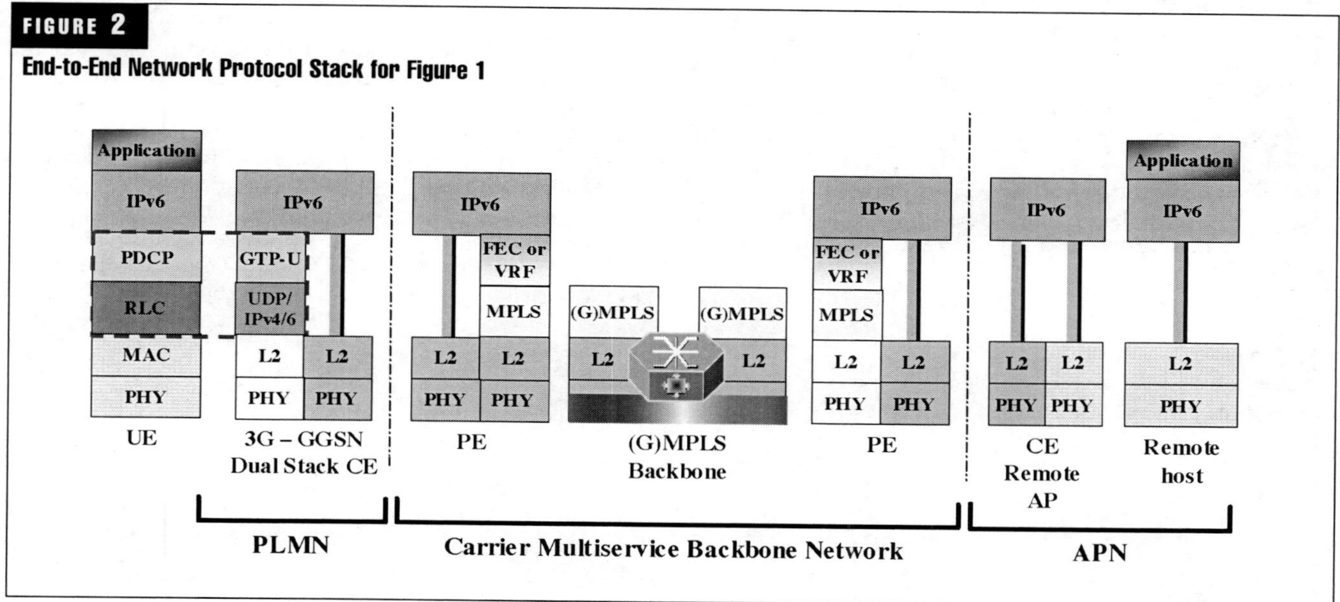

- A mobile operator network or site IP infrastructure tier that extends IP connectivity to the GGSN and customer edge (CE) router.

Figure 2 illustrates the protocol stack for the network architecture shown in *Figure 1*. Each mobile network infrastructure is attached to the backbone network through an edge router, in this case a CE router, which serves as a traffic aggregation point and demarcation point between the mobile operator IP core network and the carrier's multiservice backbone IP network. The service providers' multiservice backbone network is a common backbone network that also interconnects sites belonging to the other non-wireless network operators (such as Internet service providers [ISPs], xDSL, cable, fiber-to-the-home [FTTH], etc.). The backbone tier provides wide-area IP connectivity between sites or between mobile operator networks and corporate networks that have an APN identifier for GGSN [6] to direct the traffic over the backbone network as shown in *Figure 2*. It is designed for high-speed packet transport and is typically built with gigabit or terabit backbone routers along with long optical transmission consisting of DWDM equipment. Recently, GMPLS has emerged as an NG networking protocol for providing intelligent optical control planes for long-haul optical networks. GMPLS is considered to be superset of MPLS, as it extends forwarding plane and control plane to include not only packet-based and cell-based, but also SONET–, DWDM–, and fiber-based network elements. In this paper, references to GMPLS cover both MPLS multiservice backbone networks and NG GMPLS multiservice backbone networks, as the techniques discussed here apply to both.

The Role of IPv6 in Next-Generation WCDMA/GPRS Mobile Wireless Networks

There are two roles that IPv6 plays in mobile networks, as shown in *Figure 3*. *Figure 3* shows an end-to-end protocol stack for an IPv6 application running on top of a WCDMA/GPRS network. The traffic leaves mobile core network at the GGSN. There are other scenarios in which traffic from the GGSN remains in the core network. However, for the sake of describing the roles of IPv6 technology in the mobile core network, a simplified scenario is used.

Two separate IPv6 layers can be seen in *Figure 3*. The upper layer (in green) denotes the application layer, which runs between the UE and an external entity with which the UE is communicating. This external entity is typically an IPv6 application server or another IPv6–enabled UE. As shown in *Figure 4*, IPv6 is mandated for the IMS domain of 3G network architecture, whereas it is an optional protocol for GPRS network architecture.

The lower layer (in purple) denotes the IPv6 transport layer, which has only local significance in the public land mobile network (PLMN). The IPv6 transport layer is needed to transport subscriber data (control and user plane) traffic within the mobile network in maintaining the relationship between the UE and the GGSN, anchor point for the APN access. This uses the GPRS tunnel protocol (GTP). In this particular case, the GTP transport layer is terminated at the GGSN, where the traffic leaves the PLMN; routing for the application traffic is performed directly on the IP application layer. In case of R5 UTRAN, IPv6 support is mandated when IP is used as a transport protocol (still IPv4 deployment is possible), whereas in GPRS, it is optional, as shown in *Figure 4*.

Interworking Scenarios between Mobile Wireless and Wireline Networks

This section discusses some of the internetworking scenarios between mobile operator networks and multiservice backbone networks. These scenarios include the following:

- Connecting IPv6–enabled mobile operator networks or sites using GMPLS backbone network
- ISP and PLMN roaming scenarios

Scenario 1: Connecting IPv6 enabled mobile operator networks or sites using GMPLS backbone network. As shown in *Figure 1* and *Figure 2*, an IPv6 MS, running an IMS application, opened a connection to a remote IPv6 peer node. IPv6 peer node can be an IPv6 application server or another IPv6 MS. As shown in *Figure 1*, it first set up an IPv6 PDP context to the IMS APN. The GGSN terminates the PDP context and directs the IPv6 traffic (both SIP control plane and data) into IMS. The SIP proxies, part of the IMS architecture, have first set up the SIP session with the remote end point via the GMPLS network. When the session is set up, the IPv6 traffic goes through the GMPLS network.

The GGSN is the first place in the MSP network where the MS traffic is routable. However, any IPv6 traffic (from control or user plane) that needs to be able to reach a remote network thus to leave the MSP's IMS will have to use an external backbone. In the figure, the MSP's CE is connected to a GMPLS backbone. The GMPLS operator PE will take care of the IPv6 traffic up to a remote PE that connects to the remote APN's CE.

Scenario 2: ISP and PLMN roaming scenarios over GMPLS backbone networks. *Figure 5* below summarizes the ISP and PLMN roaming scenarios in an IMS context.

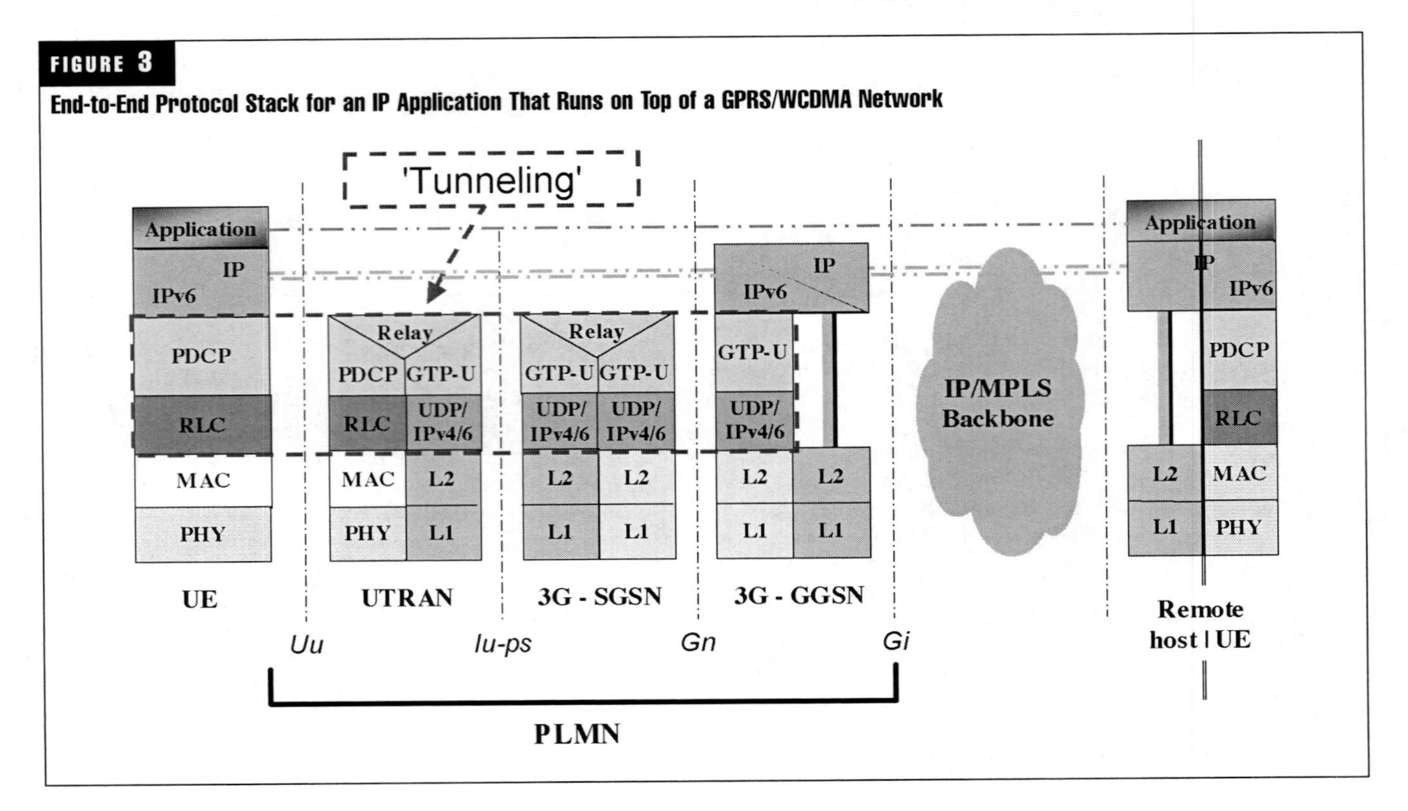

FIGURE 3

End-to-End Protocol Stack for an IP Application That Runs on Top of a GPRS/WCDMA Network

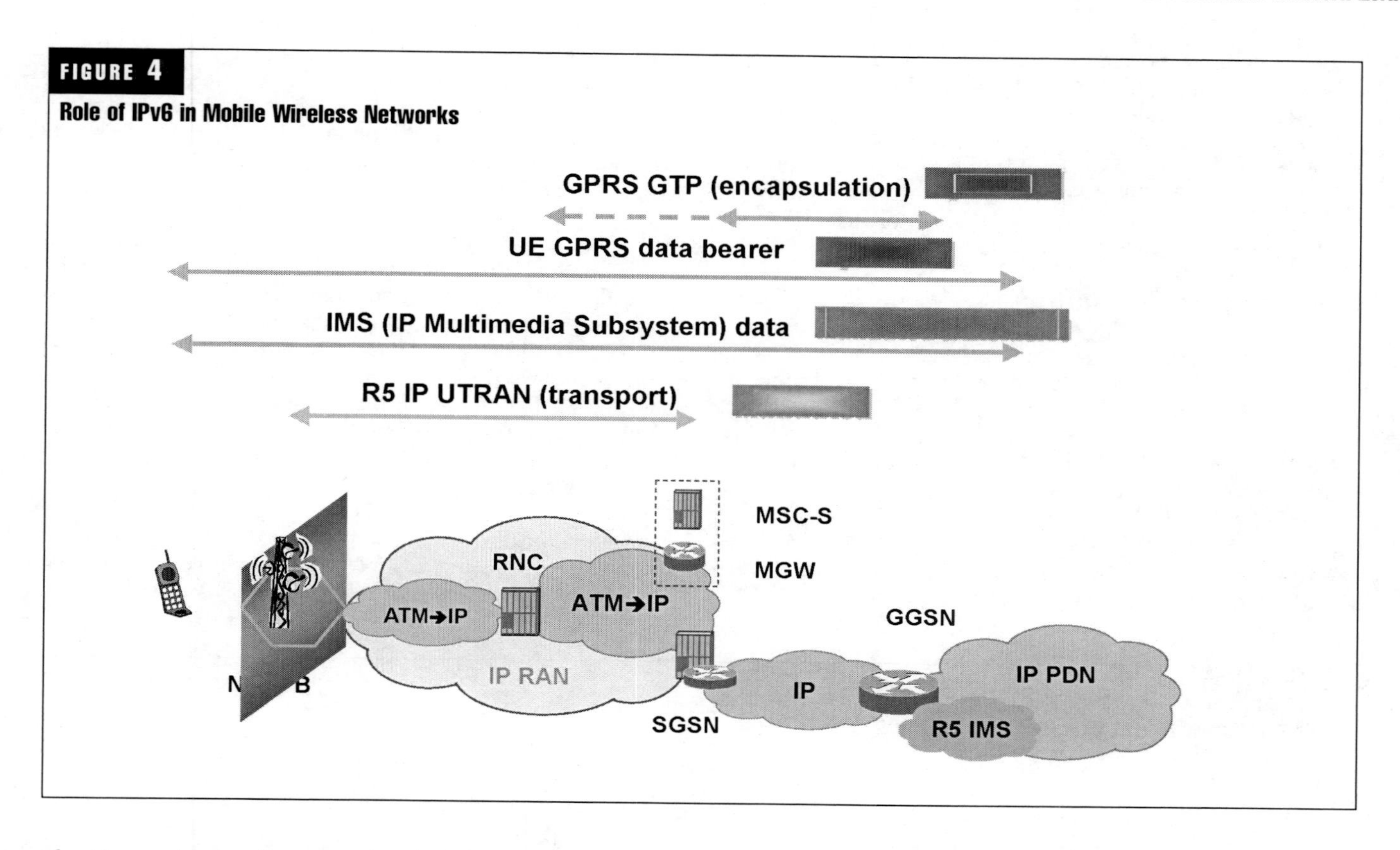

FIGURE 4

Role of IPv6 in Mobile Wireless Networks

The blue and green lines are showing the user plane only. The control planes are not drawn, in particular the SIP traffic between the CSCF (SIP proxies).

The lines in blue reflect the GRX model where GTP tunneling is used between the two PLMNs whatever traffic is (GPRS or IMS). Signaling for PDP context and SIP are also using this path. The green lines show the ISP roaming mode. Please note that in that case it is not necessary for the traffic from visited GGSN (the bottom network) to go through the home GGSN (top network), but it needs to reach the PDN. If the traffic is sourced at subscriber A to subscriber B, the two GGSNs will have to forward the traffic within the respective PDP contexts.

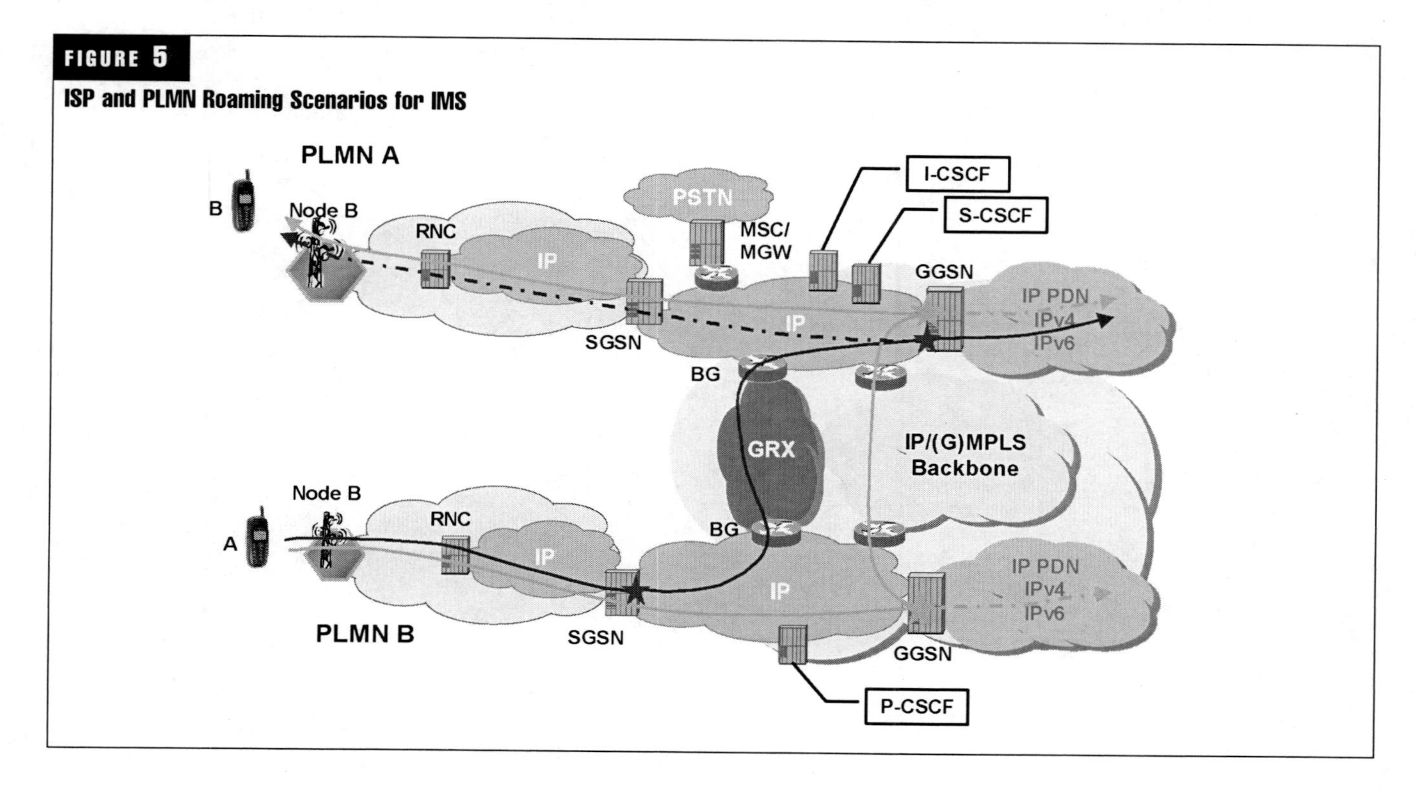

FIGURE 5

ISP and PLMN Roaming Scenarios for IMS

In this paper, we consider these two scenarios for defining internetworking techniques of IPv6 mobile networks and GMPLS–based multiservice backbone network. We present IPv6 packet flow in this paper for these scenarios. Section 5 presents the IPv6 packet flow within the mobile network, and section 6 presents internetworking techniques for transporting IPv6 packets from mobile operator networks over GMPLS–based backbone network.

IPv6 Packet Flow within PLMN

While the scope of this section is to discuss the IPv6 packet flow within PLMN, refer to [6] for basics of GPRS concepts.

The GGSN informs the MS that it shall perform stateful address auto-configuration by means of the router advertisements as defined in RFC2461. For this purpose, the GGSN shall automatically and periodically send router advertisement toward the MS after an IPv6 PDP context is activated. The use of stateless or stateful address auto-configuration is configured per APN.

GGSN shall assign a prefix that is unique within its scope to each PDP context while applying IPv6 stateless auto-configuration. The size of the prefix is determined according to the maximum prefix length of a global IPv6 address. This avoids having to perform duplicate address detection at the network level for every address built by the MS.

To ensure that the link-local address generated by the MS does not collide with the link-local address of the GGSN, the GGSN will provide an interface identifier to the MS and the MS will use this interface identifier to configure its link-local address. In case of stateless auto-configuration, the MS can choose any interface identifier to generate address other than link-local, without involving the network. The SGSN and the GGSN are not updated with the actual address used by the MS, as the prefix alone identifies the PDP context.

As shown in *Figure 6*, in step 3, upon reception of the PDP context request, the GGSN creates an IPv6 address composed of the prefix allocated to the PDP context and an

FIGURE 6

Session Establishment with PDP Type IPv6

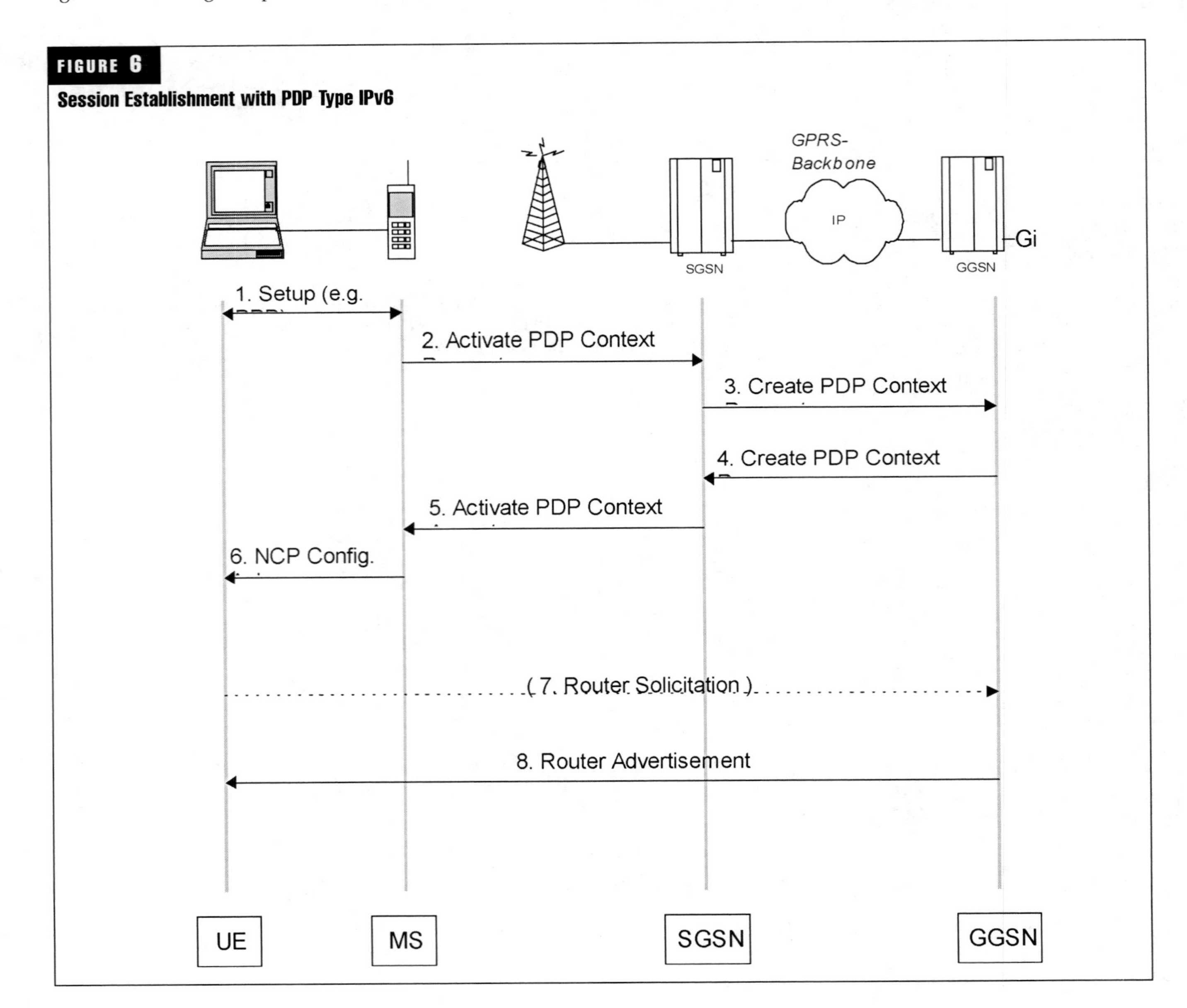

interface identifier generated by the GGSN. This address then returns in the create PDP context response message. Since the MS is considered to be alone on its link toward the GGSN, the interface identifier does not need to be unique across all PDP contexts on any APN. The MS extracts and stores the interface identifier from the address received and shall use it to build its link-local address as well as its full IPv6 address. The MS will ignore the prefix part.

After the GGSN has sent a create PDP context response message to the SGSN, it will start sending router advertisements periodically on the new MS–GGSN link established by the PDP context. As shown in step 7, the MS will issue a router solicitation directly after the user plane establishment (it is not an optional step). This will trigger the GGSN to send a router advertisement immediately. It will contain the same prefix as the one provided in step 3. After the MS received the router advertisement message, it will construct its full IPv6 address by concatenating the interface identifier received in step 3 or a locally generated interface identifier, and the prefix in the router advertisement. If the router advertisement contains more than one prefix option, the MS will only consider the first one and silently discard the others.

Because any prefix that the GGSN advertises in a PDP context is unique within the scope of the prefix, there is no need for the MS to perform duplicate address detection. Therefore, the GGSN will silently discard neighbor solicitation messages that the MS may send to perform duplicate address detection. It is possible for the MS to perform neighbor unreachability detection toward the GGSN, therefore if the GGSN receives a neighbor solicitation as part of this procedure, the GGSN shall reply with neighbor advertisement as defined in RFC2461.

IPv6 prefixes in a GGSN internal prefix pool shall be configurable and structured per APN.

After the PDP session is setup, when GGSN received an uplink packet from MS, GGSN will remove the GTP tunnel encapsulation, which include IP/UDP/GTP, then switch the internal payload IPv6 packet on corresponding interface based on IPv6 destination prefix.

When GGSN receives a downlink packet aim to an MS, GGSN does a lookup on destination address prefixes in forward information base. The IP routing table would point to the GPRS shared access interface. This will do an additional search IPv6 PDP tree by destination prefix. If the PDP is found, the packet will be encapsulated inside the correct GTP tunnel and send it to the mobile.

The following section discusses various internetworking scenarios and techniques for transporting IPv6 beyond GGSN over GMPLS backbone network to IPv6 destination host.

Internetworking Techniques for Transporting IPv6 over a GMPLS Backbone Network

IPv6 transport over GMPLS backbone enables isolated IPv6–based mobile operator networks or sites (in this section, we refer to them as IMS domains) to communicate with each other over GMPLS multiservice backbone network. A variety of deployment strategies are available, including the following:

- IPv6 using tunnels on the CE routers
- IPv6 over a circuit transport over GMPLS
- IPv6 on the provider edge (PE) routers (known as 6PE)
- IPv6 virtual private networks (VPNs)

The first of these strategies has no impact on and requires no changes to the GMPLS provider or PE routers because the strategy uses IPv4 tunnels to encapsulate the IPv6 traffic on CE routers, thus appearing as IPv4 traffic within the network.

The second of these strategies also requires no change to the core routing mechanisms. The last two strategies require changes to the PE routers to support a dual-stack implementation, but all the core functions remain IPv4. This implementation requires far fewer backbone infrastructure upgrades and lesser reconfiguration of core routers because forwarding is based on labels rather than the IP header itself, providing a very cost-effective strategy for the deployment of IPv6. Additionally, the inherent VPN and traffic engineering services available in a GMPLS environment allow IPv6 networks to be combined into VPNs or extranets over an infrastructure supporting IPv4 VPNs and GMPLS–TE.

Another strategy would be to run IPv6–based GMPLS control plane in the core, but this strategy would require a full network upgrade of all P and PE routers, with dual control planes for IPv4 and IPv6.

We describe the following mechanisms in more detail in this section:

- IPv6 using tunnels on the CE routers
- IPv6 over a circuit transport over GMPLS
- IPv6 on the PE routers
- IPv6 GMPLS VPNs

IPv6 Using Tunnels on the Customer Edge Routers

Using tunnels on the CE routers is the simplest way of deploying IPv6 over GMPLS networks because it has no impact on the operation or infrastructure of GMPLS and requires no changes to either the provider routers in the core or the PE routers connected to the customer sites.

Communication between the remote IPv6 IMS domains uses standard tunneling mechanisms, running IPv6 over IPv4 tunnels in a similar way that MPLS VPNs support native IPv4 tunnels on CE routers. The CE routers need to be upgraded to be dual-stack, and configured for IPv4–compatible or 6to4 tunnels, but communication with the PE routers is IPv4, and the traffic appears to the GMPLS domain to be IPv4. The dual-stack CE routers use the IPv4–compatible or 6to4 address, rather than an IPv6 address supplied by the service provider. *Figure 7* shows the configuration using tunnels on the CE routers.

IMS Domain Is Part of IPv6–Enabled Mobile Operator Network

IPv6 over a Circuit Transport over GMPLS

In some cases, mobile operator IP core network is adapted to low-cost, high-capacity traffic capabilities within the same operator network, typically using FastEthernet, giga-

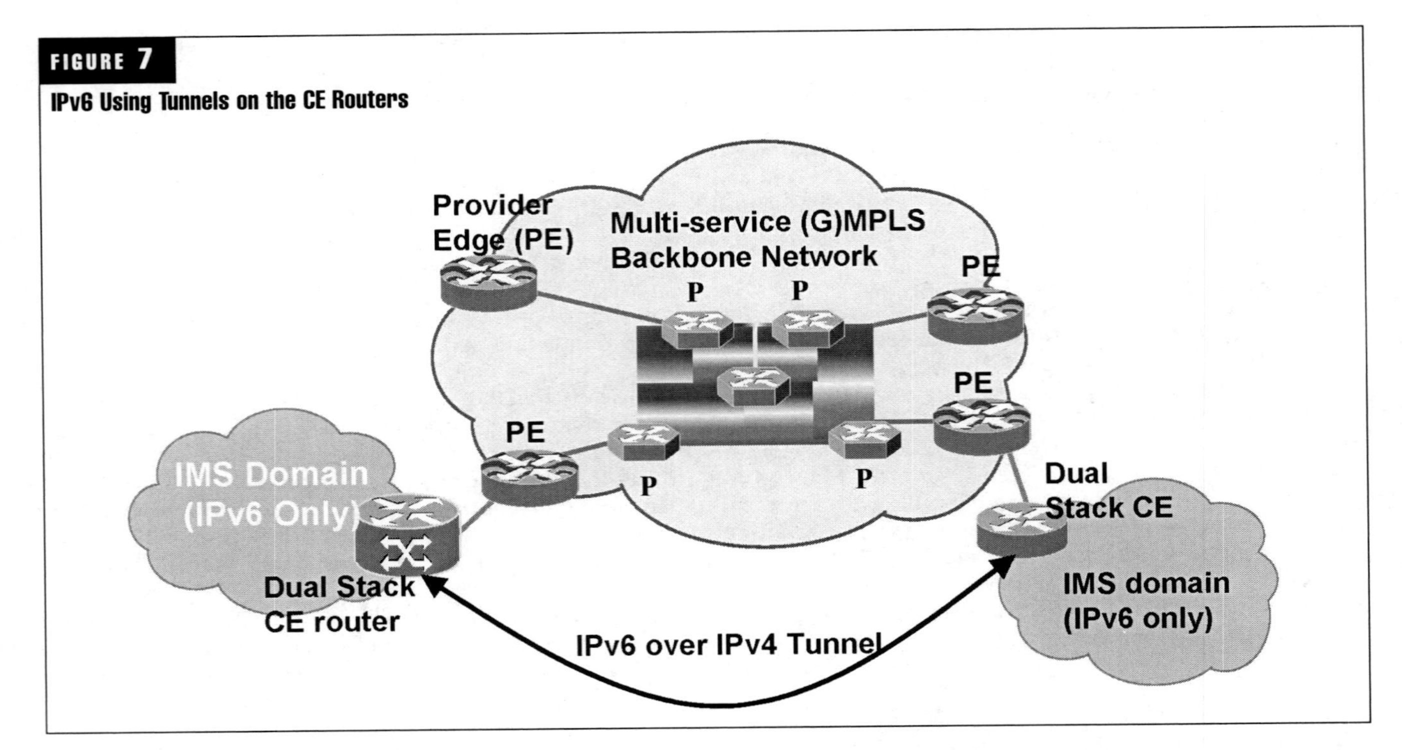

FIGURE 7

IPv6 Using Tunnels on the CE Routers

bit Ethernet, and local-area network (LAN) switches that can make use of virtual LAN techniques, as shown in *Figure 8*. In this scenario, using any circuit transport for deploying IPv6 over GMPLS networks has no impact on the operation or infrastructure of GMPLS. It requires no changes to either the provider routers in the core or the PE routers connected to the customers.

Communication between the remote IPv6 domains runs native IPv6 protocols over a dedicated link, where the underlying mechanisms are fully transparent to IPv6. The IPv6 traffic is tunneled using any transport over GMPLS (MPLS/AToM) or Ethernet over GMPLS (EoMPLS), with the IPv6 routers connected through an ATM OC–3 or Ethernet interface, respectively. *Figure 8* shows the configuration for IPv6 over any circuit transport over GMPLS.

IPv6 on the Provider Edge Routers

A further deployment technique is to configure IPv6 on the GMPLS PE routers [11]. This technique has a major advantage for service providers in that there is no need to upgrade either the hardware or software of the core network consist-

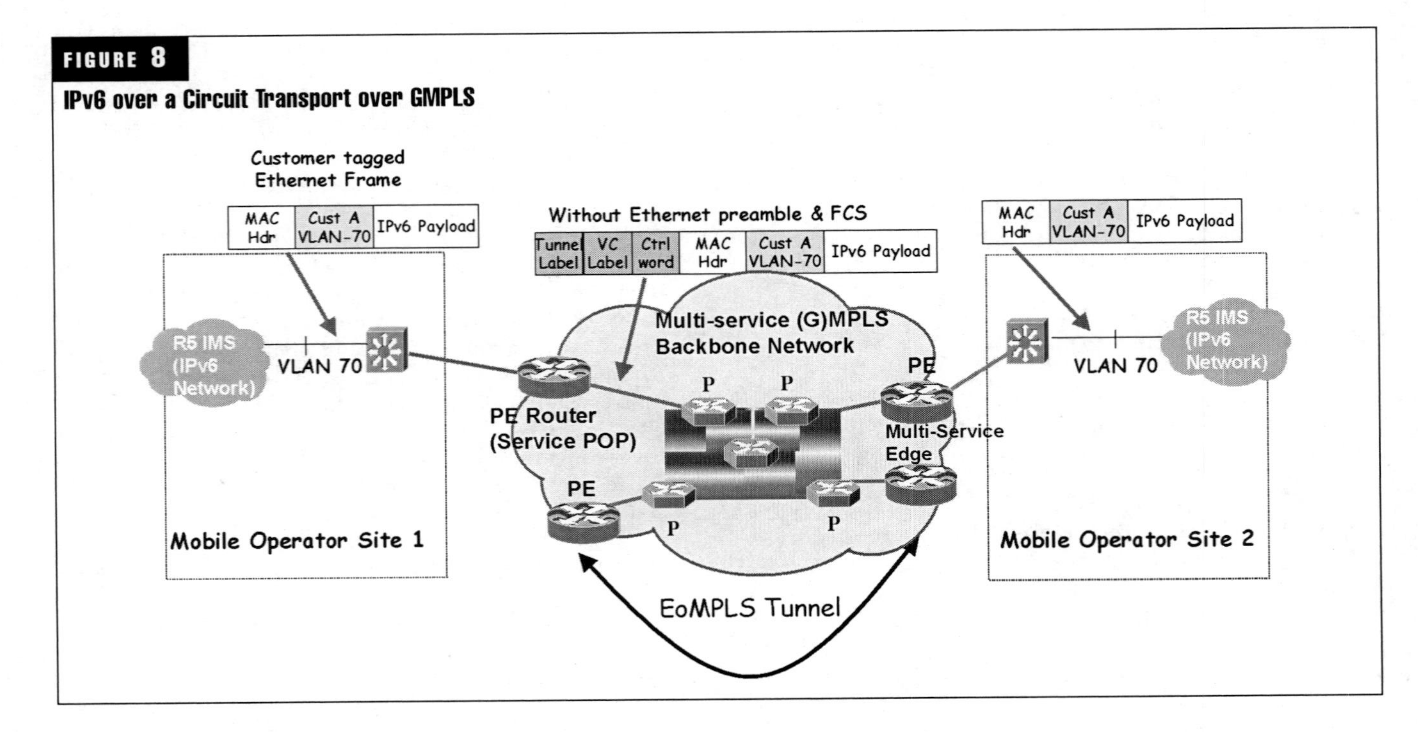

FIGURE 8

IPv6 over a Circuit Transport over GMPLS

ing of provider routers, and it thus eliminates the impact on the operation of or the revenue generated from the existing IPv4 traffic. The technique maintains the benefits of the current MPLS features (for example, GMPLS or VPN services for IPv4) while appearing to provide a native IPv6 service for enterprise customers (using ISP–supplied IPv6 prefixes). *Figure 9* shows the configuration for IPv6 on the PE routers. The IPv6 forwarding within the core is done by label switching, eliminating the need for either IPv6 over IPv4 tunnels or for an additional Layer-2 (L2) encapsulation, allowing the appearance of a native IPv6 service to be offered across the network. The core network continues to run GMPLS and any of the IPv4 interior routing protocols, eliminating the requirement for upgrades to the hardware for native IPv6 forwarding and allowing the network to continue without any upgrade.

Each PE router that must support IPv6 connectivity needs to be upgraded to be dual-stack (becoming a 6PE router) and configured to run GMPLS on the interfaces connected to the core. Depending on the site requirements, each router can be configured to forward IPv6 or IPv6 and IPv4 traffic on the interfaces to the CE routers, thus providing the ability to offer only IPv6 or both IPv6 and native IPv4 services. The 6PE router exchanges either IPv4 or IPv6 routing information through any of the supported routing protocols, depending on the connection, and switches IPv4 and IPv6 traffic using the respective fast switching path over the native IPv4 and IPv6 interfaces not running GMPLS.

The 6PE router exchanges reachability information with the other 6PE routers in the GMPLS domain using multiprotocol border gateway protocol (BGP) and shares a common IPv4 routing protocol (such as open shortest path first [OSPF] or intermediate system to intermediate system [IS–IS]) with the other provider and PE devices in the domain.

The 6PE routers encapsulate IPv6 traffic using two levels of GMPLS labels. The top label is distributed by the label distribution protocol (LDP) used by the devices in the core to carry the packet to the destination 6PE using IPv4 routing information. The second or bottom label is associated with the IPv6 prefix of the destination through multiprotocol BGP–4.

The 6PE architecture allows support for IPv6 VPNs.

IPv6 VPNs

A VPN is said to be an IPv6 VPN [12], when each mobile operators' network or site is IPv6–capable (an example IMS domain), as explained by 3GPP R5 specification, and is natively connected over an IPv6 interface or sub-interface to the PE router. Similar to IPv4 VPN routes distribution, BGP and its extensions are used to distribute routes from an IMS IPv6 VPN site to all the other PE routers connected to a site of the same IPv6 VPN. PEs use VPN routing and forwarding (VRF) to separately maintain the reachability information and forwarding information of each IPv6 VPN, as shown in *Figure 10*.

When a PE1 router receives IPv6 packet from CE A, it looks up the packet's IPv6 destination address in the VRF A. This enables it to find a VPN–IPv6 route. The VPN–IPv6 route will have an associated GMPLS label and an associated BGP next hop. This GMPLS label is imposed on the IPv6 packet. As 6PE1 directly pushes another label, the top label is bound by LDP/IGPv4 to the IPv4 address of BGP next hop to reach 6PE2 through GMPLS cloud, on the label stack of labeled IPv6 VPN packet. This topmost label imposed corresponds to the LSP starting on 6PE1 and ending on 6PE2. As mentioned above, the bottom label is bound to the IPv6 VPN prefix via BGP.

All the provider routers in the backbone network switch the VPN packet based only on the top label in the stack, which points toward the PE2 router. Because of the normal MPLS forwarding rules, the provider routers never look beyond the first label and are thus completely unaware of the second label or the IPv6 VPN packet carried across the backbone network.

The egress PE router, PE2, receives the labeled IPv6 VPN packet, drops the first label, and performs a lookup on the

FIGURE 9

IPv6 over GMPLS Network Using 6PE Architecture

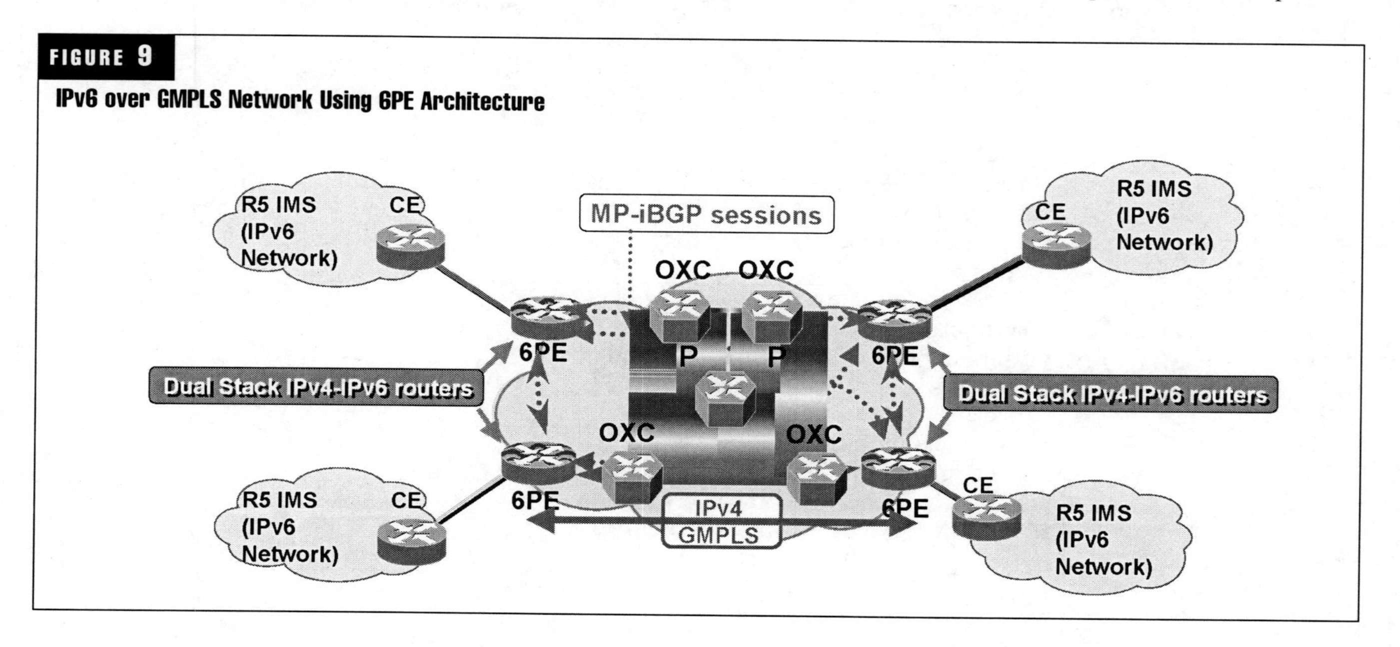

FIGURE 10

IPv6 VPN Network Architecture

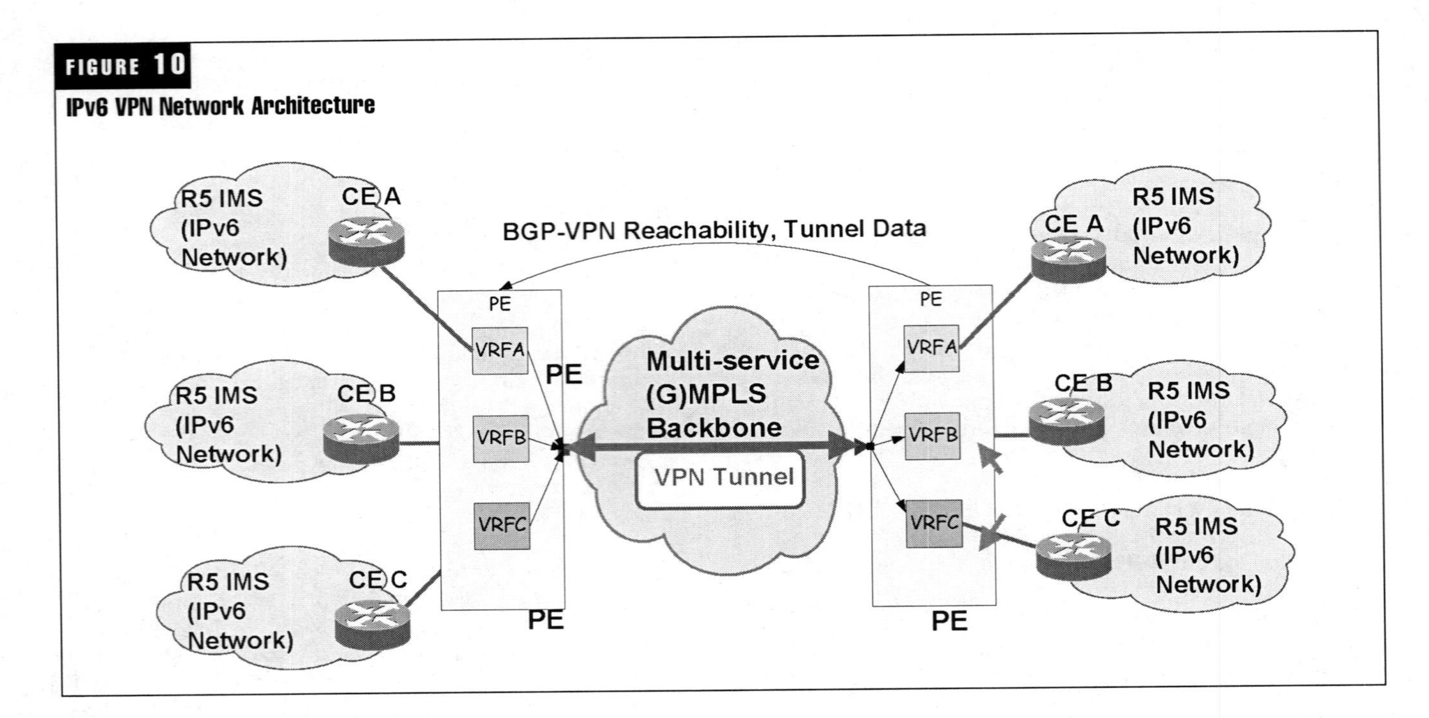

second label, which uniquely identifies the target VRF A and sometimes even the outgoing interface on the 6PE2. A lookup is performed in the target VRF A, and the IPv6 packet is sent to the proper CE router in IMS domain or site.

Comparison of IPv6 over GMPLS Network Deployment Strategies

Using tunnels on CE routers is the simplest way to deploy an IPv6–over–GMPLS network, as it has no impact on the operation or infrastructure of GMPLS and requires no changes to either the provider routers in the core or the PE routers connected to the customers' sites. However, a difficult-to-scale tunneling meshing is required as the number of CE routers to connect increases. As far as scalability, it is also difficult to delegate a global IPv6 prefix for an ISP.

The use of IPv6 over a circuit transport over GMPLS for deploying IPv6–over–GMPLS networks does not affect the operation or infrastructure of GMPLS. However, PE routers

TABLE 1

Comparison of Various IPv6 over GMPLS Backbone Transition Mechanisms

Mechanism	Primary use	Benefits	Limitations	Requirements
IPv6 using tunnels on CE routers	Enterprise customers wanting to use IPv6 over existing GMPLS services	No impact on GMPLS infrastructure	Routers use IPv4–compatible or 6to4 addresses	Dual-stack CE routers
IPv6 over a circuit transport over GMPLS	Service providers with ATM or Ethernet links to CE routers	Fully transparent IPv6 communication	No mix of IPv4 and IPv6 traffic	Need Layer-2 transport layer GMPLS
IPv6 on PE routers (6PE)	Internet and mobile service providers waning to offer an IPv6 service	Low cost and low risk upgrade to the PE routers, and no impact on GMPLS core	No VPN or VRF support	Software upgrade for PE routers
IPv6 GMPLS VPNs	Internet and mobile service providers wanting to offer IPv6 VPN services	Low-cost and low-risk upgrade to the PE routers and no impact on GMPLS core		VPN or VRF support

connected to need to support the "Ethernet over MPLS" transport mechanism. In this case, it is difficult to delegate a global IPv6 prefix for an ISP.

The 6PE feature is particularly applicable to service providers who already run a GMPLS network or plan to do it. One of the 6PE advantages is that there is no need to upgrade the core network. Thus it eliminates the impact on the operations and the revenues generated by the existing IPv4 traffic. GMPLS has been chosen by many service providers as a vehicle to deliver services to customers. GMPLS as a multiservice infrastructure technology is able to provide Layer-3 (L3) VPN, QoS, traffic engineering, fast re-routing, and integration of ATM and IP switching. GMPLS eases IPv6 introduction in existing production networks. From an operational standpoint, new CE introduction is straightforward and painless as it uses the L3 VPN scalability.

In addition to IPv6 global connectivity services, service providers need to offer IPv6 VPN services. The primary reason for such IPv6 VPN services is the same as the reason for isolation of end users' intranets as sought with IPv4 VPN services. The potential use of non-global site local IPv6 prefixes in enterprise networks is another driver for such IPv6 VPNs.

Extensions to the 6PE approach are being standardized in the IETF [12] for also supporting such IPv6 VPN services. This can be seen as combining the "IPv6 handling" of the 6PE approach described above with the "VPN handling" of BGP/MPLS IPv4 VPNs [14]. The major extensions to the 6PE approach described above are as follows:

- Use of a different address family in the MP–BGP advertisement, which is the labeled IPv6 VPN family (which allows disambiguation of non-global—and potentially overlapping—IPv6 addresses across VPNs)

- Use of the VRF concept of BGP/MPLS IPv4 VPNs [14], whereby each VPN has a separate set of routing and forwarding tables

Again, when deploying such IPv6 VPN services, only PE devices that actually connect IPv6 VPN services will need to be upgraded. There will be no impact on the core. End users will benefit from the exact same set of features for IPv6 VPN services as for existing IPv4 VPN services while service providers will benefit from the same set of operations, administration, and maintenance (OA&M) tools across IPv4 VPN services and IPv6 VPN services.

References

1. J. Wiljakka, "Analysis on IPv6 Transition in 3GPP Networks," IETF RFC4215.
2. J. Soininen, "Transition Scenarios for 3GPP Networks," IETF RFC3574.
3. J. Wiljakka, "Transition to IPv6 in GPRS and WCDMA Mobile Networks," IEEE Communication Magazine, April 2002.
4. Daniel G. Waddington et al., "Realizing the transition to IPv6," IEEE communication magazine, June 2002.
5. M. Wasserman, "Recommendations for IPv6 in 3GPP Standards," IETF Informational draft, rfc3314, September 2002.
6. 3GPP TS 23.060 V5.3.0, "General Packet Radio Service (GPRS); Service description; Stage 2 (Release 5)," September 2002.
7. 3GPP TS 29.060 V5.3.0, "General Packet Radio Service (GPRS); GPRS Tunnelling Protocol (GTP) across the Gn and Gp Interface (Release 5)," September 2002.
8. 3GPP TS 23.228 V5.5.0, "IP Multimedia Subsystem (IMS); Stage 2 (Release 5)," June 2002.
9. Luca Martini et al., "Layer2 circuit transport over MPLS," IETF draft, draft-martini-l2circuit-trans-mpls-08.txt.
10. Luca Martini et al., "Transport of Layer 2 Frames over MPLS," IETF draft, draft-martini-l2circuit-trans-mpls-16.txt, Feb. 2005.
11. J. De Clercq et al., "Connecting IPv6 Islands across IPv4 Clouds with BGP," IETF draft, draft-ooms-v6ops-bgp-tunnel-05.txt, November 2005.
12. J. De Clercq, et al., "BGP-MPLS VPN extension for IPv6 VPN," IETF draft, draft-ietf-l3vpn-bgp-ipv6-07.txt, January 2006.
13. Hamid Ould-Brahim et al., "Network based IP VPN Architecture using Virtual Routers," IETF draft, draft-ietf-l3vpn-vpn-vr-02.txt, April 2004.
14. Eric C. Rosen et al., "BGP/MPLS VPNs," IETF draft, draft-ietf-l3vpn-rfc2547bis-03.txt, April 2005.
15. Jari Arkko et al., "IPv6 for Some Second and Third Generation Cellular Hosts," IETF RFC3316.
16. Eric Mannie, "Generalized Multi-Protocol Label Switching Architecture," IETF RFC3945.

Acronym Guide

2B1Q	two binary, one quaternary
2B1Q	two binary, one quaternary
2G	second generation
3DES	triple data encryption standard
3G	third generation
3GPP	third-generation partnership project
3R	regeneration, reshaping, and retraining
4B3T	four binary, three ternary
4F/BDPR	four-fiber bidirectional dedicated protection ring
4F/BSPR	four-fiber bidirectional shared protection ring
4G	fourth generation
AAA	authentication, authorization, and accounting
AAL–[x]	ATM adaptation layer–x
ABC	activity-based costing
ABR	available bit rate
AC	alternating current OR authentication code
ACD	automatic call distributor
ACF	admission confirmation
ACH	automated clearinghouse
ACL	access control list
ACLEP	adaptive code excited linear prediction
ACM	address complete message
ACR	alternate carrier routing OR anonymous call rejection
ADM	add/drop multiplexer OR asymmetric digital multiplexer
ADPCM	adaptive differential pulse code modulation
ADS	add/drop switch
ADSI	analog display services interface
ADSL	asymmetric digital subscriber line
AES	advanced encryption standard
AFE	analog front end
AGW	agent gateway
AIM	advanced intelligent messaging
AIN	advanced intelligent network
ALI	automatic location identification
AM	amplitude modulation
AMA	automatic messaging account
AMI	alternate mark inversion
AMPS	advanced mobile phone service
AN	access network
ANI	automatic number identification
ANM	answer message
ANSI	American National Standards Institute
AOL	America Online
AON	all-optical network
AP	access point OR access provider
APC	automatic power control
API	application programming interface
APON	ATM passive optical network
APS	automatic protection switching
ARCNET	attached resource computer network
ARI	assist request instruction
ARM	asynchronous response mode
ARMS	authentication, rating, mediation, and settlement
ARP	address resolution protocol
ARPANET	Advanced Research Projects Agency Network
ARPU	average revenue per customer
A-Rx	analog receiver
AS	application server OR autonomous system
ASAM	ATM subscriber access multiplexer
ASC	Accredited Standards Committee
ASCII	American Standard Code for Information Interchange
ASE	amplified spontaneous emission
ASIC	application-specific integrated circuit
ASIP	application-specific instruction processor
ASON	automatically switched optical network
ASP	application service provider
ASR	access service request OR answer-seizure rate OR automatic service request OR automatic speech recognition
ASSP	application-specific standard part
ASTN	automatically switched transport network OR analog switched telephone network
ATC	automatic temperature control
ATIS	Alliance for Telecommunications Industry Solutions
ATM	asynchronous transfer mode OR automated teller machine
ATMF	ATM Forum
ATP	analog twisted pair
ATU-C	ADSL transmission unit-CO
ATU-R	ADSL transmission unit-remote
A-Tx	analog transceiver
AUI	attachment unit interface
AVI	audio video interleaved
AWG	American Wire Gauge OR arrayed waveguide grating
AYUTOS	as-yet-unthought-of services
B2B	business-to-business
B2C	business-to-consumer
BCSM	basic call state model
BDCS	broadband digital cross-connect system
BDPR	bidirectional dedicated protection ring
BE	border element
BER	bit-error rate
BERT	bit error–rate test
BGP	border gateway protocol
BH	busy hour
BHCA	busy hour call attempt
BI	bit rate independent
BICC	bearer independent call control
BID	bit rate identification
BIP	bit interactive parity
B–ISDN	broadband ISDN
BLEC	broadband local-exchange carrier OR building local-exchange carrier

BLES	broadband loop emulation services
BLSR	bidirectional line-switched ring
BML	business management layer
BOC	Bell operating company
BOF	business operations framework
BOND	back-office network development
BOSS	broadband operating system software
BPON	broadband passive optical network
BPSK	binary phase shift keying
B–RAS	broadband–remote access server
BRI	basic rate interface
BSA	business services architecture
BSPR	bidirectional shared protection ring
BSS	base-station system OR business support system
BTS	base transceiver station
BVR	best-value routing
BW	bandwidth
CA	call agent
CAC	call admission control OR carrier access code OR connection admission control
CAD	computer-aided design
CAGR	compound annual growth rate
CALEA	Communications Assistance for Law Enforcement Act
CAM	computer-aided manufacture
CAMEL	customized application of mobile enhanced logic
CAP	competitive access provider OR carrierless amplitude and phase modulation OR CAMEL application part
CAPEX	capital expenditures/expenses
CAR	committed access rate
CARE	customer account record exchange
CAS	channel-associated signaling OR communications applications specification
CAT	conditional access table OR computer-aided telephony
CATV	cable television
C-band	conventional band
CBDS	connectionless broadband data service
CBR	constant bit rate
CBT	core-based tree
CC	control component
CCB	customer care and billing
CCF	call-control function
CCI	call clarity index
CCITT	Consultative Committee on International Telegraphy and Telephony
CCK	complementary code keying
CCR	call-completion ratio
CCS	common channel signaling
CD	chromatic dispersion OR compact disc
cDCF	conventional dispersion compensation fiber
CDD	content delivery and distribution
CDDI	copper-distributed data interface
CDMA	code division multiple access
CDMP	cellular digital messaging protocol
CDMS	configuration and data management server
CDN	control directory number
CDPD	cellular digital packet data
CDR	call detail record OR clock and data recovery
CD–ROM	compact disc–read-only memory
CWDM	coarse wavelength division multiplexing
CE	customer edge
CEI	comparable efficient interface
CEO	chief executive officer
CER	customer edge router
CERT	computer emergency response team
CES	circuit emulation service
CES	circuit emulation service
CESID	caller emergency service identification
CEV	controlled environment vault
CFB/NA	call forward busy/not available
CFO	chief financial officer
CGI	common gateway interface
CHN	centralized hierarchical network
C–HTML	compressed HTML
CIC	circuit identification code
CID	caller identification
CIM	common information model
CIMD2	computer interface message distribution 2
CIO	chief information officer
CIP	classical IP over ATM
CIR	committed information rate
CIT	computer integrated telephone
CLASS	custom local-area signaling services
CLE	customer-located equipment
CLEC	competitive local-exchange carrier
CLI	command-line interface OR call-line identifier
CLID	calling-line identification
CLLI	common language location identifier
CLR	circuit layout record
CM	cable modem
CM&B	customer management and billing
CMIP	common management information protocol
CMISE	common management information service element
CMOS	complementary metal oxide semiconductor
CMRS	commercial mobile radio service
CMTS	cable modem termination system
CNAM	calling name (also defined as "caller identification with name" and simply "caller identification")
CNAP	CNAM presentation
CNS	customer negotiation system
CO	central office
CODEC	coder-decoder OR compression/decompression
COI	community of interest
COO	chief operations officer
COPS	common open policy service
CORBA	common object request broker architecture
CoS	class of service

COT	central office terminal
COTS	commercial off-the-shelf
COW	cell site on wheels
CP	connection point
CPAS	cellular priority access service
CPC	calling-party category (also calling-party control OR calling-party connected)
CPE	customer-premises equipment
CPI	continual process improvement
CPL	call-processing language
CPLD	complex programmable logic device
CPN	calling-party number
CPU	central processing unit
CR	constraint-based routing
CRC	cyclic redundancy check OR cyclic redundancy code
CRIS	customer records information system
CR–LDP	constraint-based routed–label distribution protocol
CRM	customer-relationship management
CRTP	compressed real-time transport protocol
CRV	call reference value
CS	client signal
CS–[x]	capability set [x]
CSA	carrier serving area
CSCE	converged service-creation and execution
CSCF	call-state control function
CSE	CAMEL service environment
CS–IWF	control signal interworking function
CSM	customer-service manager
CSMA/CA	carrier sense multiple access with collision avoidance
CSMA/CD	carrier sense multiple access with collision detection
CSN	circuit-switched network
CSP	communications service provider OR content service provider
CSR	customer-service representative
CSU	channel service unit
CSV	circuit-switched voice
CT	computer telephony
CT–2	cordless telephony generation 2
CTI	computer telephony integration
CTIA	Cellular Telecommunications & Internet Association
CTO	chief technology officer
CWD	centralized wavelength distribution
CWDM	coarse wavelength division multiplexing
CWIX	cable and wireless Internet exchange
DAC	digital access carrier
DACS	digital access cross-connect system
DAM	DECT authentication module
DAMA	demand assigned multiple access
DAML	digital added main line
DARPA	Defense Advanced Research Projects Agency
DAVIC	Digital Audio Video Council
DB	database
dB	decibel(s)
DBMS	database management system
dBrn	decibels above reference noise
DBS	direct broadcast satellite
DC	direct current
DCC	data communications channel
DCF	discounted cash flow OR dispersion compensation fiber
DCLEC	data competitive local-exchange carrier
DCM	dispersion compensation module
DCN	data communications network
DCOM	distributed component object model
DCS	digital cross-connect system OR distributed call signaling
DCT	discrete cosine transform
DDN	defense data network
DDS	dataphone digital service
DECT	Digital European Cordless Telecommunication
demarc	demarcation point
DEMS	digital electronic messaging service
DES	data encryption standard
DFB	distributed feedback
DFC	dedicated fiber/coax
DGD	differentiated group delay
DGFF	dynamic gain flattening filter
DHCP	dynamic host configuration protocol
DiffServ	differentiated services
DIN	digital information network
DIS	distributed interactive simulation
DITF	Disaster Information Task Force
DLC	digital loop carrier
DLCI	data-link connection identifier
DLE	digital loop electronics
DLEC	data local-exchange carrier
DLR	design layout report
DM	dense mode
DMD	dispersion management device
DMS	digital multiplex system
DMT	discrete multitone
DN	distinguished name
DNS	domain name server OR domain naming system
DOC	department of communications
DOCSIS	data over cable service interface specifications
DOD	Department of Defense
DOJ	Department of Justice
DoS	denial of service
DOS	disk operating system
DOSA	distributed open signaling architecture
DOT	Department of Transportation
DP	detection point
DPC	destination point code
DPE	distributed processing environment
DPT	dial pulse terminate
DQoS	dynamic quality of service
D-Rx	digital receiver
DS–[x]	digital signal [level x]
DSAA	DECT standard authentication algorithm
DSC	DECT standard cipher
DSCP	DiffServ code point
DSF	dispersion-shifted fiber
DSL	digital subscriber line [also xDSL]
DSLAM	digital subscriber line access multiplexer

DSLAS — DSL–ATM switch
DSP — digital signal processor OR digital service provider
DSS — decision support system
DSSS — direct sequence spread spectrum
DSU — data service unit OR digital service unit
DTH — direct-to-home
DTMF — dual-tone multifrequency
DTV — digital television
D-Tx — digital transceiver
DVB — digital video broadcast
DVC — dynamic virtual circuit
DVD — digital video disc
DVMRP — distance vector multicast routing protocol
DVoD — digital video on demand
DVR — digital video recording
DWDM — dense wavelength division multiplexing
DXC — digital cross-connect
E911 — enhanced 911
EAI — enterprise application integration
EAP — extensible authentication protocol
EBITDA — earnings before interest, taxes, depreciation, and amortization
EC — electronic commerce
ECD — echo-cancelled full-duplex
ECRM — echo canceller resource module
ECTF — Enterprise Computer Telephony Forum
EDA — electronic design automation
EDF — electronic distribution frame OR erbium-doped fiber
EDFA — erbium-doped fiber amplifier
EDGE — enhanced data rates for GSM evolution
EDI — electronic data interchange
EDSX — electronic digital signal cross-connect
EFM — Ethernet in the first mile
EFT — electronic funds transfer
EJB — enterprise Java beans
ELAN — emulated local-area network
ELEC — enterprise local-exchange carrier
EM — element manager
EMI — electromagnetic interference
EML — element-management layer
EMS — element-management system OR enterprise messaging server
E–NNI — external network-to-network interface
ENUM — telephone number mapping
E–O — electrical-to-optical
EO — end office
EoA — Ethernet over ATM
EOC — embedded operations channel
EoVDSL — Ethernet over VDSL
EPD — early packet discard
EPON — Ethernet PON
EPROM — erasable programmable read-only memory
ERP — enterprise resource planning
ESCON — enterprise systems connectivity
ESS — electronic switching system
ETC — establish temporary connection
EtherLEC — Ethernet local-exchange carrier
ETL — extraction, transformation, and load
eTOM — enhanced telecom operations map
ETSI — European Telecommunications Standards Institute
EU — European Union
EURESCOM — European Institute for Research and Strategic Studies in Telecommunications
EXC — electrical cross-connect
FAB — fulfillment, assurance, and billing
FAQ — frequently asked question
FBG — fiber Bragg grating
FCAPS — fault, configuration, accounting, performance, and security
FCC — Federal Communications Commission
FCI — furnish charging information
FCIF — flexible computer-information format
FDA — Food and Drug Administration
FDD — frequency division duplex
FDDI — fiber distributed data interface
FDF — fiber distribution frame
FDM — frequency division multiplexing
FDMA — frequency division multiple access
FDS–1 — fractional DS–1
FE — extended framing
FEC — forward error correction
FEPS — facility and equipment planning system
FEXT — far-end crosstalk
FHSS — frequency hopping spread spectrum
FICON — fiber connection
FITL — fiber-in-the-loop
FM — fault management OR frequency modulation
FOC — firm order confirmation
FOT — fiber-optic terminal
FOTS — fiber-optic transmission system
FP — Fabry-Perot [laser]
FPB — flex parameter block
FPGA — field programmable gate array
FPLMTS — future public land mobile telephone system
FPP — fast-packet processor
FR — frame relay
FRAD — frame-relay access device
FSAN — full-service access network
FSC — framework services component
FSN — full-service network
FT — fixed-radio termination
FT1 — fractional T1
FTC — Federal Trade Commission
FTE — full-time equivalent
FTP — file transfer protocol
FTP3 — file transfer protocol 3
FTTB — fiber-to-the-building
FTTC — fiber to the curb
FTTCab — fiber-to-the-cabinet
FTTEx — fiber-to-the-exchange
FTTH — fiber-to-the-home
FTTN — fiber-to-the-neighborhood
FTTS — fiber-to-the-subscriber
FTTx — fiber-to-the-x
FWM — four-wave mixing
FX — foreign exchange
GA — genetic algorithm

Gb	gigabit
GbE	gigabit Ethernet [also GE]
GBIC	gigabit interface converter
Gbps	gigabits per second
GCRA	generic cell rate algorithm
GDIN	global disaster information network
GDMO	guidelines for the definition of managed objects
GE	[see GbE]
GEO	geosynchronous Earth orbit
GETS	government emergency telecommunications service
GFF	gain flattening filter
GFR	guaranteed frame rate
Ghz	gigahertz
GIF	graphics interface format
GIS	geographic information services
GKMP	group key management protocol
GMII	gigabit media independent interface
GMLC	gateway mobile location center
GMPCS	global mobile personal communications services
GMPLS	generalized MPLS
GNP	gross national product
GOCC	ground operations control center
GPIB	general-purpose interface bus
GPRS	general packet radio service
GPS	global positioning system
GR	generic requirement
GRASP	greedy randomized adaptive search procedure
GSA	Global Mobile Suppliers Association
GSM	Global System for Mobile Communications
GSMP	generic switch management protocol
GSR	gigabit switch router
GTT	global title translation
GUI	graphical user interface
GVD	group velocity dispersion
GW	gateway
HCC	host call control
HD	home domain
HDLC	high-level data-link control
HDML	handheld device markup language
HDSL	high-bit-rate DSL
HDT	host digital terminal
HDTV	high-definition television
HDVMRP	hierarchical distance vector multicast routing protocol
HEC	head error control OR header error check
HEPA	high-efficiency particulate arresting
HFC	hybrid fiber/coax
HIDS	host intrusion detection system
HLR	home location register
HN	home network
HOM	high-order mode
HomePNA	Home Phoneline Networking Alliance [also HomePNA2]
HomeRF	Home Radio Frequency Working Group
HQ	headquarters
HSCSD	high-speed circuit-switched data
HSD	high-speed data

HSIA	high-speed Internet access
HSP	hosting service provider
HTML	hypertext markup language
HTTP	hypertext transfer protocol
HVAC	heating, ventilating, and air-conditioning
HW	hardware
IAD	integrated access device
IAM	initial address message
IAS	integrated access service OR Internet access server
IAST	integrated access, switching, and transport
IAT	inter-arrival time
IBC	integrated broadband communications
IC	integrated circuit
ICD	Internet call diversion
ICDR	Internet call detail record
ICL	intercell linking
ICMP	Internet control message protocol
ICP	integrated communications provider OR intelligent communications platform
ICS	integrated communications system
ICW	Internet call waiting
IDC	Internet data center OR International Data Corporation
IDE	integrated development environment
IDES	Internet data exchange system
IDF	intermediate distribution frame
IDL	interface definition language
IDLC	integrated digital loop carrier
IDS	intrusion detection system
IDSL	integrated services digital network DSL
IEC	International Electrotechnical Commission OR International Engineering Consortium
IEEE	Institute of Electrical and Electronics Engineers
I-ERP	integrated enterprise resource planning
IETF	Internet Engineering Task Force
IFITL	integrated [services over] fiber-in-the-loop
IFMA	International Facility Managers Association
IFMP	Ipsilon flow management protocol
IGMP	Internet group management protocol
IGP	interior gateway protocol
IGRP	interior gateway routing protocol
IGSP	independent gateway service provider
IHL	Internet header length
IIOP	Internet inter–ORB protocol
IIS	Internet Information Server
IKE	Internet key exchange
ILA	in-line amplifier
ILEC	incumbent local-exchange carrier
ILMI	interim link management interface
IM	instant messaging
IMA	inverse multiplexing over ATM
IMAP	Internet message access protocol
IMRP	Internet multicast routing protocol
IMSI	International Mobile Subscriber Identification
IMT	intermachine trunk OR International Mobile Telecommunications

IMTC	International Multimedia Teleconferencing Consortium
IN	intelligent network
INAP AU	INAP adaptation unit
INAP	intelligent network application part
INE	intelligent network element
InfoCom	information communication
INM	integrated network management
INMD	in-service, nonintrusive measurement device
I–NNI	internal network-to-network interface
INT	[point-to-point] interrupt
InterNIC	Internet Network Information Center
IntServ	integrated services
IOF	interoffice facility
IOS	intelligent optical switch
IP	Internet protocol
IPBX	Internet protocol private branch exchange
IPcoms	Internet protocol communications
IPDC	Internet protocol device control
IPDR	Internet protocol data record
IPe	intelligent peripheral
IPG	intelligent premises gateway
IPO	initial public offering OR Internet protocol over optical
IPoA	Internet protocol over ATM
IPQoS	Internet protocol quality of service
IPSec	Internet protocol security
IPTel	IP telephony
IPv6	Internet protocol version 6
IPX	Internet package exchange
IR	infrared
IRU	indefeasible right to user
IS	information service OR interim standard
IS-IS	intermediate system to intermediate system
ISA	industry standard architecture
ISAPI	Internet server application programmer interface
ISC	integrated service carrier OR International Softswitch Consortium
ISDF	integrated service development framework
ISDN	integrated services digital network
ISDN–BA	ISDN basic access
ISDN–PRA	ISDN primary rate access
ISEP	intelligent signaling endpoint
ISM	industrial, scientific, and medical OR integrated service manager
ISO	International Organization for Standardization
ISOS	integrated software on silicon
ISP	Internet service provider
ISUP	ISDN user part
ISV	independent software vendor
IT	information technology OR Internet telephony
ITSP	Internet telephony service provider
ITTP	information technology infrastructure library
ITU	International Telecommunication Union
ITU–T	ITU–Telecommunication Standardization Sector
ITV	Internet television
IVR	interactive voice response
IVRU	interactive voice-response unit
IWF	interworking function
IWG	interworking gateway
IWU	interworking unit
IXC	interexchange carrier
J2EE	Java Enterprise Edition
J2ME	Java Micro Edition
J2SE	Java Standard Edition
JAIN	Java APIs for integrated networks
JCAT	Java coordination and transactions
JCC	JAIN call control
JDBC	Java database connectivity
JDMK	Java dynamic management kit
JMAPI	Java management application programming interface
JMX	Java management extension
JPEG	Joint Photographic Experts Group
JSCE	JAIN service-creation environment
JSIP	Java session initiation protocol
JSLEE	JAIN service logic execution environment
JTAPI	Java telephony application programming interface
JVM	Java virtual machine
kbps	kilobits per second
kHz	kilohertz
km	kilometer
L2F	Layer-2 forwarding
L2TP	Layer-2 tunneling protocol
LAC	L2TP access concentrator
LAI	location-area identity
LAN	local-area network
LANE	local-area network emulation
LATA	local access and transport area
LB311	location-based 311
L-band	long band
LBS	location-based services
LC	local convergence
LCD	liquid crystal display
LCP	link control protocol
LD	laser diode OR long distance
LDAP	lightweight directory access protocol
LD–CELP	low delay–code excited linear prediction
LDP	label distribution protocol
LDS	local digital service
LE	line equipment OR local exchange
LEAF®	large-effective-area fiber
LEC	local-exchange carrier
LED	light-emitting diode
LEO	low Earth orbit
LEOS	low Earth-orbiting satellite
LER	label edge router
LES	loop emulation service
LIDB	line information database
LL	long line
LLC	logical link control
LMDS	local multipoint distribution system
LMN	local network management

LMOS	loop maintenance operation system
LMP	link management protocol
LMS	loop-management system OR loop-monitoring system OR link-monitoring system
LNNI	LANE network-to-network interface
LNP	local number portability
LNS	L2TP network server
LOL	loss of lock
LOS	line of sight OR loss of signal
LPF	low-pass filter
LQ	listening quality
LRN	local routing number
LRQ	location request
LSA	label switch assignment OR link state advertisement
LSB	location-sensitive billing
LSMS	local service management system
LSO	local service office
LSP	label-switched path
LSR	label-switched router OR leaf setup request OR local service request
LT	line terminator OR logical terminal
LTE	lite terminating equipment
LUNI	LANE user network interface
LX	local exchange
M2PA	message transfer protocol 2 peer-to-peer adaptation
M2UA	message transfer protocol 2–user adaptation layer
M3UA	message transfer protocol 3–user adaptation layer
MAC	media access control
MADU	multiwave add/drop unit
MAN	metropolitan-area network
MAP	mobile applications part
MAS	multiple-application selection
Mb	megabit
MB	megabyte
MBAC	measurement-based admission control
MBGP	multicast border gateway protocol
MBone	multicast backbone
Mbps	megabits per second
MC	multipoint controller
MCC	mobile country code
MCU	multipoint control unit
MDF	main distribution frame
MDSL	multiple DSL
MDTP	media device transport protocol
MDU	multiple-dwelling unit
MEGACO	media gateway control
MEMS	micro-electromechanical system
MExE	mobile execution environment
MF	multifrequency
MFJ	modified final judgment
MG	media gateway
MGC	media gateway controller
MGCF	media gateway control function
MGCP	media gateway control protocol
MHz	megahertz
MIB	management information base
MII	media independent interface
MIME	multipurpose Internet mail extensions
MIMO	multiple inputs, multiple outputs
MIN	mobile identification number
MIPS	millions of instructions per second
MIS	management information system
MITI	Ministry of International Trade and Industry (in Japan)
MLT	mechanized loop testing
MM	mobility management
MMDS	multichannel multipoint distribution system
MMPP	Markov-Modulated Poisson Process
MMS	multimedia message service
MMUSIC	Multiparty Multimedia Session Control [working group]
MNC	mobile network code
MOM	message-oriented middleware
MON	metropolitan optical network
MOP	method of procedure
MOS	mean opinion score
MOSFP	multicast open shortest path first
MOU	minutes of use OR memorandum of understanding
MPC	mobile positioning center
MPEG	Moving Pictures Experts Group
MPI	message passing interface
MPLambdaS	multiprotocol lambda switching
MPLS	multiprotocol label switching
MPOA	multiprotocol over ATM
MPoE	multiple point of entry
MPoP	metropolitan point of presence
MPP	massively parallel processor
MPx	MPEG–Layer x
MRC	monthly recurring charge
MRS	menu routing system
MRSP	mobile radio service provider
ms	millisecond
MSC	mobile switching center
MSF	Multiservice Switch Forum
MSIN	mobile station identification number
MSNAP	multiple services network access point
MSO	multiple-system operator
MSP	management service provider
MSPP	multiservice provisioning platform
MSS	multiple-services switching system
MSSP	mobile satellite service provider
MTA	message transfer agent
MTBF	mean time between failures
MTP [x]	message transfer part [x]
MTTR	mean time to repair
MTU	multiple-tenant unit
MVL	multiple virtual line
MWIF	Mobile Wireless Internet Forum
MZI	Mach-Zender Interferometer
N11	(refers to FCC–managed dialable service codes such as 311, 411, and 911)
NA	network adapter
NAFTA	North America Free Trade Agreement
NANC	North American Numbering Council
NANP	North American Numbering Plan

NAP	network access point
NARUC	National Association of Regulatory Utility Commissioners
NAS	network access server
NASA	National Aeronautics and Space Administration
NAT	network address translation
NATA	North American Telecommunications Association
NBN	node-based network
NCP	network control protocol
NCS	national communications system OR network connected server
NDA	national directory assistance
NDM–U	network data management–usage
NDSF	non-dispersion-shifted fiber
NE	network element
NEAP	non-emergency answering point
NEBS	network-equipment building standards
NEL	network-element layer
NEXT	near-end crosstalk
NFS	network file system
NG	next generation
NGCN	next-generation converged network
NGDLC	next-generation digital loop carrier
NGF	next-generation fiber
NGN	next-generation network
NGOSS	next-generation operations system and software OR next-generation OSS
NHRP	next-hop resolution protocol
NI	network interface
NIC	network interface card
NID	network interface device
NIDS	network intrusion detection system
NIIF	Network Interconnection Interoperability Forum
NIS	network information service
NIU	network interface unit
nm	nanometer
NML	network-management layer
NMS	network-management system
NND	name and number delivery
NNI	network-to-network interface
NNTP	network news transport protocol
NOC	network operations center
NOMAD	national ownership, mobile access, and disaster communications
NP	number portability
NPA	numbering plan area
NPAC	Number Portability Administration Center
NPN	new public network
NP–REQ	number-portable request query
NPV	net present value
NRC	Network Reliability Council OR nonrecurring charge
NRIC	Network Reliability and Interoperability Council
NRSC	Network Reliability Steering Committee
NRZ	non–return to zero
NS/EP	national security and emergency preparedness
NSAP	network service access point
NSAPI	Netscape server application programming interface
NSCC	network surveillance and control center
NSDB	network and services database
NSP	network service provider OR network and service performance
NSTAC	National Security Telecommunications Advisory Committee
NT	network termination OR new technology
NTN	network terminal number
NTSC	National Television Standards Committee
NVP	network voice protocol
NZ–DSF	nonzero dispersion-shifted fiber
O&M	operations and maintenance
OA&M	operations, administration, and maintenance
OADM	optical add/drop multiplexer
OAM&P	operations, administration, maintenance, and provisioning
OBF	Ordering and Billing Forum
OBLSR	optical bidirectional line-switched ring
OC–[x]	optical carrier–[level x]
OCBT	ordered core-based protocol
OCD	optical concentration device
OCh	optical channel
OCR	optical character recognition
OCS	original call screening
OCU	office channel unit
OCX	open compact exchange
OD	origin-destination
ODBC	open database connectivity
ODSI	optical domain services interface
O–E	optical-to-electrical
O–EC	optical–electrical converter
OECD	Organization for Economic Cooperation and Development
OEM	original equipment manufacturer
O–E–O	optical-to-electrical-to-optical
OEXC	opto-electrical cross-connect
OFDM	orthogonal frequency division multiplexing
OIF	Optical Internetworking Forum
OLA	optical line amplifier
OLAP	on-line analytical processing
OLI	optical link interface
OLT	optical line termination OR optical line terminal
OLTP	on-line transaction processing
OMC	Operations and Maintenance Center
OMG	Object Management Group
OMS SW	optical multiplex section switch
OMS	optical multiplex section
OMSSPRING	optical multiplex section shared protection ring
ONA	open network architecture
ONE	optical network element
ONI	optical network interface
ONMS	optical network-management system
ONT	optical network termination
ONTAS	optical network test access system
ONU	optical network unit

OP optical path
OPEX operational expenditures/expenses
OPS operator provisioning station
OPTIS overlapped PAM transmission with interlocking spectra
OPXC optical path cross-connect
ORB object request broker
ORT operational readiness test
OS operating system
OSA open service architecture
OSC optical supervisory panel
OSD on-screen display
OSGI open services gateway initiative
OSI open systems interconnection
OSMINE operations systems modification of intelligent network elements
OSN optical-service network
OSNR optical signal-to-noise ratio
OSP outside plant OR open settlement protocol
OSPF open shortest path first
OSS operations support system
OSS/J OSS through Java
OSU optical subscriber unit
OTM optical terminal multiplexer
OTN optical transport network
OUI optical user interface
O-UNI optical user-to-network interface
OUSP optical utility services platform
OVPN optical virtual petabits network OR optical virtual private network
OWSR optical wavelength switching router
OXC optical cross-connect
P&L profit and loss
PABX private automatic branch exchange
PACA priority access channel assignment
PACS picture archiving communications system
PAL phase alternate line
PAM Presence and Availability Management [Forum] OR pulse amplitude modulation
PAMS perceptual analysis measurement system
PAN personal access network
PBCC packet binary convolutional codes
PBN point-to-point–based network OR policy-based networking
PBX private branch exchange
PC personal computer
PCF physical control field
PCI peripheral component interconnect
PCM pulse code modulation
PCN personal communications network
PCR peak cell rate
PCS personal communications service
PDA personal digital assistant
PDC personal digital cellular
PDD post-dial delay
PDE position determination equipment
PDH plesiochronous digital hierarchy
PDN public data network
PDP policy decision point
PDSN packet data serving node
PDU protocol data unit
PE provider edge
PER packed encoding rules
PERL practical extraction and report language
PESQ perceptual evolution of speech quality
PFD phase-frequency detector
PHB per-hop behavior
PHY physical layer
PIC point-in-call OR predesignated interexchange carrier OR primary interexchange carrier
PICS plug-in inventory control system
PIM personal information manager OR protocol-independent multicast
PIN personal identification number
PINT PSTN and Internet Networking [IETF working group]
PINTG PINT gateway
PKI public key infrastructure
PLA performance-level agreement
PLC planar lightwave circuit OR product life cycle
PLCP physical layer convergence protocol
PLL phase locked loop
PLMN public land mobile network
PLOA protocol layers over ATM
PM performance monitoring
PMD physical-medium dependent OR polarization mode dispersion
PMDC polarization mode dispersion compensator
PMO present method of operation
PMP point-to-multipoint
PN personal number
PNNI private network-to-network interface
PnP plug and play
PO purchase order
PODP public office dialing plan
POET partially overlapped echo-cancelled transmission
POF plastic optic fiber
POH path overhead
POIS packet optical interworking system
PON passive optical network
PoP point of presence
POP3 post office protocol 3
POS packet over SONET OR point of service
PosReq position request
POT point of termination
POTS plain old telephone service
PP point-to-point
PPD partial packet discard
PPP point-to-point protocol
PPPoA point-to-point protocol over ATM
PPPoE point-to-point protocol over Ethernet
PPTP point-to-point tunneling protocol
PP–WDM point-to-point–wavelength division multiplexing
PQ priority queuing
PRI primary rate interface
ps picosecond
PSAP public safety answering point
PSC Public Service Commission

PSD	power spectral density
PSDN	public switched data network
PSID	private system identifier
PSN	public switched network
PSPDN	packet-switched public data network
PSQM	perceptual speech quality measure
PSTN	public switched telephone network
PTE	path terminating equipment
PTN	personal telecommunications number service
PTP	point-to-point
PTT	Post Telephone and Telegraph Administration
PUC	public utility commission
PVC	permanent virtual circuit
PVM	parallel virtual machine
PVN	private virtual network
PWS	planning workstation
PXC	photonic cross-connect
QAM	quadrature amplitude modulation
QoE	quality of experience
QoS	quality of service
QPSK	quaternary phase shift keying
QSDG	QoS Development Group
RAD	rapid application development
RADIUS	remote authentication dial-in user service
RADSL	rate-adaptive DSL
RAM	remote access multiplexer
RAN	regional-area network
RAP	resource allocation protocol
RAS	remote access server
RBOC	regional Bell operating company
RCP	remote call procedure
RCU	remote control unit
RDBMS	relational database management system
RDC	regional distribution center
RDSLAM	remote DSLAM
REL	release
RF	radio frequency
RFC	request for comment
RFI	request for information
RFP	request for proposal
RFPON	radio frequency optical network
RFQ	request for quotation
RGU	revenue-generating unit
RGW	residential gateway
RHC	regional holding company
RIAC	remote instrumentation and control
RIP	routing information protocol
RISC	reduced instruction set computing
RJ	registered jack
RLL	radio in the loop
RM	resource management
RMA	request for manual assistance
RMI	remote method invocation
RMON	remote monitoring
ROADM	reconfigurable optical add/drop multiplexer
ROBO	remote office/branch office
ROI	return on investment
RPC	remote procedure call
RPF	reverse path forwarding
RPR	resilient packet ring
RPRA	Resilient Packet Ring Alliance
RPT	resilient packet transport
RQMS	requirements and quality measurement system
RRQ	round-robin queuing or registration request
RSU	remote service unit
RSVP	resource reservation protocol
RSVP–TE	resource reservation protocol–traffic engineering
RT	remote terminal
RTCP	real-time conferencing protocol
RTOS	real-time operating system
RTP	real-time transport protocol
RTSP	real-time streaming protocol
RTU	remote test unit
RxTx	receiver/transmitter
RZ	return to zero
SAM	service access multiplexer
SAN	storage-area network
SAP	service access point OR session announcement protocol
SAR	segmentation and reassembly
S-band	short band
SBS	stimulated Brillouin scattering
SCAN	switched-circuit automatic network
SCCP	signaling connection control part
SCCS	switching control center system
SCE	service-creation environment
SCF	service control function
SCL	service control language
SCM	service combination manager OR station class mark OR subscriber carrier mark
SCN	service circuit node OR switched-circuit network
SCP	service control point
SCR	sustainable cell rate
SCSI	small computer system interface
SCSP	server cache synchronization protocol
SCTP	simple computer telephony protocol OR simple control transport protocol OR stream control transmission protocol
SD	selective discard
SD&O	service development and operations
SDA	separate data affiliate
SDB	service design bureau
SDC	service design center
SDF	service data function
SDH	synchronous digital hierarchy
SDM	service-delivery management OR shared data model
SDN	software-defined network
SDP	session description protocol
SDRP	source demand routing protocol
SDSL	symmetric DSL
SDTV	synchronous digital hierarchy
SDV	switched digital video
SE	service element
SEC	Securities and Exchange Commission
SEE	service-execution environment

SEP	signaling endpoint
ServReq	service request
SET	secure electronic transaction
SFA	sales force automation
SFD	start frame delimiter
SFF	small form-factor
SFGF	supplier-funded generic element
SG	signaling gateway
SG&A	selling, goods, and administration OR sales, goods, and administration
SGCP	simple gateway control protocol
SGSN	serving GPRS support node
SHDSL	single-pair high-bit-rate DSL
SHLR	standalone home location register
SHV	shareholder value
SI	systems integrator
SIBB	service-independent building block
SIC	service initiation charge
SICL	standard interface control library
SID	silence indicator description
SIF	SONET Interoperability Forum
sigtran	Signaling Transport [working group]
SIM	subscriber identity module OR service interaction manager
SIP CPL	SIP call processing language
SIP	session initiation protocol
SIP–T	session initiation protocol for telephony
SISO	single input, single output
SIU	service interface unit
SIVR	speaker-independent voice recognition
SKU	stock-keeping unit
SL	service logic
SLA	service-level agreement
SLC	subscriber line carrier
SLEE	service logic execution environment
SLIC	subscriber line interface circuit
SLO	service-level objective
SM	sparse mode
SMC	service management center
SMDI	simplified message desk interface
SMDS	switched multimegabit data service
SME	small-to-medium enterprise
SMF	single-mode fiber
SML	service management layer
SMP	service management point
SMPP	short message peer-to-peer protocol
SMS	service-management system OR short message service
SMSC	short messaging service center
SMTP	simple mail transfer protocol
SN	service node
SNA	service node architecture OR service network architecture
SNAP	subnetwork access protocol
SNMP	simple network-management protocol
SNPP	simple network paging protocol
SNR	signal-to-noise ratio
SO	service objective
SOA	service order activation
SOAC	service order analysis and control
SOAP	simple object access protocol
SOCC	satellite operations control center
SOE	standard operating environment
SOHO	small office/home office
SON	service order number
SONET	synchronous optical network
SOP	service order processor
SP	service provider OR signaling point
SPC	stored program control
SPE	synchronous payload envelope
SPF	shortest path first
SPIRITS	Service in the PSTN/IN Requesting Internet Service [working group]
SPIRITSG	SPIRITS gateway
SPM	self-phase modulation OR subscriber private meter
SPoP	service point of presence
SPX	sequence packet exchange
SQL	structured query language
SQM	service quality management
SRF	special resource function
SRP	source routing protocol
SRS	stimulated Raman scattering
srTCM	single-rate tri-color marker
SS	softswitch
SS7	signaling system 7
SSE	service subscriber element
SSF	service switching function
SSG	service selection gateway
SSL	secure sockets layer
SSM	service and sales management
SSMF	standard single-mode fiber
SSP	service switching point
STE	section terminating equipment
STM	synchronous transfer mode
STN	service transport node
STP	shielded twisted pair OR signal transfer point OR spanning tree protocol
STR	signal-to-resource
STS	synchronous transport signal
SUA	SCCP user adaptation
SVC	switched virtual circuit
SW	software
SWAN	storage wide-area network
SWAP	shared wireless access protocol
SWOT	strengths, weaknesses, opportunities, and threats
SYN	IN synchronous transmission
TALI	transport adapter layer interface
TAPI	telephony application programming interface
TAT	terminating access trigger OR termination attempt trigger OR transatlantic telephone cable
Tb	terabit
TBD	to be determined
Tbps	terabits per second
TC	tandem connect
TCAP	transactional capabilities application part
TCB	transfer control block
TCIF	Telecommunications Industry Forum
TCL	tool command language
TCM	time compression multiplexing
TCO	total cost of ownership

TCP	transmission control protocol
TCP/IP	transmission control protocol/Internet protocol
TC–PAM	trellis coded–pulse amplitude modulation
TDD	time division duplex
TDM	time division multiplex
TDMA	time division multiple access
TDMDSL	time division multiplex digital subscriber line
TDR	time domain reflectometer OR transaction detail record
TE	traffic engineering
TEAM	transport element activation manager
TED	traffic engineering database
TEM	telecommunications equipment manufacturer
TFD	toll-free dialing
THz	terahertz
TIA	Telecommunications Industry Association
TIMS	transmission impairment measurement set
TINA	Telecommunications Information Networking Architecture
TINA-C	Telecommunications Information Networking Architecture Consortium
TIPHON	Telecommunications and Internet Protocol Harmonization over Networks
TIWF	trunk interworking function
TKIP	temporal key integrity protocol
TL1	transaction language 1
TLDN	temporary local directory number
TLS	transparent LAN service OR transport-layer security
TLV	tag length value
TMF	TeleManagement Forum
TMN	telecommunications management network
TMO	trans-metro optical
TN	telephone number
TNO	telecommunications network operator
TO&E	table of organization and equipment
TOM	telecom operations map
ToS	type of service
TP	twisted pair
TPM	transaction processing monitor
TPS–TC	transmission control specific–transmission convergence
TR	technical requirement OR tip and ring
TRA	technology readiness assessment
TRIP	telephony routing over Internet protocol
trTCM	two-rate tri-color marker
TSB	telecommunication system bulletin
TSC	terminating call screening
TSI	time slot interchange
TSP	telecommunications service provider
TSS	Telecommunications Standardization Section
TTC	Telecommunications Technology Committee
TTCP	test TCP
TTL	transistor-transistor logic
TTS	text-to-speech OR TIRKS® table system
TUI	telephone user interface
TUP	telephone user part
TV	television
UA	user agent
UADSL	universal ADSL
UAK	user-authentication key
UAWG	Universal ADSL Working Group
UBR	unspecified bit rate
UBT	ubiquitous bus technology
UCP	universal computer protocol
UCS	uniform communication standard
UDDI	universal description, discovery, and integration
UDP	user datagram protocol
UDR	usage detail record
UI	user interface
ULH	ultra-long-haul
UM	unified messaging
UML	unified modeling language
UMTS	Universal Mobile Telecommunications System
UN	United Nations
UNE	unbundled network element
UNI	user network interface
UOL	unbundled optical loop
UPC	usage parameter control
UPI	user personal identification
UPS	uninterruptible power supply
UPSR	unidirectional path-switched ring
URI	uniform resource identifier
URL	universal resource locator
USB	universal serial bus
USTA	United States Telecom Association
UTOPIA	Universal Test and Operations Interface for ATM
UTS	universal telephone service
UWB	ultra wideband
UWDM	ultra-dense WDM
V&H	vertical and horizontal
VAD	voice activity detection
VAN	value-added network
VAR	value-added reseller
VAS	value-added service
VASP	value-added service provider
VBNS	very–high-speed backbone network service
VBR	variable bit rate
VBR–nrt	variable bit rate–non–real-time
VBR–rt	variable bit rate–real time
VC	virtual circuit OR virtual channel
VCC	virtual channel connection
VCI	virtual channel identifier
VCLEC	voice CLEC
VCO	voltage-controlled oscillator
VCR	videocassette recorder
VCSEL	vertical cavity surface emitting laser
VD	visited domain
VDM	value delivery model
VDSL	very-high–data-rate DSL
VeDSL	voice-enabled DSL
VGW	voice gateway
VHE	virtual home environment
VHS	video home system
VITA	virtual integrated transport and access

VLAN virtual local-area network OR voice local-area network
VLR visitor location register
VLSI very-large-scale integration
VM virtual machine
VMS voice-mail system
VoADSL voice over ADSL
VoATM voice over ATM
VoB voice over broadband
VoD video on demand
VoDSL voice over DSL
VoFR voice over frame relay
VoIP voice over IP
VON voice on the Net
VoP voice over packet
VOQ virtual output queuing
VoT1 voice over T1
VP virtual path
VPDN virtual private dial network
VPI virtual path identifier
VPIM voice protocol for Internet messaging
VPN virtual private network
VPR virtual path ring
VPRN virtual private routed network
VRU voice response unit
VSAT very-small–aperture terminal
VSI virtual switch interface
VSM virtual services management
VSN virtual service network
VSR very short reach
VT virtual tributary
VTN virtual transport network
VToA voice traffic over ATM
VVPN voice virtual private network
VXML voice extensible markup language
W3C World Wide Web Consortium
WAN wide-area network
WAP wireless application protocol
WATS wide-area telecommunications service
WB DCS wideband DCS
WCDMA wideband CDMA
WCT wavelength converting transponder
WDCS wideband digital cross-connect
WDM wavelength division multiplexing
WECA wireless Ethernet compatibility alliance
WEP wired equivalent privacy
WFA work and force administration
WFQ weighted fair queuing
Wi-Fi wireless fidelity
WIM wireless instant messaging
WiMAX worldwide interoperability for microwave access
WIN wireless intelligent network
WLAN wireless local-area network
WLL wireless local loop
WMAP wireless messaging application programming interface
WML wireless markup language
WNP wireless local number portability
WRED weighted random early discard
WS work station
WSP wireless session protocol
WTA wireless telephony application
WUI Web user interface
WVPN wireless VPN
WWCUG wireless/wireline closed user group
WWW World Wide Web
XA transaction management protocol
XC cross-connect
XD extended distance
xDSL [see DSL]
XML extensible markup language
XPM cross-phase modulation
XPS cross-point switch
xSP specialized service provider
XT crosstalk
XTP express transport protocol
Y2K year 2000